Such News of the Land

Such News of the Land

U.S. Women Nature Writers

Edited by
Thomas S. Edwards and
Elizabeth A. De Wolfe

University Press of New England

Hanover and London

University Press of New England, Hanover, NH 03755

© 2001 by University Press of New England

All rights reserved

Printed in the United States of America

5 4 3 2 1

Library of Congress Cataloging-in-Publication Data

Such news of the land : U.S. women nature writers / Thomas S. Edwards and Elizabeth A. De Wolfe, editors.

   p. cm.

Includes bibliographical references and index.

ISBN 1–58465–097–4 (alk. paper)— ISBN 1–58465–098–2 (pbk. : alk. paper)

1. American literature—Women authors—History and criticism.
2. Nature in literature. 3. Women and literature—United States—History.
4. Natural history literature—United States—History. I. De Wolfe, Elizabeth A., 1961–   II. Title.

   PS163 .E39 2001

   810.9'36—dc21   00–010815

Tom: to Barbara, Emily, and Liam, and to Cecelia Edwards, for giving books, love, and the love of books: "So we always keep the same hearts, though our outer framework fails and shows the touch of time" (Sarah Orne Jewett, *The Country of the Pointed Firs*).

Beth: to Helen Grant Kenneth (1904–1999) and Susie Nigrelli Ferrigno (1906–1998), who wrote of nature with paint and canvas.

Contents

I. Laying the Foundation

II. Expanding the Genre

III. Nature Writing in the Twentieth Century

Foreword

Vera Norwood

In 1994, Vice President Al Gore wrote a new introduction to Rachel Carson's environmental classic *Silent Spring,* in which he commented that he and Carson shared a passion for protecting nature from the assaults of an industrial age and that both had been subjected to unfair, personalized attacks as a result of their environmental advocacy. She was branded a "hysterical" "priestess of nature" while he was labeled "Ozone Man." Carson was thus cast as a woman marginalized by female biological destiny while Gore represented a weak link, an aberrant branch of the masculine family tree. Gore salvages both their reputations by inviting his readers into his office, a seat of unquestionable masculine power, filled with photographs of "political leaders, the presidents, and prime ministers," to view a picture of Rachel Carson that "has been there for years—and it belongs there."

Carson resides among the heads of state not only because her book forced policy makers to pass legislation and regulations and create new federal agencies aimed at addressing environmental pollution and threats to America's last natural places, but because it engaged a wide public "and put our democracy itself on the side of saving the Earth." Gore counters attempts to marginalize himself and Carson with appeals to gender stereotypes by locating their shared values at the heart of American national culture. He casts her accomplishments as "powerful proof of the difference that one individual can make" if the moment is ripe, the moment here being the rise of environmental concern in need of a flash point.

Gore's move to salvage Carson's reputation from the wages of sexism by stressing her individual achievement demonstrates feminism's impact on public discourse in late twentieth-century America. And yes, such a move also isolates Carson, overstates the uniqueness of her concerns for nature, and begs the question of why it was that Gore first read Carson at his mother's urging and held lively dinner table discussions about her book with his sister. We must further examine why it was a woman who reminded Americans of their identity as nature's nation and why gender-role stereotyping remains a key tactic in attacks on environmental advocates if we are to understand how Rachel Carson and the moment came together to spark the modern environmental movement.

These critical issues are taken up in *Such News of the Land: U.S. Women Nature Writers.* As the title of the collection makes clear, Rachel Carson was not the lone woman in a portrait gallery of the masculine elite. She was preceded by the likes of Mary Hunter Austin, who was indebted to Sarah Orne Jewett, who had a precursor in Susan Fenimore Cooper. Each of these women was widely influential in defining the meanings of nature in America, but much of what they valued about the plants and animals of their homes, neighborhoods, and country arose from gendered social roles. We need to know more about the impact of their nature values on the developing environmental narratives of the nineteenth and twentieth centuries. The essays in this collection address such critical questions as the definitions of wilderness as seen from the confines of home and the values of local, native plants and animals. Just as Rachel Carson turned a wary eye on the man with the spray gun in suburban yards, women throughout American history have taken a critical look at their fathers, husbands, sons, and colleagues, and used their influence to temper and modify dominant values.

Gore's introduction to *Silent Spring* concludes with an extended discussion of the Clinton-Gore environmental agenda, with the vice president convinced of a popular mandate on environmental issues and consumed with policy implementation questions and regulatory pragmatics. While *Silent Spring* set the stage for broadly supported assaults on air and water pollution, on threats to natural heritage sites and wild lands, not every American who supports such initiatives has felt empowered by the movement. One of the most brilliant aspects of *Silent Spring* was Carson's call to lay people, to ordinary citizens, to wrest control of home and neighborhood spaces from those who threatened their health and recreation. Yet those who followed in her footsteps quickly discovered that continued inequities of gender, complicated by class and race bias, made it difficult for their voices to be heard, even within the environmental movement of the 1970s and 1980s. Here too, women have played a key cautionary role in redefining the issues and changing the agendas. Environmental historians and ecocritics are just beginning to consider how the different nature values of African Americans, Native Americans, and Hispanics have led people in these groups to question what is meant by nature's nation, who is denied access to such spaces, and off of whose backs they have been preserved and created. In including essays on such nature writers as Zora Neale Hurston and Leslie Marmon Silko and the writings of women in environmental justice movements, *Such News of the Land* engages the current debates about the differences among those who feel strongly about the need to preserve the green places of home, neighborhood, and country, and encourages us to consider the ways in which gender plays a part in all these discussions.

Currently, many American environmentalists worry that in the current moment, forces against nature conservation and preservation may reassert themselves. At a time when landmark legislation—such as the Clear Air Act, the Clean Water Act, the Wilderness Act and the Endangered Species Act—all seem threatened, those concerned with the future of environmentalism in the twenty-first century are looking for ways to engage a broader public. Doing so requires that a tradition usually perceived as the domain of a white, male elite recognizes the contributions of women and people of color to the national discourse on the meaning and value of nature in America. *Such News of the Land* offers an excellent place to begin the search for the next wave of environmentalism.

Acknowledgments

This collection had its roots in the first American Women Nature Writers conference in Portland, Maine. Beth would like to thank the many individuals at the University of New England on both the Biddeford and Westbrook College campuses who were instrumental in making that gathering a success, especially Roberta Gray and Pat Milligan and the staff at the libraries on both campuses. For guiding her through the conference planning and introducing her to the topic of women nature writers, thanks go to Thomas Edwards; for continued support on projects big and small, to Scott De Wolfe. At Castleton State College, thanks go to Michèle Thomson, for her excellent editorial assistance, to Nancy Stearns, for her help in all areas, and to Dean Joseph T. Mark and President Martha K. Farmer for their support. Thanks also to Susan McKelvie for the index. Karen Kilcup at the University of North Carolina was generous with her time and her wisdom, as was Michael P. Branch from the University of Nevada/Reno, and Daniel J. Philippon, from the University of Minnesota/Twin Cities.

T. S. E. and E. A. D.

Such News of the Land

Introduction

> But if you ever come beyond the borders as far as the town that lies in a hill
> dimple at the foot of Kearsage, never leave it until you have knocked at the
> door of the brown house under the willow-tree at the end of the village street,
> and there you shall have such news of the land, of its trails and what is astir in
> them, as one lover of it can give to another.
>
> —Mary Austin, *The Land of Little Rain*[1]

"I went to the woods because I wanted to live deliberately," wrote Henry David Thoreau in *Walden* in 1854. In the process, he helped craft a powerfully romantic image of a division between nature and human society that has entranced readers and writers ever since. Thoreau's call to live "as deliberately as Nature" is often identified with his journey into the woods, as if one necessarily required the other. In an opposition that has loomed large in the American consciousness, the woods—nature—are set against our life in society and our human constructions on, around, and in nature.[2] The transcendentalists, of course, aspired to overcome this dichotomy of nature and culture. Feeling the "currents of the Universal Being" circulating through him, Emerson could claim that "I am part or parcel of God."[3] In a quieter moment, Thoreau enjoyed fishing at night on Walden Pond, where "It seemed as if I might next cast my line upward into the air, as well as downward into this element which was scarcely more dense."[4] Neither image, however, carried the power of the simple and compelling idea that *Walden* communicated to many of its readers: that in order to live as deliberately *as* nature one must live *in* nature.[5] From Thoreau to John Muir and Edward Abbey, the formula has been an enduring one, in part because it has functioned so well. But while this romantic formula has been undeniably powerful, it has also been debilitating, for this division between nature and culture has ultimately made our view of what constitutes a contribution to the examination of the "deliberate life" unnecessarily narrow and restricted.

Part of the problem lies in our equating "nature" with "wilderness," as Thoreau himself discovered on his journeys into the Maine woods. The top of Katahdin (or even, for that matter, the potential violence of the sea that he touches on in *Cape*

I

Cod) was quite a different place from the more pastoral woods surrounding Concord. And if in the popular consciousness nature and wilderness were roughly equivalent, it could be expected that "wilderness writing" was what one meant when the topic of "nature writing" was introduced. Wilderness, the province of (often-solitary) men, was no place for a lady, whether that wilderness be threatening or simply sublime.[6] Like the travel narrative, much nature writing in America has followed a pattern of separation, whether through travel, or isolation, or both. In many travel narratives, American authors "discover" the "true" America by traveling its length and breadth, rediscovering in local, regional flavors the distinctive character that they cannot—or choose not to—recognize in their own backyards. In the same way, much nature writing in America consists of attempts to encounter "true" nature somehow separate and isolated from our everyday existence.[7] Both travel narratives and nature writing endorse the need for separation and distance in order to provide for true contemplation, study, and reflection. The philosophical assumptions of the nature/culture dichotomy have led to a genre that favors a specific type of isolated study of and interaction with the natural world.

Yet, as much as this formula has contributed to the construction of a distinctly American literature, it has contained within itself the seeds of a potentially fatal flaw. "Wilderness" writing leaves out the rich and fertile literature of a broader, more inclusive "nature" writing, which at its best is no less observant, no less insightful, and no less valuable in our examination of the interaction of the human and the physical environment. By focusing on those writers (predominately white males) who have found success within this tradition, we have ignored the work of countless women authors who have contributed so much to the American investigation of how our interaction with the natural landscape helps form and shape our lives. The record that has been presented heretofore has been incomplete: important pages have been missing, neglected, or ignored. Appearing in letters, notebooks, popular periodicals, fiction, market bulletins, and a wide variety of other venues, women's nature writings provide keen insight both into women's relationship to nature and into nature writing's role in the construction of American identity.

In his book *This Incomperable Lande: A Book of American Nature Writing,* Thomas Lyon defines nature writing as a genre that stresses "the awakening of perception to an ecological way of seeing." For Lyon, this way of seeing encompasses "the capacity to notice pattern in nature, and community, and to recognize that the patterns observed ultimately radiate outward to include the human observer."[8] Nothing in this definition would, on its own, exclude the kinds of writing represented here. Nevertheless, the practice among scholars has been to consider in great detail the contributions of male writers while neglecting the range of women writers whose work is so crucial to fully understanding the patterns in nature and community that this genre can represent. Known and unknown, trained and untrained, these women authors experienced nature as intimately connected to culture, and within a culture that failed to take their writings as seriously as those of their male counterparts.[9]

Lawrence Buell argues that "environmental interpretation requires us to re-think our assumptions about the nature of representation, reference, metaphor, characterization, personae, and canonicity."[10] When we begin to dismantle the foundation of the nature/culture dichotomy found in much traditional nature writing, we are invited at the same time to dismantle the genre assumptions that went along with it. Scholars ranging from Roderick Nash (1967) to Buell (1996) have laid the crucial foundation for studying the links between nature writing and American thought and culture. John Elder's collection *American Nature Writers* (1996) marks an important step in documenting the vast treasure house of American nature writers, their work, and their influence.[11] Parallel to their efforts, other writers from Annette Kolodny (1975) to Vera Norwood (1993) have been engaged in the process of examining the gendered nature of this branch of American litera-ture.[12] In Norwood's *Made from this Earth: American Women and Nature,* she states, "American women who took part in nature study recognized that they en-tered terrain controlled by men." At the same time, however, her study makes a powerful case for a "distinctly female tradition in American nature study."[13] Lor-raine Anderson, editor of the anthology *Sisters of the Earth,* writes, "Although I've concluded that there is no such thing as a woman's view of nature, I do think there is a feminine way of being in relationship to nature. This way is caring rather than controlling; it seeks harmony rather than mastery; it is characterized by hu-mility rather than arrogance, by appreciation rather than acquisitiveness. It is available to both men and women, but it hasn't been exercised much in the history of Western civilization."[14]

The essays in this collection reveal how enriching Lyon's "ecological way of seeing" can be when we examine how American women have molded their work to the literary, social, and biological community around them. In exploring these women's lives and writings, these authors make a powerful argument for a re-newed focus on the specific contributions of American women to nature writing. Our understanding of preeminent nature writers such as Henry David Thoreau or John Muir, as well as Rachel Carson or Annie Dillard, is greatly enhanced when we widen the context in which they operated. We need to consider what authors as diverse as Susan Fenimore Cooper, Mary Austin, or Gene Stratton-Porter have brought us from their unique perspectives as women writers over the past century and a half.

Growing both within and away from the male-dominated literary culture of the nineteenth century, women writers found niches of their own in nature essays, re-gional sketches, fiction, and science. Publishing in periodicals, pamphlets, and books of their own, women nature writers enlarged the audience for whom nature mattered in defining an element of the self and of the nation. While critics often la-beled women's nature writing, especially in the nineteenth century, as sentimental, the essays in this collection demonstrate that overt sentimentality often masked a foundation of careful observation and ardent advocacy. With approaches drawn from literary criticism, history, and anthropology, the authors here collectively argue for a recognition of the tremendous diversity and richness of American nature writing, in both fiction and the more traditional nonfiction essay, than we have previously acknowledged.

Many women writers have enjoyed success within the format of the traditional nonfiction nature essay. Lyon's anthology, for example, contains entries from Rachel Carson and Annie Dillard. But as Vera Norwood indicates in her foreword to this work, between and beyond the best-known practitioners in the field, male and female, were a host of lesser-known women authors who wrote movingly and insightfully of the natural world around them. Social historians have already made great strides in this direction by demonstrating how much fuller, richer—and more accurate—our picture is when we are alert to the variety of sources women have employed in their contributions to the historical record.[15] In the realm of nature writing, we now recognize that in addition to the individual journey, the isolated study, and the division of a world into the cultured and the wild, lies a wealth of materials written by women where the natural world is an integral part of everyday existence, where the garden outside the front door supplies an experience as immediate and direct as the mountains in the distance.

And it is this wealth of material that makes structuring a collection such as this a difficult, but rewarding task. The process of reexamining genre and genre expectations brings with it new problems of classification. While many women nature writers have written within (or alongside) the established nature writing tradition, many others, especially contemporary authors, have found alternative formats and modes of expression. So while a number of possible organizational schemes suggest themselves, we have chosen to divide this collection into three parts. Part one lays the foundation for looking at women and the nature writing genre in essays that introduce some of its earliest practitioners as well as the descendants of this literary tradition. Part two looks beyond the traditional venues for nature writing and expands—and challenges—our understanding of the boundaries of the genre. Part three explores the variety afforded by modern women nature writers, and reveals the richness, the complexity, and the scope of modern contributors to a field that is as expansive as it is provocative. In the process, this collection demonstrates both the chronological and the philosophical development of these nature writers and the tradition they have created.

From the carefully wrought drawings of Jane Colden to the environmental activism of Barbara Kingsolver, American women writers have crafted a literature of the natural world in which nature and culture are not separate entities available for separate and "deliberate" living. Rather, they have created a corpus of material that emphasizes how intricately intertwined the natural and the cultural landscape are. In turn, they have provided a rich and varied landscape to "write." If in many ways the writing of these women paralleled the work of their male counterparts, there is also much to suggest that at many key junctures they have been remarkably prescient in identifying issues of gender, ethnicity, and class as intimate factors in our relationship with the land. Offering such news of the land to audiences public and private, trained and untrained, women nature writers wrote of nature in their botanical notes, market bulletins, sentimental fiction, educational pamphlets, and more. Indeed, as the essays in this volume suggest, we must look even further afield to comprehend and benefit from the rich literature of the environment, looking even to the land itself as women write upon the altered earth.

Recognizing that even the act of mere observation changes the meaning of nature, women nature writers have argued tirelessly for preserving the environment, and for decentering the human experience within the natural world. Following their lead, when we expand our "ecological way of seeing" and look more broadly and deeply at women writers, women's writings, and the places they are found, we find that the unique perspective of women offers a rich reward. As the essays in this volume demonstrate, this reward is gained not by isolating nature from human society, but from understanding the natural world as simultaneously beyond, around, and within each of us. Thus when women writers and their readers share such news of the land, we share a part of our selves. Sarah Orne Jewett and Willa Cather, two authors with an intimate knowledge of the landscapes that were inseparable from their stories, recognized this link. In her introduction to *The Country of the Pointed Firs and Other Stories,* Cather wrote that Jewett's stories "melt into the land and the life of the land until they are not stories at all but life itself."[16] And as these women nature writers demonstrate, the stories of their lives and the stories of their land are one and the same. Their lives melt into the land, and the result *is* life itself.

Part I

Laying the Foundation

Although Thoreau's *Walden* (1854) remains the best-known touchstone of the genre, most chronologies of American writing will point to Bartram's *Travels* (1791) as an important precursor and link to an earlier, European tradition. Susan Fenimore Cooper's *Rural Hours,* now enjoying somewhat of a renaissance, pre-dated Thoreau's work by four years and is generally cited as the first fully realized piece of nature writing by an American woman. Few scholars, however, have focused on the role of women nature writers in the period prior to 1850. In his investigation of the eighteenth-century nature writers Eliza Lucas Pinckney, Martha Daniell Logan, and Jane Colden Farquhar, Dan Philippon establishes the important context for the later development of nineteenth-century nature writing. Philippon argues that Pickney, Logan, and Farquhar wrote works that were simultaneously "literary" and "scientific," living liminal lives on the edges of a new continent, a new nation, new sciences, and new roles for women, producing works remarkable in their science and their artistry, a confluence of nature and culture. By tracing the connections of these colonial writers to the more established nature writing tradition in Europe, Philippon underscores the important international aspect of American colonial literature. At the same time, as the title of his essay suggests, these women challenged traditional concepts of gender as well as of genre.

Rick Magee explores Susan Fenimore Cooper's *Rural Hours.* In his examination of Cooper's "sentimental ecology," Magee argues that Thoreau's *Walden,* firmly separating nature and culture, arrived on a national scene prepared, in part, by authors such as Cooper. As Magee demonstrates, Cooper's nature is a tool (used largely by women) that provides a guide to and lesson in domestic values, and her *Rural Hours* emphasizes the interconnectedness of all life on earth. Cooper's contribution to American nature writing, however, has been undervalued compared to that of Thoreau because of the supposedly more "sentimental" nature of her writing. Magee examines the overlap between Cooper's earlier attempts at sentimental fiction and *Rural Hours,* and argues that her nature writing cannot be adequately understood separate from a context that takes into account her knowledge of genre and gender expectation.

Matt Bolinder approaches the question of the response of female nature writers to Thoreau's dominating influence in American nature writing. In his essay, Bolinder analyzes how female authors such as Mary Austin "appropriated" Thoreau's literary legacy and adapted it to a more unforgiving, though no less beautiful, desert landscape in a way distinct from Thoreau or from Austin's contemporary, John Muir. In his reading of Austin's *The Land of Little Rain,* Bolinder places Austin as Thoreau's literary descendent who promoted the significance of *Walden* as a model for future nature writers. At the same time, however, Austin was careful to address its limitations. *Walden* may have provided a touchstone for Austin's *The Land of Little Rain,* but her work demonstrates that she was keenly aware that gender was an important factor that Thoreau had not adequately addressed.

Carol Dickson examines the literary relationship between Mary Austin and her contemporary, John Muir. While Austin and Muir responded to the same shifting cultural attitudes in their writing and shared certain ideas about the relationship between man and nature, as Dickson demonstrates, Muir and Austin exhibit key differences in their use and in their understanding of narrative and language to represent nature. In her reading of Austin's *The Land of Little Rain,* Dickson finds a gendered critique of "patriarchal attitudes towards nature and of narratives." Austin, like many of the women nature writers here, demonstrated an awareness that the language of nature writing itself was problematic. Dickson shows that this awareness distinguishes Austin for anticipating later poststructuralist and ecofeminists concerns for the role of language in constructing the environment.

Daniel J. Philippon

Gender, Genus, and Genre
Women, Science, and Nature Writing
in Early America

In *Sisters of the Earth* (1991), Lorraine Anderson observed that "[t]he canon of nature writing, like the literary canon in general, has been male-centered, both a reflection and a reinforcement of a culture that once defined only men's experience as important and only men's voices as authoritative."[1] While scholars have worked assiduously in the last decade to call attention to women's nature writing from the nineteenth and twentieth centuries, writing about nature by early American women has gone largely unnoticed.

There are several reasons for this oversight. First and foremost, the genre of nature writing has traditionally emphasized nonfiction essays in the Thoreauvian mode, which excludes almost all texts written before the mid-nineteenth century. Even though it predates *Walden* (1854) by four years, the first fully developed work of essayistic nature writing written by a woman—Susan Fenimore Cooper's *Rural Hours* (1850)—is undeniably a mid-nineteenth-century text.[2] And while some critics consider the late eighteenth-century natural history essayists Hector St. John de Crèvecoeur, Gilbert White, and William Bartram to be continuous with the Thoreauvian tradition, most still see these writers as a tradition unto themselves, more as precursors to Thoreau than as his thematic and stylistic contemporaries.[3]

Another reason early American women have been overlooked by scholars of nature writing is that they wrote about nature in a variety of nontraditional literary forms, addressing private as well as public audiences. Native American women, for instance, created oral songs and stories; African-American women composed poems, letters, and petitions; and European-American women wrote diaries, letters, poems, captivity narratives, travel narratives, botanical journals, and garden books, among other kinds of texts.[4]

9

In addition to writing in nontraditional forms, these women often wrote about unconventional subjects, at least from the perspective of conventional definitions of "nature." Cookbooks and narratives of food preparation, for instance, are not usually considered to be "nature writing," yet the human consumption of plants and animals must surely rank among the most complex of human-nature interactions. Likewise, the intimacies of pregnancy and childbirth, including the roles of midwives, are hardly subjects most readers would identify with nature writing, yet what could be more distinctively "natural" than the mysteries of human reproduction?[5]

The frequency and consistency with which early American women referred to the nonhuman world in their writings has been another complicating factor for scholars. How much writing about nature need a text contain, one might ask, before it can be called nature writing? How much writing about nature need a writer produce before he or she can be called a nature writer?

Finally, writing about nature by early American women is also difficult to locate, since many of the books, newspapers, and almanacs in which these women published are now quite rare, and many of the letters and diaries they created now reside—often unpublished and unheralded—in restricted archives and manuscript collections.

Scholarship on early American women and nature therefore has not been extensive. What little of it does exist has been published mainly by literary scholars and historians interested in the environment, natural history, science, and agriculture. This neglect is clearly unjustified, and in this essay I want to suggest that early American women's nature writing deserves our attention because, among other things, it raises questions about the three fundamental subjects of my title: gender, genus, and genre. (All three terms refer to the attempt to categorize objects according to some common attribute—whether they be sexual, scientific, or literary—and all three originate in the Latin *genus,* or "kind.") In what follows, I examine the nature writings of three early American women whose work could be considered both "literary" and "scientific": Eliza Lucas Pinckney (c. 1722–1793), Martha Daniell Logan (1704–1779), and Jane Colden Farquhar (1724–1766).[6] Pinckney, a prolific letter-writer, was the first person to cultivate indigo successfully in South Carolina; Logan, who wrote the first gardening treatise published in America, was a highly respected horticulturist; and Colden, the first woman to write a local flora in America, was the nation's first female botanist.[7] In looking at the writings of these three women, I want to explore further Joan Hoff Wilson's assertion that "[t]he neglect by historians of literature of the books and letters written by Jane Colden, Martha Logan, and others reflects double banishment—banishment of women from science and science from literature."[8]

One of the major tasks of Thoreauvian nature writing, as conventionally defined, is the reconciliation of science and literature, or what C. P. Snow has called the "two cultures."[9] John Elder, for instance, describes the genre as composed of "personal, reflective essays grounded in appreciation of the natural world and of science, but also open to the spiritual meaning and value of the physical creation."[10] This definition, however, is a product of the genre's roots in the mid-nineteenth century, when literature and science had just begun to assume their distinct, modern identities. As Howard Mumford Jones has observed, "During the decades before, say, 1830, science meant, vaguely, any form of organized learning, the 'sciences' slid easily into the penumbra of the 'arts,' and the arts included everything from invention and manufacture to epic poetry and historical painting. The word 'literature' itself, it is too often forgotten, meant virtually anything that took shape as the printed word."[11] Writers in the eighteenth century, in other words, witnessed not only the transformation of amateur natural history into professional science, but also the gradual distinction of professional science from the arts and humanities. The writings of Pinckney, Logan, and Colden—all of whom lived and worked during the mid-eighteenth century—thus straddle the same boundary between science and literature as do the better-known writings of Crèvecoeur, White, and Bartram—all of whom published their work a few decades later in the century.[12]

Unlike these male natural history writers, Pinckney, Logan, and Colden also found themselves engaged in another balancing act that pitted their scientific interests against eighteenth-century norms of female behavior. All three women were the daughters of provincial governors, and as a result they received privileged introductions to the prominent scientific and literary ideas of their day. These introductions, however, were mediated by male authorities.[13] In the case of Pinckney and Colden, the middlemen were their fathers; in the case of Logan, it was William Bartram.[14] The subsequent scientific and literary activities of these women were also heavily influenced by their roles as wives and mothers. For Pinckney and Colden, their marriages brought an end to their agronomical and botanical investigations; for Logan, the need to support her eight children drove her to experiment with plant breeding as well as with writing.

Despite the limitations placed on these women, all of them managed to defy convention and distinguish themselves as both effective scientists and expert rhetoricians. Their investigations into the natural world advanced the state of knowledge in their respective fields, and their written words exhibit a perceptive awareness of both their rhetorical context and their competing interests—as women, as scientists, and as writers. In addition, their attempts to make textual as well as experimental sense of the natural world can also been seen as efforts, comparable to those of other women in other places and at other times, to make sense of themselves—to be among the imposers of taxonomic order, in other words, rather than the ones being imposed upon.[15]

"I love the vegitable world extremly" (Eliza Lucas Pinckney)

The first of these women, Elizabeth Lucas Pinckney, known as Eliza, was born in Antigua, in the West Indies, on 28 December, probably in 1722. Her father, George Lucas, was a major in the British army, stationed in Antigua, where Eliza spent the first few years of her life.[16] She was sent to England to be educated, but in 1738 Major Lucas moved his wife and two daughters to South Carolina, where he intended to settle permanently.[17] Lucas had inherited three Carolina plantations, all located near Charles Town:[18] Wappoo, a six-hundred-acre tract overlooking Wappoo Creek, located seventeen miles outside of Charles Town; Garden Hill, a fifteen-hundred-acre plantation on the Combahee River, located forty miles southwest of Wappoo; and another three thousand acres on the Waccamaw River, located eighty miles to the northeast. Lucas's plans for residence in South Carolina were short-lived, however, because when political conflicts between England and Spain broke out in 1739, he was forced to return to Antigua, leaving his family behind. For the next five years, while her father remained in Antigua (where he was eventually appointed lieutenant governor in 1742), sixteen-year-old Eliza lived at Wappoo, managed its twenty working slaves, and supervised the operation of her father's other two plantations—all while simultaneously caring for her ailing mother, looking after her younger sister Polly, and attempting to remain active in the social affairs of the fledgling colony.

Founded in 1670, Charles Town was the first permanent settlement in colonial South Carolina. Its major cash crop was rice, which dominated the economic life of the low country until the mid-nineteenth century, when exports peaked at about 160 million pounds. Rice was the central crop on the Lucas plantations, as well, but the plantations' estates and slaves were also heavily mortgaged, and so Eliza's father sent several different types of seed to her, in the hope that she might generate some additional income. Other colonists had previously experimented with these seeds, but none had met with commercial success. Eliza succeeded with indigo, however, and her struggles to make the crop productive, though well known to colonial historians, merit discussion here.

As the source for one of the few deep blue dyes available in the eighteenth century, indigo was in great demand by Europe's rapidly expanding textile industries. Demand in England was especially large, and it grew even larger after the outbreak of King George's War in 1740, when English trade routes to the French West Indies were severed, cutting off manufacturers from their major suppliers of indigo. Because the war also disrupted the exportation of rice from South Carolina, the 1740s were a particularly opportune period for Eliza to attempt to master the complicated process of growing and manufacturing indigo.

Around 1739, Major Lucas first sent his daughter the seed of the Hispaniola strain of indigo, *Indigofera tinctoria,* from Antigua, and Eliza began experimenting almost immediately. In July 1740 she wrote to her father of "the pains I had taken to bring the Indigo, Ginger, Cotton and Lucerne and Casada to perfection, and had greater hopes from the Indigo (if I could have the seed earlier next year from the West India's) than any of the rest of the things I had tryd" (*LEP* 8).[19] The next year's crop was damaged by frost, but Eliza was undaunted, telling her father that she had "no doubt [that] Indigo will prove a very valuable Commodity in time" (*LEP* 16). In 1741, she was able to report a small production of "20 weight" of indigo (*LEP* 22), but not until 1744 was she able to make a large enough quantity of the dye to send six pounds to England for evaluation. That year, she also gave her fellow planters some of the seed she had grown, to ensure that a local supply would be available in the future. Only three years later, in 1747, South Carolina exported more than 135,000 pounds of indigo. After 1748, when the British Parliament approved a bounty of six pence per pound on indigo raised in its American colonies, indigo became South Carolina's second most important crop. Exports peaked at about one million pounds in 1775, when the Revolution ended the bounty, but planters were said to have doubled their capital every three or four years during the thirty years when indigo was a major crop. The historian David Ramsay, writing in 1809, indicated the significance of Pinckney's achievement when he noted that indigo "proved more really beneficial to Carolina than the mines of Mexico or Peru are or ever have been either to old or new Spain."[20]

If Pinckney's agricultural legacy can be found in her successful cultivation of indigo, her literary legacy lies in her letterbook, which is one of the most important collections of personal writing ever assembled by a mid-eighteenth-century woman. In this parchment-bound book, Pinckney drafted and recorded the family and business correspondence she sent to her father; the personal letters she sent to her brothers, sons, and former associates in England; and the social notes she sent to her circle of friends in colonial Charleston. Not surprisingly, the letters that most concern the natural world appear early in her letterbook, during the period from 1739 to 1743 when she was managing her father's plantations. Of particular interest are the fourteen letters she sent in 1742 to her friend Mary Bartlett, who was visiting Charleston from England. In these, Eliza's tone is particularly playful and intimate, a fact that may reflect not only her fondness for Bartlett, but also her affection for Charles Pinckney, Bartlett's uncle, who subsequently became Eliza's husband, following the death of his first wife in 1744.[21]

Eliza's likely awareness that whatever she wrote to Bartlett would also find its way to Charles imparts many of these letters with a lively double meaning. This is especially the case when Eliza discusses her various agricultural schemes, which she characterizes as nothing more than romantic mischief. In so doing, she manages to maintain her readers' interest while also deflecting any potential judgment against her unladylike behavior. In one letter, for instance, Eliza dismisses her creation of a new orchard: "I have planted a large figg orchard with design to dry and export them. I have reckoned my expence and the prophets to arise from these figgs, but was I to tell you how great an Estate I am to make this way, and how 'tis to be laid out you would think me far gone in romance. Your good Uncle I know has long thought I have a fertile brain at schemeing. I only confirm him in his opinion; but I own I love the vegitable world extremly. I think it an innocent and useful amusement" (*LEP* 35). A more ambitious scheme to plant and harvest oaks for the Royal Navy—and to keep the profits for herself—receives a similar treatment from her pen: "Wont you laugh at me if I tell you I am so busey in providing for Posterity I hardly allow my self time to Eat or sleep and can but just snatch a minnet to write to you and a friend or two now. I am making a large plantation of Oaks which I look upon as my own property, whether my father gives me the land or not; and therefore I design many years hence when oaks are more valueable than they are now—which you know they will be when we come to build fleets" (*LEP* 38).

The combination of agricultural and literary pursuits that marked Eliza's life also appears in her correspondence, particularly when she comments on a volume of Virgil that she recently received from Charles:

I have got no further than the first volume of Virgil but was most agreeably disapointed to find my self instructed in agriculture as well as entertained by his charming penn; for I am persuaded tho' he wrote in and for Italy, it will in many instances suit Carolina. I had never perused those books before and imagined I should imediately enter upon battles, storms and tempest that puts one in a maze and makes one shudder while one reads. But the calm and pleasing diction of pastoral and gardening agreeably presented themselves, not unsuitably to this charming season of the year, with which I am so much delighted that had I but the fine soft language of our poet to paint it properly, I should give you but little respite till you came into the country and attended to the beauties of pure nature unassisted by art. The majestick pine imperceptably puts on a fresher green; the young mirtle joyning its fragrance to that of the Jesamin of golden hue perfumes all the woods and regales therural wander[er] with its sweets; the daiseys, the honeysuckles and a thousand nameless beauties of the woods invite you to partake the pleasures the country affords. (LEP 35–36)

She continues, describing how Virgil has influenced not only the content of her correspondence, but also the products of her plantation: "You may wonder how I could in this gay season think of planting a Cedar grove, which rather reflects an Autumnal gloom and solemnity than the freshness and gayty of spring. But . . . I intend . . . to connect in my grove the solemnity (not the solidity) of summer or autumn with the cheerfulness and pleasures of spring, for it shall be filled with all kind of flowers, as well wild as Garden flowers, with seats of Camomoil and here and there a fruit tree—oranges, nectrons, Plumbs, &c., &c." (*LEP* 36).

The interrelationship of gender, genus, and genre in Eliza's letters is especially evident when, in another letter to Bartlett, she describes the song of the mockingbird in a poem she wrote while lacing her corset:

I promised to tell you when the mocking bird began to sing. The little warbler has done wonders; the first time he opened his soft pipe this spring, he inspired me with the spirit of Rymeing, and [I] produced the 3 following lines while laceing my stays:

> Sing on thou charming mimick of the feathered kind
> and let the rational a lesson learn from thee,
> to Mimick (not defects) but harmony.

If you let any mortal besides your self see this exquisite piece of poetry, you shall never have a line more than this specimen; and how great will be your loss you who have seen the above may jud[g]e. (LEP 38–39)

Perhaps most revealing, though, is a letter in which Eliza responds to Bartlett's query about the appearance of a comet in the pre-dawn sky. In it, she not only declares that such a "glorious" celestial body could hardly be male, but she also reveals her latent sexuality, almost to the point of inviting male pursuit: "The brightness of the Committ was too dazleing for me to give you the information you require. I could not see whether it had petticoats on or not, but I am inclined to think by its modest appearance so early in the morning it wont permitt every Idle gazer to behold its splendour, a favour it will only grant to such as take pains for it—from hence I conclude if I could have discovered any clothing it would have been the female garb. Besides if it is any mortal transformed to this glorious luminary, why not a woman" (*LEP* 31).

Although the extent of Eliza's awareness of her audience can never be fully known, Charles Pinckney may well have been taken in by the teasing mixture of revelation and concealment that seems to have been Pinckney's trademark. When Major Lucas sent his oldest son, George, to escort his family back to Antigua in 1744, Charles Pinckney, then a widower, proposed to Eliza, and the two were married on 27 May 1744. He was forty-five, she twenty-one.

After her marriage, Eliza continued her agricultural experiments (trying her hand at the cultivation of flax, hemp, and silk), but her letters turned increasingly to the social realm and to the moral education of her two sons.[22] Her achievements with indigo, however, and the subsequent services she and her children rendered to South Carolina during the Revolution, were never forgotten by the leaders of the new nation. When she died on 26 May 1793 in Philadelphia, where she had gone to seek treatment for breast cancer, George Washington himself asked to be a pallbearer at her funeral.

"Her garden is her delight" (Martha Daniell Logan)

Born one generation earlier than Eliza Pinckney, Martha Daniell Logan was another prominent resident of colonial Charleston, which was quickly becoming a center for scientific investigation in the South.[23] Much less is known about Logan's life than Pinckney's, but their biographies do share a few characteristics. Like Pinckney's father, George Lucas, Logan's father, Robert Daniell, traveled from the West Indies to take up residence in South Carolina. In 1679, he moved from Barbados to Charleston, where he received the title of "Landgrave" from the Lords Proprietors, which allowed him to acquire forty-eight thousand acres. Martha, who was born in Charleston, probably on 29 December 1704, was the second of the four children and three daughters born to Daniell and his second wife, Martha Wainwright. Richard Beale Davis has speculated that Martha may have been educated in England, and this speculation is not unfounded, considering the important role her father played in the early life of the colony.[24] Daniell helped to revise the Fundamental Constitutions in 1698, was second in command of the South

Carolina expedition against St. Augustine in 1702, and served two terms as lieu-tenant governor of South Carolina from 1715 to 1717. When he died in May 1718, however, Martha was only thirteen. The next year, she married George Logan, Jr., on 30 July 1719—only two months after her mother married George Logan, Sr.

The younger Logans spent the early years of their marriage on a plantation on the Wando River, about ten miles from Charleston, on land that Martha had inher-ited from her father. Between 1720 and 1736 they had eight children, six of whom lived to adulthood. Such a large family must have strained the couple's finances, because advertisements in the *South Carolina Gazette* show that Martha was working for at least twenty years prior to her husband's death on 1 July 1764. (His occupation is unknown.) A notice on 20 March 1742, for instance, announced that "Any Persons desirous to board their Children . . . may have them taught to read and write, also work plain Work, Embroidery, tent and cut work" by Logan.[25] By the end of the 1740s, she and her family must have moved back to Charleston from their Wando River plantation, because on 6 March 1750 Logan advertised "that she is pleasantly situated near Mrs. Trotts point, and assigns on April next to open a school."[26] She continued to advertise her teaching at this location through at least August 1754.[27]

In addition to teaching, Logan's other occupation was gardening, which she seems to have begun in earnest following her return to Charleston. A notice in the *Gazette* for 12 November 1753 announces: "Just imported from London and to be sold by [Robert] Daniel Logan, at his Mother's house on the Green, near Trotts point, a parcel of very good seeds, flower roots, and fruit stones of several kinds."[28] Whether Robert was the sole proprietor of this plant nursery or was working as an agent for his mother is unclear, but at some point Martha assumed full control of the business, perhaps after her son's death. Ann Manigault wrote in her diary in 1763 that she "went to Mrs. Logan's to buy roots," and an advertise-ment in the *Gazette* for 15 February 1768 reports:

Just imported from Capt. Lloyd from London and to be sold very reasonably by Martha Logan at the house in Meeting-street, three doors without the gate:
    A fresh assortment of very good garden seeds and flower roots: also many other sorts of flowering shrubs and box edging beds now growing in her garden.[29]

More than just a business venture, Logan's garden was also an important source of plants and seeds for John Bartram, one of the eighteenth century's most industri-ous botanists. Logan first met Bartram in Charleston in the spring of 1760, while Bartram was on one of his many collecting expeditions south of his home in Kingsessing, Pennsylvania, outside Philadelphia. Logan gave him a tour of her garden, and soon thereafter the two began exchanging letters, seeds, and plants. Bartram described their initial encounter, and subsequent correspondence, in a 22 May 1761 letter to the English plant collector Peter Collinson. Bartram identified Logan as "an elderly widow Lady," saying that she:

spares no pains or cost to oblige me. Her garden is her delight & she hath a fine [one]. I was with her about five minits in much company yet we contracted much mutual Correspondence that one silk bag hath past & repast full of seeds several times last fall. I desired her last march to send me some seeds of ye hors sugar or yellow leaf. She directly sent me A box with 2 fine growing plants mixt with several other sorts that she thought would pleas & paid freight with promises to send any vegitable in her power to procure & thay thrive finely.[30]

Collinson replied on 1 August 1761, noting that he, too, had observed the eagerness of many women to embark on botanical studies. At the same time, though, he displayed the common male belief (mocked by Eliza Pinckney) that such studies are but "amusements" to women, who are fated to be nothing more than mere "assistants" to real botanists: "I plainly see thou knowest how to fascinate the Longing Widow by so close a Correspondence—When the Women enter into these amusements I ever found them the best assistants, now I shall not wonder if thy Garden abounds with all the Rarities of Carolina" (*CJB* 530).

Logan's own dedication to botany is evident from Bartram's statement in a 30 May 1763 letter to Collinson, in which he noted that Logan "will pass thro fire or water to get any curiosity for mee alltho our personal acquaintance was but A few minits & with much company" (*CJB* 594).

One of the more interesting aspects of the eight extant letters Logan sent to Bartram between 1760 and 1765 is Logan's reference to the many male *and* female gardeners working in Charleston, including not only the well-known botanist and physician Alexander Garden, but also a Mrs. Wragg, Mr. Raper, Mr. Glen, Mrs. Hopton, Mr. Ratlive, and Mrs. Bee (*CJB* 519, 547–48). Clearly, Logan was part of a network of Carolina gardeners in regular contact with one another. With the exception of Alexander Garden, however, she alone corresponded with Bartram, which indicates her prominent position in this horticultural community.

The appearance of Logan's "Gardener's Calendar" in John Tobler's 1752 *South Carolina Almanack*—the first almanac printed in South Carolina—must have both reflected and helped to enhance Logan's identity as an expert in the garden.[31] Logan revised and enlarged her calendar throughout her life, and versions of it appeared intermittently in various almanacs for more than fifty years. The earliest extant copy is from the *South Carolina Almanack* for 1756, where it appeared as "Directions for managing a Kitchen-Garden every month in the Year. Done by a Lady." Although the calendar did not appear under Logan's own name until seventeen years after her death—when it was published as "Gardener's Calendar, by Mrs. Logan; Known to succeed in Charleston, and its Vicinity, for many Years" in the 1796 *Palladium of Knowledge: or, the Carolina and Georgia Almanac*—Logan's identity as the author was probably common knowledge, at least among Charleston residents.[32] According to David Ramsay, Logan's calendar regulated the practice of gardeners in and around the city as late as 1809.[33] The calendar may also, therefore, have helped to boost the sale of plants, roots, and seeds from Logan's nursery—if not by solidifying her own identity as a master gardener, then at least by encouraging increased horticultural activity throughout colonial South Carolina.

Sarah Pattee Stetson observes that most of the gardening manuals published in eighteenth-century America "were not only the condensation and adaption of English books to American needs; they were also the result of practical experience and of long and painstaking observation."[34] Such is certainly the case with Logan's "Gardener's Calendar," which must have been an extremely useful tool in its time. A practical manual, concerned largely with the kitchen garden (although it also mentions roses, flowering shrubs, and ornamental trees), Logan's calendar describes the common eighteenth-century practices of planting and fertilizing— including several references to the supposed influence of lunar cycles on vegetation.[35] In it, Logan demonstrates both a sensitivity to the local conditions in and around Charleston and an intimate knowledge of the plants and agricultural techniques needed to succeed in such an environment.

If Logan's identity as the "Lady" who wrote the "Gardener's Calendar" was in fact common knowledge in Carolina, then the calendar's assertive style, in addition to its practical content, may have helped Logan to refine her identity as an authority on Southern gardening. Logan's directions brook no disagreement, and only once does she cite another authority—"a gentlemen [sic] in Carolina"—but even his expertise is filtered through her own. Some of her gardening instructions echo Benjamin Franklin's moralistic maxims in *Poor Richard's Almanac:* "[I]f your trees have not been pruned," she writes in February, "they must not be neglected now."[36] Others, however, are more lenient. In March she advises: "What was neglected last month, may be successfully done in this." And in June: "If you have lost the last seasons for sowing carrots and parsnips, you may now sow them." Interestingly, Logan's use of the second person throughout her calendar obviates her need to specify the gender of her readers, a circumstance that not only may have helped both men and women feel welcome in Logan's garden, but also may have helped Logan to solidify her authority. More than a woman speaking only to other women, she instead became a gardener speaking to other gardeners.

"This young lady merits your esteem" (Jane Colden Farquhar)

In 1755, three years after Martha Logan began publishing her "Gardener's Calendar," her friend and colleague Alexander Garden was on his way to Philadelphia when he discovered what he believed to be a new plant about a mile outside of New York City. He sent a description of the plant to Jane Colden, whom he had met the previous summer at Coldengham, her father's home, in what is now the town of Montgomery, in Orange County, New York. Colden, who was assembling a flora of New York, replied that she had already catalogued the plant as number 153 of her collection, but that, "using the privilege of a first discoverer," she would name it *Gardenia* in his honor. Hoping to make the designation official, Garden sent Colden's description of the plant to the European botanists Robert Whytt in Edinburgh and John Frederic Gronovius in Leiden. That description became Colden's only known publication when it appeared, under Garden's name, as "The Description of a New Plant" in the Philosophical Society of Edinburgh's *Essays and Observations, Physical and Literary* in 1756.[37] Unfortunately for Garden, Colden's name for the plant was not accepted, since the plant itself turned out not to be new. It was a *Hypericum,* a St. Johnswort, probably *Hypericum virginicum.*[38]

That Colden's description was acknowledged at all by the all-male international botanical community was no small feat, considering the views of such men as Peter Collinson. What impressed Collinson and others, however, was not Colden's interest in identifying plants, nor her familiarity with their common names, but rather her knowledge of the new binomial system of taxonomy created by the Swedish botanist Carolus Linnaeus.

More than just an important development in science, the appearance of the Linnaean system was also a significant moment in the history of nature writing. As David Rains Wallace has observed, before Linnaeus, "nature was generally seen in two dimensions. It was a backdrop to a historical cosmos, or a veneer over a religious one." After Linnaeus, however, "nature rapidly acquired a new substantiality, and became a subject as well as a setting."[39] Pamela Regis agrees, arguing that the appearance of the Linnaean method changed the rhetorical context of early American natural history, providing its practitioners with the means to identify and celebrate the uniqueness of the new nation by distinguishing its flora and fauna, as well as its scientific pursuits, from those occurring in Europe.[40] And Robert Finch and John Elder trace Linnaeus's influence into the present, noting that all nature writers are in fact "the children of Linnaeus."[41]

Jane Colden was able to master the Linnaean system thanks to the unique situation in which she found herself, as the fourth child and second daughter of the accomplished philosopher-politician Cadwallader Colden and his wife, Alice Chrystie. Cadwallader Colden, who served as Surveyor General, Lieutenant Governor, and—on several occasions—Acting Governor of the Province of New York, was also a physician, a physical scientist, and an authority on American Indian history. But it was his work as a botanist that most influenced his daughter's choice of occupation.

Four years after Jane (or Jenny, as she was often called) was born on 27 March 1724 in New York City, her father moved his family north to Coldengham in 1728. Although her mother was responsible for educating Jane in the social and domestic realms, her father took it upon himself to instruct her in the sciences. He recounted the circumstances of her scientific education in an 18 October 1757 letter to John Fothergill of London: "When I removed my family into the country & thereby my children were deprived of all those amusements in which young people take delight I thought the putting them at some research which would fix their attention & at the same time please their fancy might remove that disgust to their present situation which I apprehended otherwise could not be avoided."[42]

Jane's education in botany probably began sometime in the early 1750s, following her father's own initiation into the new Linnaean system. Cadwallader had studied botany at the University of Edinburgh, but admitted that he understood only its rudiments. Around 1742, however, when he encountered a copy of Linnaeus's *Genera Plantarum* (1737), he began compiling a list of the plants growing around Coldengham. After cataloging 141 native species in the Linnaean method, he sent the list to Gronovius, who then forwarded it to Linnaeus himself. The catalog so impressed the Swede that he published it in Uppsala as *Plantae Coldenghamiae* (1749), the first work of its kind to be completed by an American.[43] Soon thereafter, politics took Colden away from botany, but by the time he found the leisure to pursue it again in the 1750s, he was nearing seventy and no longer able to botanize. As he wrote to Gronovius on 1 October 1755, "My eyes so far fail me that I cannot now with sufficient accuracy examine the parts of fructification nor can I bear the fatigue which accompanies Botanical researches" (*LPCC* 5: 29).[44]

It was thus fortuitous that Jane, still unmarried in her late twenties, was as desirous of occupation as she was capable of scientific pursuits. Her father memorably described the process by which she learned the Linnaean system in the same 1 October 1755 letter to Gronovius:

I thought that Botany is an Amusement which may be made agreable for the Ladies who are often at a loss to fill up their time if it could be made agreable to them. Their natural curiosity & the pleasure they take in the beauty & variety of dress seems to fit them for it. The chief reason that few or none of them have hitherto applied themselves to this study I believe is because all the books of any value are wrote in Latin & so filled with technical words that the obtaining the necessary previous knowledge is so tiresome & disagreable that they are discouraged at the first setting out & give it over before they can receive any pleasure in the pursuit.

I have a daughter who has an inclination to reading & a curiosity for natural phylosophy or natural History & a sufficient capacity for attaining a competent knowledge. I took the pains to explain Linnaeus's system & to put it in English for her use by fre[e]ing it from the Technical terms which was easily don[e] by useing two or three words in place of one. She is now grown very fond of the study and has made such progress in it as I believe would please you if you saw her performance. Tho' perhaps she could not have been persuaded to learn the terms at first she now understands in some degree Linnaeus's characters notwithstanding that she does not understand Latin. (*LPCC* 5: 29–30)[45]

As a result of Jane's progress, when John and William Bartram visited Coldengham in the summer of 1753, they were able to look over "some of ye Dr daughters botanical curious observations," as John Bartram wrote to Collinson later that fall (*CJB* 360).

Jane Colden's achievement clearly impressed every botanist who learned of it, but none more than Peter Collinson. For two years, Collinson and the English botanist John Ellis attempted to gain recognition for Colden from Linnaeus, a task at which they were only partly successful. In a 12 May 1756 letter, Collinson wrote Linnaeus that Cadwallader Colden was well, "but, what is marvellous, his daughter is perhaps the first lady that has so perfectly studied your system. She deserves to be celebrated."[46] The following year, on 1 April 1757, he informed Linnaeus about the publication of Colden's plant description in the Edinburgh *Essays,* noting that she was "perhaps the only lady that makes profession of the Linnaean system; of which you may be proud" (*SCL* 1: 40). Five days later, on 6 April, Collinson wrote to her father, noting that Linnaeus was "not a Little proud" of "the Only Lady that I have yett heard off that is a proffessor [of] the Linnaean System" (*LPCC* 5: 139). The next year, on 25 April 1758, Ellis attempted to gain a more lasting form of recognition for her, sending Jane's description of what she thought might be another new plant to Linnaeus and asking him to name it after her:

Mr. Colden of New York has sent Dr. Fothergill a new plant described by his daughter; I shall send you the characters as near as I can translate them. It is called Fibraurea, Gold thread . . .

This young lady merits your esteem, and does honour to your System. She has drawn and described 400 plants in your method only: she uses English terms. Her father has a plant called after him Coldenia, suppose you should call this Coldenella or any other name that might distinguish her among your Genera. (*SCL* 1: 95)

Five days later, on 30 April, Collinson sent a follow-up letter to Linnaeus, citing Ellis's correspondence and continuing his lobbying campaign on behalf of Jane: "As this accomplished lady is the only one of the fair sex that I have heard of, who is scientifically skilful in the Linnaean system, you, no doubt, will distinguish her merits, and recommend her example to the ladies of every country" (*SCL* 1: 45). Unfortunately, as Linnaeus later informed Ellis, Colden's *Fibraurea* had already been described. Nevertheless, Linnaeus seems to have been complimentary of Colden, as Ellis's 21 July 1758 reply to him suggests: "I shall let her know what civil things you say of her—her Christian name is Jane" (*SCL* 1: 98).[47]

For seven or more years, from her late twenties to her marriage at thirty-five, Jane Colden remained an active member of the scientific community, exchanging specimens and seeds with the Bartrams, Garden, Collinson, and Gronovius. John Bartram himself considered her on equal terms with his European counterparts, telling her in his 24 January 1757 letter: "I received thine of october ye 26th 1756 & read it several times with agreeable sattisfaction. Indeed I am very carefull of it & it keeps company with ye choicest correspondents, ye european letters" (*LJB* 414).[48] In 1756, she taught botany to the fourteen-year-old Samuel Bard, who later became a prominent New York physician.[49] That year she also kept a "Memorandum of Cheese" in which she methodically described the ingredients and conditions under which she produced a large variety of cheeses—itself a kind of controlled scientific experiment (*LPCC* 5: 55–63).[50] After she married Dr. William Farquhar, a forty-year-old Scotsman and widower, on 12 March 1759, however, Jane's scientific studies appear to have ceased. Like her contemporary, Eliza Pinckney, Jane must have found her life transformed by the obligations that accompanied a traditional eighteenth-century marriage. After seven years of an apparently happy domestic life, Colden died, probably in New York City, on 10 March 1766, a few weeks shy of her forty-second birthday, and her only child died later that same year.[51]

Despite what these scattered references to her in the letters of family members and scientific correspondents might suggest, Colden's literary legacy is in fact quite substantial. Her botanic manuscript, which now resides in the British Museum, contains drawings of 340 native plants, many of which are accompanied by descriptions in English, rendered in the Linnaean style.[52] What distinguishes Colden's manuscript most, however, is not its employment of the new Linnaean taxonomy (impressive though that achievement is), but the many characteristics it shares with the herbals of the English Renaissance.

In the earliest western form of the herbal, ancient Greek and Roman writers described and catalogued the plants of Mediterranean for their uses in medicine, cookery, and manufacture. In most cases, these herbalists saw plants mainly as "simples," or simple ingredients for compound recipes, and thus were not concerned with the question of botanical classification. In the Middle Ages, when these ancient texts all but disappeared from view, plant lore was preserved in manuscript form by male physicians and in oral form by lay women healers. In the Renaissance, however, when humanistic scholars rediscovered the classical herbals of Theophrastus, Dioscorides, and other writers, the folk wisdom of women increasingly came into conflict with the alphabetic literacy of men, as male physicians and apothecaries increasingly sought to document the good and ill effects of their own native plants. Systematic observation and identification of local plants for their own sake, improved illustration and description, the search for new medicines, and the discovery of new plants as a result of global exploration all contributed to the growth of a rational, mathematical, observational view of nature from which modern medicine and botany gradually emerged. "As time went on," notes Agnes Arber, "the *herbal,* with its characteristic mixture of medical and botanical lore, gave way before the exclusively medical *pharmacopoeia* on the one hand, and the exclusively botanical *flora* on the other. As the use of home-made remedies declined, and the chemist's shop took the place of the housewife's herb-garden and still-room, the practical value of the herbal diminished almost to [the] vanishing point."[53]

Despite its principal purpose as a botanical flora, Colden's manuscript resembles the Renaissance herbals in several important ways. First, like these earlier herbals, Colden's manuscript reflects the influence of texts from other lands, which offered Colden models for her collection and illustration of the native plants of Coldengham. Around October 1755, her father attempted to acquire several eighteenth-century herbals for her from England, telling Peter Collinson: "As she cannot have the opportunity of seeing plants in a Botanical Garden I think the next best is to see the best cuts or pictures of them for which purpose I would buy for her Tourneforts Institutes & Morison's Historia plantarum, or if you know any better books for this purpose as you are a better judge than I am I will be obliged to you in making the choice" (*LPCC* 5: 37). Eighteen months later, on 6 April 1757, Collinson finally responded, informing Cadwallader, "I have att last been So luckky to gett you a fine Tournforts Herbal & his History of Plants . . . [translated by] Martin in Excellent preservation" (*LPCC* 5: 139).[54]

Jane had already been "shewn a method of takeing the impression of the leaves on paper with printers ink by a simple kind of rolling press" as her father's 1 October 1755 letter to Gronovius makes clear (*LPCC* 5: 30). That she was also appreciative of botanical illustration can be surmised from a 24 January 1757 letter from John Bartram to Jane herself, in which Bartram notes that his son William was "so well pleased" with her last letter "that he presently made A pockit of very fine drawings for the[e] far beyond Catesbys, took them to town & tould me he would send them very soon"(*CJB* 414).[55] It was not until after her receipt of Tournefort's *Herbal* and *History of Plants,* however, that her father began to sing her praises as an illustrator more widely. "When it is considered that she has no instructor in drawing has few or no good copies & was only shewn how to use china ink with a pencil you will easily pardon where she has failed in the art & yet allow her some genius for that kind of drawing," he wrote to John Fothergill on 18 October 1757 (*LPCC* 5: 203). And later, in a 15 February 1758 letter, probably to Robert Whytt, he commented: "considering that she had no instructor the proficiency she has made & the justness of her figures surprise those who have seen them" (*LPCC* 5: 217). Walter Rutherfurd, who was one of those who had seen them during his visit to Coldengham in the mid-1750s, noted that Colden "draws and colors . . . [her plants] with great beauty."[56]

Although modern writers have been less generous about Colden's drawings, these critics have used other eighteenth-century floras as their standards of evaluation, rather than the sixteenth- and seventeenth-century herbals whose illustrations Colden's drawings more closely resemble. Raymond Phineas Stearns, for example, finds Colden's figures "neat, but as they lack flowers and root structures, as well as suffient detail of stems and leaves for present-day botanical use, they are more curious than valuable."[57] James Britten more bluntly describes Colden's drawings as "very poor," Buckner Hollingsworth says they are "wholly without artistic merit," Joseph Ewan labels them "caricatures" executed in a "crude style," and H. W. Rickett calls them "poorly executed" and "of no use for identification."[58] Even the feminist critic Vera Norwood, who argues that Colden's "actual drawings were not as important than the act of making them," admits that they are "more decorative than accurate" and sees in them a demonstration of "how the art of floral reproduction became a fitting female occupation."[59] If Colden's drawings are compared to those of her more practiced contemporaries, such as

Mark Catesby and William Bartram, who drew their plants from life, they are sure to disappoint. If, however, they are compared to those of the Renaissance herbalists, who usually drew their plants from pressed specimens, as did Colden, her drawings are much more deserving of note.[60]

Colden's manuscript also resembles the herbals of the English Renaissance in its documentation of orally transmitted remedies in both rural and Native American communities. At the end of several of her plant descriptions, Colden notes the utility of these plants in medicine, a habit that reflects not only her father's occupation as a physician but also the historic relationship of medicine to botany. She observes, for instance, that Lousewort *(Pedicularis canadensis)* "is call'd by the Country People Betony[;] they make Thee [tea] of the Leaves, et use it for the Fever & Ague et for sikness of the Stomak" *(BM* 46). And she writes of the Mountain-mint *(Pycnanthemum incanum)* that "the Country people here make a Tea of the Leaves & use it for pain or sickness at their stomach" *(BM* 167).

By documenting this folk wisdom, Colden is also participating in the appropriation of these remedies by the emerging medical establishment—another characteristic that her manuscript shares with the Renaissance herbals. She writes that the root of the Butterfly-weed *(Asclepias tuberosa),* "taken in powder, is an excellent cure for the Colick, about halff a Spoonfull at a time. This cure was learn'd from a Canada Indian, & is calld in New England Canada Root. The Excellency of this Root for the Colik is confirmed by Dr. Porter of New England, and Dr. Brooks of Maryland likewise confirm'd this" *(BM* 80). Seneca snakeroot *(Polygala senega),* she notes, is also "much used by some Physicians in America, principally Long Island, in the Pleurisy, especially when it inclines to a Perip neumony, they give it either in Powder or a Decoction" *(BM* 53). And of the Virginia snakeroot *(Aristolochia serpentaria),* she says similarly, "This is the S'erpentaria used in the S'hops, the Virtues of it is generally known" *(BM* 185).[61]

The complexity of Colden's identity as a woman botanist in early America, as well as her careful negotiation of that identity, may best be illustrated by a letter she wrote to the Edinburgh botanist Charles Alston on 1 May 1756. In it, she wrestles with the expectations raised about her by Alexander Garden, minimizing her own achievements in comparison to those of her European counterparts. Alston's correspondence with her, she demurs, could not possibly be as colleague to colleague, but only as teacher to student. Nevertheless, while couching her findings in deferential language, she boldly suggests that she may in fact have discovered a new plant. Mindful of her previous experience with the gardenia, though, she begs Alston not to make her speculations public until they can be verified against those of John Clayton and Gronovius in the *Flora Virginica.*[62] In short, she demonstrates in this letter not only her considerable abilities and ambitions as a scientist, but also her equally worthy skills as a rhetorician. She writes:

Sir I have the unexpected honour of your very obliging favour of the 8th of Septr last, but blush to think how much you will be disappointed in the expectations Dr.

Garden has raised in you, if ever you should have a further knowledge of me, his fondness for incouraging the Study of Botany, made him pass the most favourable judgement on the little things I have done in that way.[63] It must be no less a desire to incourage the Study of this delightfull Science, could induce you to condescend to honour me with an invitation to correspond with you. If I can in any way serve you, either by sending you Seeds, or the Characters (such as I am able to form) of our Native Plants, I shall always be proud of obeying your Commands; And shall at this time, so far presume on the liberty you give me as to send you the Characters of a Plant which my Father tells me, he takes to be the same that Dr Gronovius in his Flora Virginica calls Panax foliis ternis tornatis Page 147. I shall be glad to know whether you think it a new Plant, if not, to what Genus you suppose it belongs; To which if you will please to add any Corrections, or be so good as to give me any In-structions in Botany, they will be most thankfully received.

You conplasantly intimate that anything that I shall communicate to you, shall not be conceald, but this I must beg as a favour of you, that you will not make any thing publick from me, till (at least) I have gained more knowledge of Plants, and then perhaps I shall be able to make some amendments to my Discriptions. I am

Sir

Your greatly Obliged &
Most Hum[bl] [e] Serv[t]
Jane Colden[64]

Living on the Edge: Early American Women Nature Writers

Greg Dening has suggested that "liminality" was a common characteristic of early American identity. "Early America was a place of thresholds, margins, boundar-ies," he writes. "It was a place of ambivalence and unset definition. The search for identity in that place was multivalent and unending . . . Edginess is the quintessen-tial feel of early America."[65] That same edginess is evident in nature writing, and the tension the genre exhibits between science and literature, and between profes-sional and amateur perspectives, shares a great deal with the tenor of eighteenth-century America. John Elder, for one, has observed the similarity of nature writing to the "ecotone," or the transitional zone between two ecological communities. These ecotones, Elder notes, "contain more species than are found in either con-stituent ecosystem, and they have a greater density of organisms within a given species."[66] It might similarly be said that the peculiar richness of early American women's nature writing exists in part because these women were "living on the edge": not only the edge of a new continent and a new nation, but also the edge of their opportunities as women, the edge of the new scientific disciplines, and the edge of the genres in which they chose to write.

While the edginess of these writers is particularly grounded in their historical moment, their texts also intermingle attention to practical and aesthetic concerns in the same way as do texts written by the next generation of American women na-ture writers. On the one hand, early American women often represent the natural world in terms of its uses: Pinckney sees it as a source of future economic prosperity, Logan as a garden for food production, and Colden as a storehouse for medicine. On the other hand, they also display an awareness of the natural world as a subject for aesthetic contemplation: Pinckney's letters feature elements of Romanticism, with their focus on the individual's emotional and intuitive experience of nature; Logan's calendar includes instructions for cultivating ornamental flowers, shrubs, and trees, as well as those for maintaining a kitchen garden; and Colden's manuscript illustrations are as much artistic creations as they are the means for plant identification. The nature writings of early nineteenth-century women—such as Anne Grant, Lucy Hooper, and Margaret Fuller—demonstrate this inextricable entanglement of nature and culture in a remarkably similar fashion. The increasing emphasis on aesthetics over science in nineteenth-century American women's nature writings, however, can be seen as evidence not only of the professionalization of natural history, but also of the marked gap between women's and men's scientific education. Not until the late nineteenth century, and the reappearance of amateur science in the writings of Florence Merriam Bailey, Neltje Blanchan Doubleday, Olive Thorne Miller, and Mabel Osgood Wright, do a group of women nature writers exceed the collective achievements of Pinckney, Logan, and Colden in the mid-eighteenth century.

Although these three women have previously received some degree of recognition, however small, for their scientific activities, their importance as nature writers has yet to be fully articulated. Just as all three women demonstrated to their contemporaries that their gender need not limit their behavior, their activities as writers can also help modern readers to recognize not only that women *were* writing about nature in the eighteenth century, but also that the manner in which they did so continues to be used today—in e-mail correspondences, in local gardening guides, and in the nature journals of amateur natural historians. To ignore these modes of nature writing, I would argue, would be to limit the genre in the same way that the restrictive social norms of the eighteenth century once hindered the full acceptance of these women as scientists.

Richard M. Magee

Sentimental Ecology
Susan Fenimore Cooper's *Rural Hours*

Susan Fenimore Cooper, the daughter of the famous novelist, was an early voice in the tradition of American nature writing, publishing her nature journal *Rural Hours* four years before Thoreau's *Walden* appeared.[1] This journal of her life in rural Cooperstown is vastly important in the tradition of American nature writing, as it was one of the first American natural histories and the first written by a woman. She acknowledges her thematic debt to male European writers such as Gilbert White and John Leonard Knapp, but claims the genre both for American writers and for women writers.[2] Before writing *Rural Hours,* Cooper tried her talents in a more traditional forum for women writers with her domestic-sentimental novel *Elinor Wyllys; or, The Young Folk of Longbridge.* Many of the concerns, themes, and tropes in this novel prefigure the passionate reverence for nature's beauties that informs much of *Rural Hours,* a foreshadowing that tells us much about the motivating forces and structure of the environmental genre. For Cooper, rural values and domestic values are congruent, if not always identical. Domestic values are brought into the wilderness by enlightened and concerned men and women and are passed on largely through the educational efforts of mothers who use nature as a lesson or guide for their children. Thus, domestic values are enhanced by the wilderness, which is moderated but not urbanized.

Susan Cooper's interest and participation in the two genres—the domestic/sentimental and the natural history/environmental—suggest that the overlap in her writings may be more than historical coincidence. Certainly, the most popular and economically successful books written during the middle of the nineteenth century were those written by and for women—Hawthorne's "d——d mob"—so Cooper's entry into the market with *Elinor Wyllys* is not unusual. On the other hand, her subsequent entry into the natural history or environmental genre was without precedent. Ironically, the novel, in spite of its firm grounding in the popular sentimental mode, did very poorly, while *Rural Hours* was successful enough to warrant numerous editions, including a "fine" gift edition with color illustrations. What makes Cooper's foray into domestic fiction important to students of environmental literature is the dramatic manner in which the language of the sentimental informs so much of her later work, so that any study of Cooper's nature writing that ignores her domestic writing is incomplete. Her domestic-influenced nature writing, in fact, indicates an important sub-genre, which I call sentimental ecology, whereby the demands of community and domestic life are intertwined, much like models of ecosystems, with the demands of the natural environment.

Susan Cooper's voice or author's persona also places *Rural Hours* firmly within the sentimental narrative of community, to use Sandra Zagarell's term. Throughout the journal, she consistently uses the "we" rather than the "I" to talk about her observations; this marked difference between Cooper and Thoreau, says Lawrence Buell, makes her eye "public, not merely idiosyncratic."[3] By creating this more public persona, Cooper offers an alternative to the narrative of self, an alternative whereby the "self exists . . . as a part of the *interdependent network* of the community rather than as an individualistic unit."[4] The narrative "we" constructs a linguistic counterpart to the ecological formulation that all life forms a complex web of interrelations dependent upon each other for survival. Arcadian ecology, the scientific notion expounded by Gilbert White at the end of the eighteenth century, states that "each organism, no matter how insignificant to human eyes or in the human economy, [has] a role to play in nature's economy."[5] White's view was still current at the time that Cooper was writing, and accurately reflects her attitude toward the natural world that surrounded her at Cooperstown. The connection between the sentimental ethos of human networks and the ecological ethos of bio-networks is reinforced by Cooper's participation in both genres, domestic or sentimental literature and nature writing.

Cooper's narrative also participates in what Buell terms the first major work of American bioregionalism, a subgenre related to local color or regional realism.[6] The "finely calibrated environmental sense" of *Rural Hours* is manifested by the variety of environmental goods that are uniquely products of the Otsego County area and include maple sugaring, rural housekeeping, and local farming.[7] Although Cooper was keenly interested in the regional products of her home, we would be wrong to mistake this interest for provincialism. On the contrary, she believes that Americans should be aware of the wider world but should use this awareness better to appreciate the "variety of the world God had given them."[8] Bioregionalism thus operates in a number of ways: It values the natural products of the land and, by extension, the land; it values the American over the foreign; and it values the community as much as the individuals within the community.

Cooper's journal was written over a two-year span but covers only one year, just as Thoreau's *Walden* did. She begins her entries on March 4 and concludes February 28 of the following year, with the entries further grouped according to the seasons. Cooper's seasonal ordering of her observations does not present any new or unique challenges; such reliance on seasonal change as a literary trope extends back at least as far as Virgil. The cyclical nature of changing seasons is, aside from the day and night cycle, the "most perceptible in everyday life."[9] *Rural Hours,* though, uses the seasonal cycle for reasons other than to denote the passage of time or to express an Ecclesiastes-like hymn to the cyclical nature of life. Instead, the book's awareness of the seasons serves to reinforce the environmental as well as nationalistic concerns that occupied a part of Susan Cooper's life. This first becomes evident in the manner in which she divides the seasons and the passages in her book. The book opens with an entry for early March that begins the "Spring" section of Cooper's observations. The "Summer" section begins on June 1 and ends the last day of August, while "Autumn" covers September through November, and "Winter," December through February. These divisions do not correspond to strict, calendrical definitions of the seasons: March 4 is actually still in winter; spring begins on March 21. Cooper's divisions, then, mark a more natural perception of the changing seasons based on observation and an empathetic feel for and understanding of the yearly cycle in her own region.

The first two entries in the book illustrate this point very well. As with many of her observations, Cooper begins with a brief note about the weather, and the first line of *Rural Hours* observes that everything "looks thoroughly wintry still, and fresh snow lies on the ground to the depth of a foot."[10] Her next entry, for Tuesday, March 7, indicates that she is eagerly anticipating spring weather when she begins, "Milder; thawing." She also points out in this entry that she has seen loons moving northward while walking near the river. The keen observer of birds' migratory patterns soon checks herself, though, saying, "It is early for loons, however, and we may have been deceived." Whether the birds are loons is not so important as what the migrating birds represent: nature moves at its own, mysterious, non-human pace, and humans may only observe but may not set constraints on the change. Furthermore, any date that we set for the advent of spring is necessarily arbitrary; the birds, who have a much better and more empathetic connection to nature, are better equipped to mark seasonal change than a date on a calendar.[11]

Another passage, again involving birds, further emphasizes this point. In the 1868 edition of *Rural Hours,* Susan Cooper added a chapter entitled "Later Hours," which takes the observations of the *Rural Hours* year and adds a greater perspective of time. Here she notes that, on June 26, a pair of robins has nested between deer antlers that hang over the door of her verandah. A year later the avian couple returns to build the nest on May 14, and a year after that on June 1. In the "New and Revised" edition of 1887, Cooper notes that the robins are still building in the same spot, having made their 1886 arrival on an unspecified day in May. Although the exact date the robins choose to build their nests never varies by much—except for the anomalous June 26, they are within two weeks—the different calendar dates of the nest-building indicates that nature will not be tied to arbitrary human demarcations.

Arbitrary human demarcations, though, are inevitable, and Cooper recognizes their importance in establishing a link between the natural and the human. According to Buell, "polite culture" and "popular culture" intersect in the almanac

tradition through the shared trope of seasonal change: both "polite" poetry such as
Thomson's "Seasons" and "popular" culture such as the almanac rely on the
trope.[12] Cooper furthers the connective qualities of the almanac by noting its importance to farmers and other country people as a literary piece and as a mediator
between humans and nature. On Tuesday, July 3, Cooper describes in detail the
visit she takes to a local Farmer B——'s house, where she notes the prominent position of the "well-fingered almanac, witty and wise as usual."[13] She goes on to say
that the almanac printers had, a year or two previously, decided to leave out the
weather prognostications, but the resulting book sold so poorly that the printers
had to resume the weather in subsequent editions. For country people, farmers especially, some knowledge of the weather is crucial, and the almanac helps mediate
between wild nature and a domestic pastoral. Farmers can use the information
about nature contained within the almanac to create a better, more comfortable
living situation working within nature.[14] In a sense, then, Cooper's book itself is
an almanac, both for its seasonal awareness and for its attempts to bridge natural
and domestic concerns.

Cooper's visit to Farmer B—— creates an extended narrative of community and
rural values that is sustained longer than any other single episode in the book. For
Cooper, the small details of daily rural life constitute a microcosm and ideal form of
her national heritage. Her opening exclamation about the farmhouse sets the tone:
"How pleasant things look about a farm-house! There is always much that is interesting and respectable connected with every better labor, every useful or harmless
occupation of man."[15] She continues to list all of the products of human labor and
nature's bounty that occupy the farm house and make it a prime model of domestic
satisfaction, and she is particularly impressed by the dairy the wife keeps.[16] While
on the visit at Farmer B——'s, Susan and her unnamed companion (her sister) admire the domestic accoutrements and comestibles, all of which are "almost wholly
the produce of their own farm."[17] The farm wife is the model of self-sufficiency and
the model of proper domestic roles; she is one who can take the goods of nature and
produce comfort, food, and clothing. Interestingly, although Susan Cooper resisted
feminism, she did feel that the domestic was a corrective and "recall[ed] her readers
to their republican roots in the unassuming country life."[18]

Later, in an entry dated Friday, September 29, Cooper describes going to town
for the local fair and market day. She again lists the various items crafted by the
men and (mostly) women of Otsego County. The goods available range from
woven and rag rugs, flannels, leather goods, woolen stockings, and table linens to
"very neat shoes and boots, on Paris patterns."[19] Cooper is not content to list these
items and let her readers draw their own conclusions, but must add her editorial,
saying "Every one must feel an interest in these fairs; and it is to be hoped they
will become more and more a source of improvement and advantage in everything
connected with farming, gardening, dairy-work, manufacturing, mechanical, and
household labors."[20] Here and elsewhere Cooper's agenda is clear: she wishes to
perpetuate a rural lifestyle that values industry as it is practiced by the agrarian

community. Such a lifestyle, performed in such a community, reflects her ideal of nationalism. A large part of this community is, of course, the industry of the farmer's wife, and such highly valued household industries as weaving, sewing, butter and cheese making, as well as other domestic duties.

One of the most important domestic duties is that of providing a moral education for the children in the family, and this task generally falls to the mother or other women in the household. In this case, the natural environment plays a crucial role, as the study of natural history provides a forum and excellent examples for teaching morality to children. Susan Cooper found herself writing in the natural history tradition, "conjoining women's roles as domesticator and the American landscape's new image as home."[21] Cooper was operating within a system of natural history writing that had been domesticated and sentimentalized already, most notably by Charlotte de la Tour's *La Langage des Fleurs,* published in France in 1819. This book purported to teach women botany, but mostly consisted of images of personified flowers and a set of symbols to communicate feelings.[22] The close scrutiny that Cooper gives to her natural surroundings had thus been mediated already by this and other quasi-scientific works that potentially undermined any serious study of botany or nature.

Cooper, however, resists being neatly pigeonholed as simply a woman writer participating in the sentimental floral tradition, as the preceding argument has already established.[23] Her own language further refutes any easy dismissal of her writings. Some of her most sustained and passionate passages in *Rural Hours* deal with complex botanical subjects, most importantly the naming of plants. She begins her June 23 diatribe with a complaint that too many country people cannot identify the plants, trees, and flowers that grow on their land, going on to say that the women sometimes know the names of herbs or simples but they often make strange mistakes. This sad state of education is troubling for two primary reasons. First, if women are to use botanical knowledge to inculcate moral knowledge in their children, they must have some understanding of what they are teaching, and the ignorant women who do not know the names of the plants are not rising to the challenge demanded and are not teaching well. Second, for a nation to grow and prosper, its citizens must understand their own landscape. The domestic pastoral ideal cannot exist when the rural inhabitants cannot identify the elements that contribute to the health and vitality of the ideal. As a corollary to this, those who do not understand the natural world are not as likely to value it, and, consequently, the stability of nature is jeopardized.

Susan Cooper's demand for a clear understanding of nature led to her great problem with the Latinate, specifically Linnaean process for naming plants. According to her domestic pastoral ideology, Latin appellations are clumsy and "very little fitted for every-day uses, just like the plants of our gardens, half of which are only known by long-winded Latin polysyllables, which timid people are afraid to pronounce."[24] One possible reason for so many of the native plants having such unfriendly nomenclature is that the American continent was settled after Linnaean taxonomy took root and names were given by botanists before the common people could give common names. More likely, though, is the possibility that Cooper, through her Linnaean critique, is defending her turf as an amateur botanist. Vernacular botanical names were in common usage, but Cooper sees the scientific language as encroachment from without the community by professionals. The danger of this is clear: professional botanists come from outside the community and thus have no stake in it. When the privilege of naming the land's bounty is removed, the inhabitants of the countryside are also removed from a crucial relationship with the land.[25] Furthermore, the spread of Latinate terms "actually perverts our common speech, and libels the helpless blossoms, turning them into so many *précieuses ridicules*."[26]

The reluctance and even antipathy that Cooper felt toward Latinate nomenclature is consistent with a sentimental ethos. As Annie Finch points out, many scholars feel that sentimental art is threatening because it becomes too close and intimate with its reader, and, "worse, sentimental art accomplishes this aim by reversing the crucial hierarchy of reason over emotion."[27] This reversal in some ways parallels Susan's balking at an overly scientific system of naming, if we take Latinate nomenclature as representative of reason and a fondness for common names as emotion. More importantly, though, Cooper is seeking, through this more emotional naming process, a closer, more vital connection between nature and the people who inhabit the bounds of nature. This real and vital connection to nature is at the heart of sentimental ecology.

When the native flowers have complicated and difficult Latinate names thrust upon them by botanists with no stake in the region, poets also suffer because of an inability to create an aesthetically pleasing poem incorporating these uncommon names. This problem, in fact, seems to be most troubling to Cooper, and she spends several pages railing against the unpoetic sound of Latinate names. She asks, "Can you picture to yourself *such* maidens, weaving in their golden tresses *Symphoricarpus vulgaris, Tricochloa, Tradescantia, Calopogon?*"[28] Her answer to this is "No, indeed!" and the reasons for this resounding denial are not as simple as aesthetic distaste. If one of the purposes of landscape and nature poetry is to help create a national American literature and, by extension, a national American identity, both of which help foster the sense of community crucial to the domestic scene, then American poets become ineffectual when their language becomes distanced and foreign.[29] Furthermore, a distinctly American literature would need to be written in a distinctly American idiolect or "ecolect" in order to be truly representative of the new country.[30] An imposed language, particularly a foreign, distant, and lofty language like Latin, undermines the nationalistic or regionalistic tendencies of poetry.

Cooper's defense of the local extends beyond nomenclature to the uniqueness of American seasons, where it takes on a fiercely patriotic tone as she unfavorably compares European nature or seasonal poetry to American. In a long entry dated Wednesday, October 11, Cooper begins by saying that "[a]utumn would appear to have received generally a dull character from the poets of the Old World." She goes on to criticize in one of her longer entries the poor presentation of autumn by various European poets, remarking on many specific passages. Shakespeare, she points out, links "*chilling* autumn" with "angry winter," while Collins calls the colorful season "*sallow*" and Wordsworth dismisses it as a "melancholy wight."[31] Not content to chastise merely the English poets, Cooper displays her European education to good advantage by quoting several French poets in the original before moving on to her critique of German and Italian poets.[32] All of these European poets lack the ability to see the majesty and beauty of autumn that Cooper spends so much time praising. Not only are the poets factually wrong—she attacks Thomson, Spenser, and Keats for wrongly placing the summer wheat harvest in the autumn—but they are also blind to the beauties of the changing colors, seeing impending winter rather than multicolored autumn.[33] Cooper does admit, however, that the European leaf season is not so colorful, saying, "[h]ad the woods of England been as rich as our own, their branches would have been interwoven among the masques of Ben Jonson and Milton; they would have had a place in more than one of Spenser's beautiful pictures. All these are wanting now."[34] The American landscape, then, is not only more "rich" but it can provide an example for American poets to create works that must by association be more rich as well, which, in turn, allows for a superior setting of domestic/didactic possibilities.

Cooper's motivation for connecting the colorful autumn season and a national literature comes almost as an afterthought following her critique of the Europeans. She notes "the march of Autumn through the land is not a silent one—it is already accompanied by song. Scarce a poet of any fame among us who has not at least some graceful verse."[35] The poets "among us," the American poets, stand in clear contrast to the Europeans who fail to write anything "graceful" about autumn, and so the Americans have an advantage in their access to more colorful and more poetic subject matter. Cooper's confidence in her fellow American writers is remarkable as she says "year after year the song must become fuller, and sweeter, and clearer."[36] This confident optimism contrasts dramatically with Emerson's vain search for his ideal poet. "We have yet had no genius in America . . . which knew the value of our incomparable materials," Emerson says.[37] However, Cooper, though she does not cite American examples to support her claim, recognizes both the "incomparable materials" and the poets who know their value. The richness of the American environment seems to make the rise of rich, sweet, and clear poetry inevitable.

The narrative of community, which crucially informs the sentimental ecology of Cooper's work, culminates in an environmental awareness that transcends aesthetic appreciation to become a conservationist ethos. Cooper's construction of her community is marked by an inclusiveness that sets her apart from her father and illustrates her attempts to link community and environment. Susan Cooper first creates a narrative of community in her domestic sentimental novel, *Elinor Wyllys; or, The Young Folk of Longbridge.*[38]

The argument for labeling this a work of nature writing can be summed up as a complex form of environmental synecdoche. Rosaly Torna Kurth dismisses the novel in extremely harsh terms, saying that it "possesses as much or, rather, as little of artistic merit as do the novels of the later domestic writers," noting that the plot is "not plausible nor . . . unified," with too much emphasis on "domestic matters."[39] These very points, however, when viewed in another light, make it an important work of environmental writing. Cooper knew from her frequent nature walks that one action or event in a field, meadow, or forest could have larger reverberations elsewhere, just as she knew from her own social contacts that no member of society lives in a vacuum. Thus when she describes in detail the social climbing vulgarity of the seemingly minor character Mr. Clapp early in the novel, she does so because this unpleasant man and his morally suspect ambitions will have implications for Hazlehurst later. At least four subplots reveal themselves throughout the novel, and each one has some significant influence on the others. Her novel, then, is a model of interrelationships, a sort of sentimental ecosystem. The few subplots are representatives or synecdoches of the larger social environment.

While James Fenimore Cooper creates a moderately favorable portrait of Native Americans, his daughter goes further in her claims of sharing communal values with the Indians.[40] Susan Cooper notes the Indians' degradation as her father does, but differs from him by noting that civilization is responsible for it. The Indian women, Susan argues, are less likely to become corrupted by the influences of civilization and remain "gentle and womanly" while the braves are marked by the "heavy, sensual, spiritless expression, the stamp of vice."[41] The gender-dependent differences that Susan Cooper notes illustrate the most important manner in which she disagrees with her father's point of view. To Susan, the "domestic qualities," which her father mentioned, are much more important, and she is better able not only to see this importance but to see the underlying causes for the disruption of these important qualities. As a woman intent on perpetuating a domestic ideal, she is able to diagnose the problem. Her diagnosis is based largely on a feminine sympathy, which her father seems to lack.

Susan Cooper's sympathy for the Indians is an important aspect of her sentimental ecology because she is sharing an emotional bond by sympathizing. The sympathetic bond is crucial in creating the interdependent links of the sentimental ecosystem. In one encounter with the Indians, she says: "The first group [of Indians] that we chance to see strike us strangely, appearing as they do in the midst of a civilized community with the characteristics of their wild race still clinging to them; and when it is remembered that the land over which they now wander as strangers in the midst of an alien race, was so lately their own—the heritage of their fathers—it is impossible to behold them without a feeling of particular interest."[42] This passage is remarkable for Susan's ability to imagine the Indian point of view, to place herself and her entire community as "alien," as other. By so doing, Susan's language is in some ways "at odds" with the dominant discourse, forming a Kristevan "semiotic discourse" which, rather than an "infantile fusion with the mother," indicates a fusion with the Indians *as* a mother.[43]

We can add another dimension to this maternal language after looking at a related passage that occurs some pages later. Still discussing the Indians, Susan Cooper says: "It is easy to wish these poor people well; but surely something more may justly be required of us—of those who have taken their country and their place on the earth. The time seems at last to have come when their own eyes are opening to the real good of civilization, the advantages of knowledge, the blessings of Christianity. Let us acknowledge the strong claim they have upon us, not in word only, but in deed also."[44] Here, Cooper is even more emphatic in her critique of white civilization. While still maintaining the supremacy of white civilization—its "blessings" and "advantages"—Cooper nevertheless acknowledges the displacement and alienation the Indians have a right to feel, as well as the debt that white Americans owe them. Significantly, this debt is not described in pecuniary terms, but in moral, specifically Christian, terms. It would not be overemphasizing the point to say that Cooper's idea of a debt is here a sentimental one: The debt may be paid, Cooper implies, by providing the means for the Indians to reap the "real good of civilization."

Cooper's next comment about the plight of the Indians points even more forcefully to the domestic or sentimental social awareness that the author brings to the observations of her community. She notes that "perhaps the days may not be distant when men of Indian blood may be numbered among the wise and the good, laboring in behalf of our common country.[45] The most striking idea expressed here is Susan Cooper's emphasis on the "common country" and her willingness—one might even say eagerness—to give a voice to the Indian. Though we might discern a note of condescension in her tone by her dismissal of any possibility that the Indians might already be wise, as well as the notion that Indians can only achieve social mobility through the generous and charitable efforts of whites, it is still notable that she stresses the importance of commonality and community. Her emphasis on community contrasts with her father's "more favorable picture of the redman than he deserves" and his understanding of the Indian as other and therefore not capable of being part of the same society that the Coopers inhabit. Susan, though, seeks to bring the Indians into her home, as she does literally in *Rural Hours,* figuratively in the national sense, and symbolically as a part of her domestic community in nature.

Perhaps the best-known entry in *Rural Hours* is a long description of a stand of old-growth pines, untouched by the threat of the axe, situated on the edge of town, which Cooper uses to illustrate the passage of natural epochs and the very short time white settlers have been on the land.[46] In this short time, the white settlers have radically altered the natural landscape in a manner that had never happened before. She concludes the passage by meditating on the fate of the forests, speculating that "[a]nother half century may find the country bleak and bare."[47] Her tone, however, is hopeful, because "as yet the woods have not all been felled" and we are left with the impression that such a calamity may be avoided. Cooper goes on to urge conservation and systematically refutes arguments in favor of clear-cutting the forests. The final value of a tree, she vehemently tells us, lies not merely in the "market price of dollars and cents," but in the "intellectual and . . . moral sense."[48] Here Cooper brings her notions of conservation back to the moral value of the landscape and nature.

Her tone as she discusses the unsparing logging in her community is very critical, even angry. She derides early colonists for looking on trees as enemies, and chastises their descendants for maintaining the same reckless spirit. "One would think," she notes wryly, "that by this time, when the forest has fallen in all the valleys . . . some forethought and care in this respect would be natural in people laying claim to common sense."[49] The brutal but understated sarcasm of her tone is consistent with other times she sees people not acting in the best interests of nature and the community. This is the same sarcasm that creeps into her voice as she chastises country women for not knowing the names of plants, and stands as a stark contrast to her warm approval of people such as Farmer B——'s wife, who recognizes the role of nature in her domestic life.

Cooper's overall tone throughout *Rural Hours,* though, is generally more positive, and she cannot sustain her anger long before she provides examples of her neighbors' care for the environment. She begins with a story of horrible degradation in the "wilds of Oregon," where a government scouting party comes across a huge tree that had been felled and left to rot. The man who cut the tree, she speculates, claimed, "no doubt, to be a civilized being."[50] Cooper compares this act to one committed by one of "the horde of Attila," and goes on to offer a counter-example that happened, "happily," near her own neighborhood. On the banks of the Susquehannah [*sic*], a large elm tree was left standing in spite of the fact that it stood in the way of the highway. Although the tree stands where a "thorough-going utilitarian would doubtless quarrel with it," its beauty rendered it safe from the axe.[51] Cooper applauds the decision of her neighbors to keep the tree and their ability to recognize that its moral and aesthetic value far outweighed any merely practical or utilitarian concerns. Environmentalism, Cooper's modern readers are led to see, is a community value that transcends mundane concerns and elevates the moral strength.

The final value of Susan Fenimore Cooper's work may be found in something other than the "market value" of her collection of natural observations, just as she insists that the final value of environmental health transcends the market. She uses her domestic voice to convince us of the intellectual and moral value of the landscape and the responsibility we have to perpetuate this value. By continuing her sentimental ethos from her novel and elaborating upon the themes of the domestic pastoral in the natural landscape, Cooper creates a sentimental ecology that illustrates the interconnectedness of all creatures on earth.

Matthew Bolinder

"Appropriated Waters"
Austin's Revision of Thoreau in *The Land of Little Rain*

In the chapter of *Walden* entitled "The Ponds," Henry Thoreau laments the utilitarian perspective of the Concord townsfolk regarding the beloved subject of his book, writing, "the villagers, who scarcely know where it lies, instead of going to the pond to bathe or drink, are thinking to bring its water, which should be as sacred as the Ganges at least, to the village in a pipe, to wash their dishes with!—to earn their Walden by the turning of a cock or drawing of a plug!"[1] In *The Land of Little Rain,* Mary Austin similarly contemplates the fate of water in her local environment, writing with a critical voice that complements Thoreau in content, if not tone, "It is the proper destiny of every considerable stream in the west to become an irrigating ditch. It would seem the streams are willing. They go as far as they can, or dare, toward the tillable lands in their own boulder-fenced gullies—but how much farther in the man-made waterways."[2]

While neither of the two authors unequivocally condemns the appropriation of water for human use, both are clearly bothered by the implications of such appropriations. What is clear in both texts is that, like most nature writers, Austin and Thoreau see their subjects as more than merely natural resources or physical commodities. They are, rather, second parties in imaginative relationships—relationships that are necessarily altered after appropriation. But while Thoreau is irked by the terms of a relationship that doesn't attempt to extend itself any further than the proximity of the closest spigot, in Austin's understanding a stream's very relational capacity is diminished when it is confined. As Austin writes of her subject, "It is difficult to come into intimate relations with appropriated waters; like very busy people they have no time to reveal themselves."[3] As opposed to a free-flowing stream, with its unregulated processes—meanderings, uneven banks, and fluctuations in level and rhythm—the direction and significance of an irrigating ditch is always already defined. It is limited to an artificial border imposed upon it, and hastily fulfills the singular purpose of its appropriation, failing to exhibit the changes that an unappropriated stream makes and displays to an interested observer.

Austin's observations are equally pertinent, however, if applied to the genre of nature writing. Nature writing is itself an act of appropriation—that is, an employment of nature for specific purposes, with a particular creative end in mind—though on an ecologically non-invasive level. As a sustainable appropriation, it is limited only by the subjective experience and the imaginative interaction between a writer and her environment. There need not be any borders to the exchange or formal structures that dictate its textual depiction save those of the sensory process itself. And yet, despite this interpretive boundlessness, in the genre one particular representation of nature (and a representation, I might add, that was a popular failure in its own time) has proved incredibly significant. A majority of critics position *Walden* not only as the first example of what is called nature writing proper, but the most historically formidable and influential literary appropriation of the environment as well. The odd result is that critical surveys of the field often make Thoreau's text look as though it functions more like a physical appropriation and barrier in its influential effect than an imaginative one. Generically speaking, we can say that *Walden* itself acts like an irrigating ditch: The very direction and shape of nature writing has, from its conception, always been predetermined. Or, to employ a different aquatic simile, *Walden* sits as a dam at the headwaters of the genre, and those writers who choose to navigate the river apparently have no choice but to grab a barrel and go over it.[4]

On the one hand, the project of tracing lines of influence from *Walden* to later examples of nature writing testifies as much to the skill of the critic in constructing literary constellations and meta-narrative as it does to Thoreau's enduring affective power or genius. And yet, on the other, in their epigraphs and textual allusions, a wide variety of nature writers, from Leopold to Ehrlich, Eiseley to Dillard, do in fact acknowledge debts—or at the very least pay lip service—to Thoreau.

My reading of *The Land of Little Rain* situates Austin as a literary descendent of Thoreau, and as such an author anticipating the growth of *Walden's* influential "wake" in the twentieth century. But, more importantly, *The Land of Little Rain* is itself a thoughtfully crafted response to certain imaginative and ecological deficiencies of this wake. Austin's text alludes to the "difficulties" of becoming "intimate" with *Walden* as a reading of nature's significance and as a literary and philosophical model for nature writing. As Mikhail Bakhtin has written, language is never "neutral or impersonal." Rather, "The word in language is half someone else's. It becomes 'one's own' only when the speaker populates it with his own intention, his own accent, when he appropriates the word."[5] Before this "moment of appropriation," the word "exists . . . in other people's contexts, serving other people's intentions: it is from there that one must take the word, and make it one's own."[6] In her first major work, Austin enacts a kind of compositional ecology, appropriating the example of her literary predecessor, placing it within her own specific context and making it her own.[7]

While Austin does not explicitly note a literary debt to Thoreau in either *The Land of Little Rain* or *Earth Horizon,* her autobiography, she does describe in the latter text that she was familiar with literature of the period in general, and the Concord Transcendentalists in particular. As a child, Austin had been given by her father unfettered access to a personal library filled with the works of American authors; in *Earth Horizon,* Austin notes (of herself, in the third person) "Mary's habit of inordinate reading," being acquainted with "everything that other people in the eighties read, and a good deal more." Of these authors, Ralph Waldo Emerson is singled out as "the only writer out of those days who affected her style. I don't know why."[8] Her description of a certain personal experience at the age of "five or six," in the presence of an old walnut tree that "reach[ed] into infinite immensities of blueness," however, would seem to illustrate an affinity between her own developing consciousness and that of Thoreau's mentor, an affinity that moves beyond "style":

Quite suddenly, after a moment of quietness there, earth and sky and tree and wind-blown grass and the child in the midst of them came al ive together with a pulsing light of consciousness. There was a wild foxglove at the child's feet and a bee dozing about it, and to this day I can recall the swift, inclusive awareness of each for the whole—I in them and they in me, and all of us enclosed in a warm lucent bubble of livingness. I remember the child looking everywhere for the source of this happy wonder, and at last questioned— God?"—because it was the only awesome word she knew. Deep inside, like the murmurous swinging of a bell, she heard the answer, "God, God . . ."[9]

Austin would refer to this experience as a formative moment in the development of a concept of self she called "I-Mary": While Austin does not declare herself to be a "transparent eyeball," her description clearly rings with an Emersonian reverberation.

Her mention of Thoreau in *Earth Horizon,* though equally brief, is worthy of note as well. Austin writes that at fourteen she found her mother "tak[ing] pains to impress upon [Mary] the childish character of her interest in nature and the inexpedience of talking about it," especially when in the presence of young men. To "exceed a ladylike appreciation of [her] surroundings," her mother exhorted, would be a grievous mistake: "You must not quote," wrote Austin, "especially poetry and Thoreau. An occasional light reference to Burroughs was permitted, but not Thoreau."[10] With due respect to Burroughs, one might assume this Thoreauvian "tabu," if taken as such, to be grounded in Thoreau's talent for and enthusiasm in describing the natural environment, certainly, as "A very little experience demonstrated . . . that outdoors as a subject was boring to most people" but more importantly in a greater enthusiasm for detailing the shortcomings of the society he distanced himself from in retreating to Walden.[11] In Thoreau, Austin found a critical thinker who not only made no apologies for his stances, but reveled in the act of making them.

Consequently, there are a number of parallels between *The Land of Little Rain* and *Walden;* Austin's description of the desert southwest oftentimes evokes Thoreau's own local depiction. At the beginning of a chapter entitled "The Basket Maker," for example, Austin writes that "To understand the fashion of any life, one must know the land it is lived in and the procession of the year."[12] It's a species of statement that, if transplanted to Thoreau's own literary landscape, would be indistinguishable from the native vegetation. The two texts are filled with impressions and motifs—large and small, general and particular—that intersect: human trails that linger after their desertion, the privileged consciousness of youth, legends of bottomless ponds. Like the mantis egg cases that another Thoreauvian learned to spot in a Virginia valley, once a single specimen is identified they seem to show up everywhere.[13] But Austin's text, like the environment it describes, is most notable in how it takes Thoreau's example and changes it. In the case of the above quote, during the "procession of the year" in the Owens Valley we encounter a "country of three seasons," rather than four: "From June on to November it lies hot, still and unbearable, sick with violent unrelieving storms," Austin writes, "then on until April, chill, quiescent, drinking its scant rain and scanter snows; from April to the hot season again, blossoming, radiant, and seductive."[14] An exploration of a few of the more noteworthy textual intersections begins to illuminate Austin's convergences with her literary predecessor and, more importantly, her divergences from his text as well.

We find a significant motif common to both writers in the text's opening pages. Like Thoreau, Austin laments the inadequacies of names bestowed upon places, what Thoreau in *Walden* describes as the "poverty of our nomenclature."[15] In "The Ponds," Thoreau harps on a certain Farmer Flint for naming a pond after himself, as in his opinion Flint had no suitable connection with it: "*Flint's Pond!* . . . What right had the unclean and stupid farmer, whose farm abutted on this sky water, whose shores he has ruthlessly laid bare, to give his name to it? Some skin-flint, who loved better the reflecting surface of a dollar, or a bright cent . . ." Instead, Thoreau suggests that its name be derived "from the fishes that swim in it, the wild fowl or quadrupeds which frequent it, the wild flowers which grow by its shores, or some wild man or child the thread of whose history is interwoven with its own."[16]

Austin moves in this same direction in her preface, refusing to divulge to her reader the proper names of the places she will write about in the pages to follow. She "confess[es] to a great liking for the Indian fashion of name-giving," in which is established a link between name and character: "For if I love a lake known by the name of the man who discovered it, which endears itself by reason of the close-locked pines it nourishes about its borders, you may look in my account to find it so described. But if the Indians have been there before me, you shall have their name, which is always beautifully fit and does not originate in the poor human desire for perpetuity."[17]

Unlike Thoreau, however, this appreciation for the natural properties of the landscape does not prevent Austin from seeing land *as property* at times, and relating to it as such: a difference reflecting a greater conceptual distinction between the two writers. While Thoreau, in describing his initial—and later abandoned—inquiries of purchasing a farm, ends up boasting of his avoiding the necessity of ownership altogether, Austin pulls no punches when noting her desire to own land adjacent to a meadow that she finds particularly pleasant. In a chapter entitled "My Neighbor's Field," Austin writes of this parcel, "I knew I should have no peace until I had bought ground and built me a house beside it, with a little wicket to go in and out at all hours, as afterward came about."[18]

My emphasis on Austin's insertion of her own desire in some ways runs counter to what other readers have found in *The Land of Little Rain*. Vera Norwood writes that Austin's text teaches a reader "how best to respond [to the land] in an interactive rather than an hierarchical mode;" Elizabeth Ammons parallels Austin's voice to that of native cultures in her espousal of an existence founded on accommodation rather than imposition, noting that *The Land of Little Rain* "constantly reiterates the simple construct of push/pull, ebb/flow, shrink/expand that Austin considered the pulse, the constant, cyclic, unaccented rhythm of the earth itself"; and Lawrence Buell notes that a "guarded impersonal voice and diffuse perceptual centers" are Austin's "hallmark," the above instance of "land-envy," according to Buell, then reads as a "caricature of romantic imperial selfhood."[19] While I would agree in general with the characterizations of Austin's text as provided by Ammons and Norwood, and grant that Austin's statement about her neighbor's field must be read as self-consciously grandiose, if not ironic, her admission is nevertheless exemplary of her own insistence on honestly assessing the place of the human within the natural environment. In providing a detailed recitation of the business transactions that comprise the fields' human story, Austin implicitly acknowledges her own participation in "all this human occupancy of greed and mischief," even though as an adjoining squatter her appropriation of the field is primarily aesthetic.[20]

This ambivalent relationship to the natural environment—Austin at one moment admonishing human possessiveness and indulging in it the next—is not indicative of a flawed narrative, an inability to weave into her textual fabric a single pattern. Rather, Austin's text alludes to what *Walden* glosses over: the fact that such a fabric can never be altogether smooth, can never be constructed without being rent at some point, in some way.

The passage cited before from "The Ponds," in which Thoreau evaluates material appropriations of Walden, aptly illustrates on an ecological level this difference between Thoreau and Austin. Thoreau seems to move in the direction of candid observation when he notes of Walden's shores that, regrettably, "the woodchoppers have still further laid them waste," and that as a result "now for many a year there will be no more rambling through the aisles of the wood, with occasional vistas through which you see the water."[21] But the passage ultimately lauds the natural realm for an intrinsic imperviousness to human degradation. Despite development on its shores, including "the railroad infring[ing] on its border," Thoreau notes that Walden "is itself unchanged," "perennially young," and that "where a forest was cut down last winter another is springing up by its shore as lustily as ever." "All the change," Thoreau writes, "is in me."[22] Austin is interested in the "retaking of old ground by the wild plants, banished by human use" as well, but does not lose her critical equilibrium in celebrating it: In the economy of nature "there is not sufficient account taken of the works of man. There is no scavenger that eats tin cans, and no wild thing that leaves a like disfigurement on the forest floor."[23]

For Thoreau, then, the significance of what he observes in his environment is often anchored in a transcendent ideal, something *other than* ecological reality and value. Take, for another example, a fairly minor, comic episode in a chapter entitled "Brute Neighbors." Thoreau writes of a "war" between two "races of ants" that "human soldiers never fought so resolutely . . . It was evident that their battle-cry was 'Conquer or die.'" While the action depicted is in fact "brutal"— the scene is filled with description of gnawed limbs and severed heads—its lesson alludes to something else. Thoreau personifies until his "brute neighbors" are no longer "brute": "I was myself excited somewhat even as if they had been men," he writes. "The more you think of it, the less the difference." What results is a "reading" of a natural phenomenon that looks well beyond the occurrence itself. Indeed, this particular skirmish proves to be a battle of classical proportion: "perchance he was some Achilles, who had nourished his wrath apart, and had now come to avenge or rescue his Patroclus"; not even the storied military history of Concord itself can compare.[24] But Austin writes of one of her own brute neighbors, rabbits "taking the [water]trail with long, light leaps, one eye and ear cocked to the hills from whence a coyote might descend upon them at any moment," simply that "They do not fight except with their own kind, nor use their paws except for feet, and appear to have no reason for existence but to furnish meals for meateaters."[25] It is not that Austin is unable or unwilling to personify her landscape, create correspondences with human experience, or attribute a spiritual significance to nature. Rather, the harsh quality of the desert requires that it always be taken at an ecological, rather than ideal, level first, where its most valuable— though often truly "brutal"—lessons lie.

This particular contrast between the two authors is perhaps most evident in a chapter in *Walden* entitled "Higher Laws," in which Thoreau writes of finding within himself "an instinct toward a higher, or, as it is named, spiritual life" that proportionately gives way to the "reptile and sensual" as it "slumbers."[26] (The aforementioned blurring of the distinction between human and animal, it seems, is no longer applicable.) Quietly residing in his cabin by Walden's temperate banks, Thoreau can finally object to the consumption of "animal food" simply because it is "not agreeable to [his] imagination."[27] But Austin's landscape grants no such privilege: as the frequently cited dictum from her book states, "Not the law, but the land sets the limit."[28] Austin interacts with an environment where "woodpeckers befriend demoniac yuccas" and meadowlarks nest "unhappily in the shelter of a very slender weed."[29] In the desert, bucolic idealism is turned on its head: domesticated sheep graze on "red, choked maw" signifying only "a kind of silly pastoral gentleness that glozes over an elemental violence."[30] In order to survive in a country where "Of all its inhabitants it has the least concern for man," a Transcendental turn to "higher laws" is an ill-afforded luxury.[31] The realities of scraping by an existence in the desert necessitate a physical presence.[32]

I would further suggest that this refusal to idealize on Austin's part, and the resulting integrity of her portrayals, is rooted not only in her interaction with a physical landscape radically different from Thoreau's pond and wood, but a social landscape that contrasted with that of her generic predecessor as well. Though Austin would later articulate more overt and complex feminist and social critiques as her literary pursuits progressed, *The Land of Little Rain* is firmly rooted in these same concerns.[33] Austin's appropriations of *Walden* often illustrate important differences in perspective between two individuals—and, in particular, writers—separated by gender. This contrast between the two authors is most fruitfully explored in a discussion of a brief but important anecdote from *Walden*'s initial pages and a chapter of *The Land of Little Rain* referred to beforehand, "The Basket Maker."

Again, we can view Austin's concern with material necessity as a direct response to Thoreau: *Walden* begins, and to an extent is sustained, not only as a work of nature writing per se, but as a sort of treatise on the depravity of work, or the "curse of trade," as Thoreau calls it. A few pages into his text, Thoreau provides a brief anecdote of a "strolling Indian" who attempts—and fails—to sell a basket to a neighborhood lawyer. According to Thoreau, the Indian had labored under the false impression that, having made his baskets, he had "done his part, and then it would be the white man's to buy them." On being turned away, we are told that the Indian departs stunned, convinced of the lawyer's intention to "starve" Indians. Thoreau reminds the reader of his personal experience in such matters, writing that he, too, "had woven a kind of basket of a delicate texture, but . . . had not made it worth any one's while to buy them." Numerous critics remind us that the "basket" referred to was *A Week on the Concord and Merrimack Rivers,* Thoreau finding himself obligated to purchase over seven hundred of the one thousand copies printed, for lack of sale.[34] This, of course, did not result in any hesitance on Thoreau's part to make more "baskets." He "studied rather how to avoid the necessity of selling them."[35]

While Thoreau's determination to evade the demands of a materially obsessed culture he found absurd is laudable, the extent of his project's success is dubious.[36] More important to my comparison, however, is the context from which his critique is constructed. Thoreau was not writing from a station of privilege, but his bachelorhood did afford him a certain level of opportunity that one finds missing from Austin's experience. In some ways, Austin's own approach to her writing co-incides with that of Thoreau. Esther Stineman notes that in 1898 Austin heard a lecture by William James on repose and relaxation, in which James encouraged artists to depart from the "strained surface tension which was the preferred intel-lectual mode of the time," a talk that Austin in turn "interpreted . . . as a personal message."[37] But, though Austin would eventually become deeply involved in and associated with the Native American culture that encouraged this "mode," domes-tic hardship at the outset of her career inevitably hindered intellectual retreat. While Thoreau regularly visited with his family for a free meal during his Walden experiment, as critics are fond of pointing out, the stove initially occupied a more prominent position in Austin's household than the writing desk. As a married woman, she found herself with obligations to meet: a husband whose income was minimal, and in 1892, the birth of a mentally retarded daughter who required a great deal of care. Such circumstances contributed to a very different conception of authorship than Thoreau's: "Caring for a hopelessly invalid child is expensive business," wrote Austin. "I had to write to make money."[38] This is not to say that Austin's works are merely a means to a monetary end, but such quotes do eluci-date the fact that Austin was all too conscious of material realities, of "baskets" as necessarily being more than avocational artifacts. In *The Land of Little Rain,* then, we find a treatment of basket making very different from Thoreau's.

In "The Basket Maker," the Paiute baskets that Austin describes are exquisitely crafted pieces, aesthetically pleasing and symbolically significant, bearing a "per-sonal note" in their pattern that reflects the pattern of their maker's life. They "are wonders of technical precision, inside and out," Austin writes, "the palm finds no fault with them," with their "subtlest appeal . . . in the sense that warns us of the humanness in the way the design spreads into the flair of the bowl." But the basket as signifier is never separated from the necessities of subsistence. While they are art objects and stories, they are commodities as well: Seyavi, the native woman central to the chapter, "made baskets for love and sold them for money, in a gener-ation that sold iron pots for utility."[39] The baskets, then, link the two spheres that Thoreau attempts to separate in *Walden*'s opening pages.

In some respects, this difference between the baskets Austin writes about and those of Thoreau may be viewed as cultural. Concepts such as wilderness and so-ciety, and art and utility, come into being only when there is a noticeable demarca-tion between them, a boundary much more difficult to trace in the culture of a peo-ple living in a Country of Lost Borders, as Austin called it, than in Concord. The synthesis that Austin documents would appear to stem from organic process rather than conscious effort: "Seyavi's baskets had a touch beyond cleverness. The weaver and the warp lived next to the earth and were saturated with the same ele-ments." But gender is central to the domestic space of Austin's sketch as well. "Every Indian woman is an artist," Austin writes—perhaps sarcastically, with Thoreau's introspective text in mind—"sees, feels creates, but does not philoso-phize about her processes."[40]

But while Austin celebrates these lives, they are not sanitized. In Thoreau's own domestic space, we find a romantically idealized version of a life lived outside the confines of walls. In a beautifully written passage in "Sounds," Thoreau clears out his cabin to wash the floor, finding great pleasure in viewing his

whole household effects out on the grass, making a little pile like a gypsy's pack, and my three-legged table . . . standing amid the pines and hickories. They seemed glad to get out themselves, and as if unwilling to be brought in. I was sometimes tempted to stretch an awning over them and take my seat there . . . A bird sits on the next bough, life-everlasting grows under, and blackberry vines run round its legs; pinecones, chestnut burs, and strawberry leaves are strewn about. It looked as if this was the way these forms came to be transformed to our furniture, to tables, chairs, and bedstead,—because they once stood in their midst.[41]

Austin's chapter on the basket maker, however, shows that such a "transformation" in fact has a human cost:

Indian women do not often live to great age, though they look incredibly steeped in years. They have the wit to win sustenance from the raw material of life without intervention, but they have not the sleek look of the women whom the social organization conspires to nourish. Seyavi had somehow squeezed out of her daily round a spiritual ichor that kept the skill in her knotted fingers long after the accustomed time, but that also failed. By all counts she would have been about sixty years old when it came her turn to sit in the dust on the sunny side of the wickiup, with little strength left for anything but looking. And in time she paid the toll of the smoky huts and became blind. This is a thing so long expected by the Paiutes that when it comes they find it neither bitter nor sweet, but tolerable because common.[42]

Austin leaves her reader with a haunting image of withered and blinded women who live on "memory and speech," calling to each other "in high, old cracked voices, gossip and reminder across the ash heaps."[43] Though Austin does not in principle oppose Thoreau's disenchantment with a mindset that he believes leads his townsfolk to "labor under a mistake," "their fingers, from excessive toil . . . too clumsy and trembl[ing] too much" to "pluck" life's "finer fruits," the lessons of her own experience, the lives of the native women she carefully portrays, and the "knotted fingers" of Seyavi suggest its lack of universal applicability.[44] Austin writes from a position that finally affirms the dignity of labor and the value of experience: "if [the Paiute women] have your speech or you theirs, and have an hour to spare, there are things to be learned of life not set down in any books," not even *Walden*.[45]

Austin continues the passage from "Other Water Borders," cited at the beginning of my discussion, by telling the story of "old Amos Judson asquat on the headgate with his gun, guarding his water right toward the end of a dry summer."[46] One day Amos finds his water supply cut short; on approaching the gate to learn why, he discovers the wife of a neighboring landowner named Diedrick sitting on the headgate "with a long-handled shovel across her lap and all the water turned into Diedrick's ditch": "It was all up with Amos; he was too much a gentleman to fight a lady—that was the way he expressed it. She was a very large lady, and a long-handled shovel is no mean weapon. The next year Judson and Diedrick put in a modern water gauge and took the summer ebb in equal inches."[47] While Austin may not have had Thoreau in mind when she relayed this story, a better statement of their literary relationship would be difficult to find. In *The Land of Little Rain,* Austin points out the significance of *Walden* as a model for nature writers, yet alludes to the proclivities and limitations of Thoreau's example. With her appropriations of *Walden,* Austin taps into Thoreau's creative channel, and makes it her own: Appropriated water is, after all, to be shared.

Carol E. Dickson

"Recounting" the Land

The Nature of Narrative in Mary Austin's Narratives of Nature

> I may not know how to write, nor how to delineate character, nor even how to tell a story. The one thing I am sure about myself is that I know the relation of letters and landscape, of life and its environment. No doubt the great literary genius of life and landscape will come after me, when the sense of man as inextricably a part of his background has become part of our social consciousness. I think it is possible it will come in time for you to say, She was the first to whom it was clear, even if she did not have the gift of making it clear to others.
>
> —Mary Austin, 1930[1]

Mary Austin (1868–1934) wrote prolifically about such diverse issues as women's rights, Native American and Hispanic cultures in the American West, environmental concerns, and the education of children, but she is known mainly for her nature writing of California and New Mexico, including her nonfiction essays in *The Land of Little Rain* (1903) and *The Land of Journey's Ending* (1924), and her short stories in *Lost Borders* (1909).[2] Representation of the natural world, in fact, remained a dominant focus for Austin in all of her work, to the extent that, reflecting upon her literary career in 1930, she was convinced that she had recognized certain truths concerning the "relation of letters and landscape, of life and its environment." Indeed, according to late twentieth-century conventional wisdom, she had. Austin's vision of "man as inextricably a part of his background," particularly in *The Land of Little Rain,* sounds familiar today.

Although such a vision was far from generally accepted at the time, especially in Western America, Austin was not alone in her focus on the relationship between humans and the land. The beginning of the twentieth century—the period of Austin's most sustained engagement with nature writing—was in fact marked by a burgeoning popular interest in nature, in the form of outdoor recreation as

47

well as of nature writing. By the end of the nineteenth century, many Americans had already begun to question the deeply ingrained attitudes of the land as existing simply to be conquered and exploited. Hunters and railroad boosters, for example, were among the first groups to talk about conserving nature (albeit in order to maintain enough wildlife to hunt or enough wild space to promote to tourists). By the turn of the century, American consumption of nature "for its own sake" was rapidly increasing through such diverse vehicles as tourist resorts, summer camps, the Audubon Society, and the American Boy Scouts.[3] Ernest Thompson Seton described the first decade of the twentieth century as "a time when the whole nation [was] turning toward the Outdoor Life," a trend that led, according to Roderick Nash, to a kind of "wilderness cult."[4]

At the same time, writers who celebrated wilderness gained popularity. One of the foremost among these writers—and perhaps the most visible—was John Muir. By the beginning of the twentieth century, Muir was already well known as a leader in the fledgling conservation movement through his founding of the Sierra Club in 1892 and for such writings as *The Mountains of California* (1894) and *Our National Parks* (1901). Although not as well known, Mary Austin responded to the same changing cultural attitudes in her writings of this period. The work of Austin and Muir, in fact, nicely complement each other in many ways. Both celebrate Western land and wilderness on its own terms and, in so doing, both insist on rethinking dominant conceptions of the natural world by decentering human presence as the measure of all things; instead, their work grants nature autonomy as a—if not *the*—central agent in human life. Ann Zwinger likewise places Muir and Austin together as "the first truly *western* nature writers . . . They celebrate landscape as animate, existing on its own, requiring no explanation, no intellectualizing, only devotion, insight and understanding."[5]

Such general comparisons, however, obscure the fundamental differences between the projects of Muir and Austin—and, by extension, the huge variety within the early twentieth-century nature movement. For while many of these currents challenged nineteenth-century, human-centered ideologies of nature (including Manifest Destiny and Romanticism), many of them also remained entangled with the implications of such thinking. The case of Muir represents well the kinds of tensions and contradictions that characterized visions of nature at this time. For example, although Muir posits the spiritual benefits of a land predominantly considered as economic resource or spectacle, and thus argues the need for its preservation, he maintains an understanding of nature as a vehicle for human benefit. Furthermore, Muir never questions his use of language in representing nature, thereby reinscribing the human-centered power dynamics that conventional narrative represents.

In addition, Zwinger's general comparison does not sufficiently recognize the uniqueness of Austin's conception of the relationship between narrative and the natural world. Although she also cannot escape the legacy of nineteenth-century thought, Austin offers a remarkably incisive critique of the implications of this legacy in *The Land of Little Rain* by anticipating more-contemporary understandings of language. For, more than simply challenging how nature is represented, Austin critiques the nature of linguistic representation itself. Realizing that any narrative of the land remains a human construct, which necessarily maintains a position of anthropocentrism, Austin highlights the inadequacy of narrative to represent fully the nonhuman world, while simultaneously acknowledging the necessity for such constructs in any human understanding of this world. In the very self-consciousness of this tension—her awareness of the difficulties as well as the importance of "narrativizing" the land—Austin suggests that the land can never be fully controlled by humans, physically or linguistically. In this way, she compellingly decenters the human role in the natural environment; but, contrary to Zwinger's assertion, Austin does indeed demand "intellectualizing," which, rather than precluding understanding, is an essential component of understanding the human relationship to nature.

Moreover, the critique of narrative convention in *The Land of Little Rain* contains an implicit gender critique of patriarchal attitudes toward nature and of narratives, like Muir's, that reveal such attitudes. Although Austin's proto-ecofeminist vision does not become explicit until such later writings as *Lost Borders,* these essays, in challenging traditional narrative structures and in recognizing the relationship between an individual female narrator and her desert environment, work to validate the experience and the voice of individual women, voices lost within Muir's broadly collective vision of "our" national parks.[6]

Austin's *The Land of Little Rain*—a text that, as tapestry of nature essay, personal essay, folk legend, and local history, defies easy generic classification—brings her readers to the eastern slope of the Sierra Nevada mountains, and invites us to experience the desert wilderness. Yet, as both place and image, the desert was unfamiliar and distant to most of her readers at the time of the book's publication in 1903, a distance made clear in contemporaneous reviews of *The Land of Little Rain. The Dial,* for example, speaks of an "atmosphere . . . of strangeness" in the desert that Austin reproduces. Similarly, *The New York Times* doubts anyone would be "likely to dispute with Mrs. Mary Austin the title she claims to all of the dreary country." *The Nation* recognizes that *The Land of Little Rain* "covers a ground rarely if ever trodden," while *Out West* credits the book's importance in part to its subject, "which is but a name to most who will hear it—and a name which has carried with it ideas of desolate loneliness at best, of tragic horror more commonly."[7]

The uniqueness of Austin's text thus lies in part in the very "kind" of nature she chooses to narrate, and her first challenge to her readers is to engage with such a landscape. To do so, Austin's narrator celebrates the land she describes, while acknowledging her reader's probable prejudices: "If one is inclined to wonder at first how so many dwellers came to be in the loneliest land that ever came out of God's hands, what they do there and why stay, one does not wonder so much after having lived there. None other than this long brown land lays such a hold on the affections. The rainbow hills, the tender bluish mists, the luminous radiance of the spring have the lotus charm."[8] Austin's alter-ego narrator here proclaims the ineffable beauty and power of the natural world in a subtle and intimate tone, at the same time that she reaches out to her audience by putting herself in the reader's position and putting the reader in hers.

The Land of Little Rain primarily works to make the "voiceless" land heard.[9] The land, not its human residents, is the central character; its cycles and patterns determine the lives of the plants, animals, and humans who live there. "The land," Austin's narrator observes, "will not be lived except in its own fashion."[10] As she decenters the role of human behavior in this world, Austin also uses simple, minimalistic language to represent it, encouraging her reader to look beyond the human words to the non-human world they depict.[11] At the same time, she calls attention to these very words.

The Land of Little Rain, in fact, opens by immediately drawing our attention to the ways in which our use of language reflects and influences our relationship with the land. The preface sets the tone for the entire narrative in its meditation on the power inherent in the act of naming. Humans impose names on the land, Austin suggests, in order to control and possess it. This naming takes different forms, however, and she draws a distinction between Anglo- and Native-American conventions. The Anglo-American convention, characterized by naming places after people, "originates in the poor human desire for perpetuity," symbolizing the impulse to own and to colonize the land. Native-American naming, on the contrary, "sets . . . well with the various natures that inhabit in us" and is "always beautifully fit." People, as well as places, are "known by that phrase which best expresses [them] to whoso names [them]," acknowledging a relationship of greater equality between namer and named.[12] Throughout this text, Austin, through the voice of her narrator, aligns herself with Native American naming and the attitudes that it represents, in order to "keep faith with the land."[13] "Desert," she claims, is the name this land "wears upon the maps, but the Indian's (the 'Country of Lost Borders') is the better word."[14]

Yet, in putting the land into narrative form at all, Austin remains complicit in the very act that she condemns; as a result, the text contains internal contradictions that parallel those of writers such as Muir. Although she attempts to let the land, in a sense, speak for itself, she necessarily remains in the position of speaking for it. These essays might be more "land-centered" than conventional narratives representing the natural world, but as human constructs they cannot describe the land in other than human-centered terms and concepts. As William Cronon puts it, stories of the land "are human inventions despite all our efforts to preserve their 'naturalness.'"[15] Although Austin wants to recognize the autonomy of the natural world, true natural autonomy—nature unmediated by human experience—is a utopian projection, impossible to imagine, much less to narrate.

Furthermore, neither Austin nor her narrator resist the empowerment that language gives them. The narrator claims this region as "her" country and retains the right to control it, and her reader's experience of it, through her words. She warns, "I am in no mind to direct you to delectable places toward which you will hold yourself less tenderly than I"; and she claims her right to "annex to my own estate a very great territory to which none has a surer title."[16] Despite her "diffusion of centers of consciousness, and her refusal to maintain an executive control over the perceptual center," as Lawrence Buell puts it, William Scheik's charge rings true: "Whenever Austin alters the figures of nature rhetorically . . . she annexes the landscape through a verbal act akin to proprietary naming by colonizers."[17]

Conveying the tensions and contradictions in her narrativizing the natural world, however, is precisely the focus of *The Land of Little Rain*. Austin realizes that appropriation is intrinsic to narrative and that she can never present a fully land-centered text through human language, so her essays explore different modes of expressing the land in order to raise the awareness of her reader about the relationships among language, the land, and human power. Through this exploration of the multiplicity of potential narratives for the natural world, Austin exposes and undermines the power dynamic implicit in conventional narratives of the desert landscape. In this country of "lost borders," where "not the law but the land sets the limit," she seeks to decenter human activity and to suspend conventional boundaries of vision.[18]

To teach us how we might do this, Austin's narrator assumes the role of a translator in this text, teaching us how to "read" the desert landscape—including where to find water, what the presence of certain flora and fauna means, and how to understand desert legends—so that we might emulate desert inhabitants in order to survive, appreciate, and tell accurate stories of this world. Yet, while such advice might help us to adapt, she also suggests that the land can only be fully understood by living there. Ideally, the narrator asserts, after having lived in the desert for an extended period of time, this land could be understood implicitly through experience rather than mediated through language. Sensory perceptions, for example, such as the desert smells that she describes at the end of "The Mesa Trail," might provide an immediate awareness; likewise symbols, such as the Native American "glyphs" that she illustrates in "Water Trails of the Ceriso," might minimize narrative intrusion. As translator and writer, however, Austin cannot adhere to a non-linguistic ideal. She cannot erase her narrative self at the same time that she challenges her readers' preconceptions of narrative. All she can offer is different narrative possibilities that point to the infinite, unfixable meanings that the land contains.

Whatever narrative strategy might be employed, Austin suggests that the land remains ultimately in control, and can never be fully appropriated nor molded by language: "[T]here are certain peaks, cañons, and clear meadow spaces which are above all compassing of words, and have a certain fame . . . to which we give no familiar names."[19] Much of nature is beyond words, inexpressible through language, resistant to human understanding. Finally, nature is inscrutable and can never be fully "read," known, or understood. Even the narrator herself, a longtime desert resident, "can never get past the glow and exhilaration of a storm, the wrestle of long dust-heavy winds, the play of live thunder on the rocks, nor past the keen fret of fatigue when the storm outlasts physical endurance."[20] Although she has lived there and considers herself something of an expert on the desert, she is still, and will always remain, a stranger in some ways. For her the land remains unfamiliar and cloaked in romanticism and sublimity.

Austin writes of a similar moment of defamiliarization in the Sierra wilderness in her diary in 1901: "getting in to camp at Kearsarge after dark. I walked along a little path among the pines, and suddenly I came to the edge of the world. Below me were clouds and stars and space. For a moment I almost fainted with shock. And then I realized I had come to the edge of a mountain lake and was looking at a reflection of the sky. But it was terrible for a moment."[21] Moments like these are important because they remind us that humans can never truly control the natural world, physically or narratively. Nor can we ever fully separate our experience of nature from culturally determined constructs of nature. For it is only when Austin "realizes" and becomes able to articulate what she sees that her experience is no longer strange and "terrible." Humans' immediate and less self-conscious reactions to the natural world, she suggests, are still mediated by convention. If, then, as William Scheik argues, *The Land of Little Rain* demonstrates Austin's "frustrated . . . desire for a transcendental encounter," that seems to be exactly the point.[22]

Austin's critique of the ways in which language reveals and extends humans' desire for control of the land seems to demonstrate a yearning for a new kind of nature writing, through which we might hear a greater variety of voices—including those of the land itself and of women's experience. For, despite the challenges to conventional thinking about the landscape in writers such as Muir, these visions of the natural world remain profoundly masculine, exemplifying what Patrick Murphy describes as the intrinsically patriarchal quality of most nature writers in their acceptance of the notion of human alienation from nature and in their monologic narrative style of "lecturing about individual perceptions" rather than "conversing."[23]

Furthermore, in contrast to the popularity of many male nature writers, Austin's voice was muffled by the patriarchal institutions that supported the main infrastructure of the nature movement, especially that of the publishing industry. Her books never sold widely, and she struggled with publishers about the best way to market them.[24] This marginalization, however, may have allowed her the perspective and freedom to forward a feminist challenge. Austin, with little public profile and no overt political aims for her text, perhaps had more leeway to critique conventional representations of nature, as well as the assumptions of the contemporaneous nature movement, than, for example, the politically outspoken Muir. She does this, as I have suggested, implicitly in her critique of patriarchal language and more explicitly in her valorization of the experiences of an individual woman in the desert.

Austin's project resonates with Patrick Murphy's notion of "ecofeminist dialogics," by which women and nature might be rendered "as speaking subjects within patriarchy in order to subvert that patriarchy not only by decentering it but also by proposing other centers"—or, as we have seen, by constantly shifting and suspending the centers that we as readers expect.[25] Drawing on the work of M. M. Bakhtin, Murphy argues that the combination of ecofeminism and dialogic narrative theory provides "a method by which we may yet effect one of the paradigm shifts necessary to break down the dualistic thinking of patriarchy that perpetuates the exploitation and oppression of nature in general and women in particular."[26] Murphy himself illustrates how we might apply such a theoretical framework to *The Land of Little Rain* by articulating her project in Bakhtinian terms. In this text, he claims nature "becomes a hero in the Bakhtinian sense, with Austin unfolding its narrative both in terms of its presenting 'a particular point of view on the world' and through continuously depicting the land's own ecosystem behavior as a viewpoint 'on oneself.' Although not a speaking subject, the land does function as a signifying agent, with its significations interpreted and represented by Austin."[27] Murphy's approach parallels my own, positing the fundamental importance Austin ascribes to an awareness of the role of language in mediating the relationship between humans and the natural world, as well as demonstrating the value of combining feminist and dialogic critiques in reading Austin. Such a lens clearly helps to frame Austin's critical impetus in *The Land of Little Rain* and is useful in considering how Austin centers the voices of women and the natural world, suggesting the truly subversive potential of her project.

Austin's feminist, dialogic critique in *The Land of Little Rain* is perhaps most pointed through the text's narrative style itself—as a conversation between the narrator and the reader—which posits the understanding of the land as a collaborative, communal, and arguably feminist process, rather than an individualistic and arguably masculine one. In so doing, the narrative structure of this text exemplifies Susan Lanser's notion of "communal voice," which valorizes female community rather than individual protagonists. Lanser defines "communal voice" as "a practice in which narrative authority is invested in a definable community and textually inscribed either through multiple, mutually authorizing voices or through the voice of a single individual who is manifestly authorized by a community"; this creates the possibility of female empowerment through "constituting a collective female voice through narrative."[28]

Austin creates such a collective narrative by constantly including the reader, repeatedly invoked through the narrator's use of "you," as an immediate and essential participant in her narrative. She concludes the preface, for example, with an image of the community she hopes to create through telling her stories and sharing her knowledge: "But if ever you come beyond the borders as far as the town that lies in a hill dimple at the foot of Kearsarge, never leave it until you have knocked at the door of the brown house under the willow-tree at the end of the village street, and there you shall have such news of the land, of its trails and what is astir in them, as one lover of it can give to another."[29] Sharing the "news of the land" is not only a logical extension of the processes of reading about and experiencing the land, it is a necessary and central part of these processes.

Furthermore, such a striving toward shared understanding emphasizes that the reader's role in creating meaning is as important as the writer's, and that the reader, by extension, must also assume responsibility for representing the landscape. The narrator declares this directly: "Guided by [the places to whom we give no familiar names] you may reach my country and find or not find, according as it lieth in you, much that is set down here. And more."[30] In qualifying what the reader might find in this land, and in this text, the narrator charges the reader with contributing to this representation. The reader is thus placed in an unfamiliar but essential position; through participating in the construction of knowledge she might more fully understand not only the desert but also the limitations to understanding and representing nature.

Interestingly, such a dialogic rather than monologic narrative structure (to use Murphy's Bakhtinian terms) assumes a certain transparency of language that Austin's critique of language seems to deny. However, the building of community through dialogue relies more on the act of talking (or reading) itself than necessarily on the words themselves. Thus casual conversation, which, in the forms of legend, gossip, and hearsay the narrator's stories validate as both useful and entertaining, is perhaps the most appropriate role of "narrativizing" nature—through creating community rather than controlling representation, and through maintaining an awareness of the continual possibility of the unreliability of language.

And once the reader has heard these stories, she ideally will experience the land herself, and then tell her own stories. It is this dynamic, these stories told and heard, that create histories of the land, and the kind of narrative history transmitted through storytelling becomes a further mode for narrativizing the land. As William Cronon argues in reference to environmental historians, "[d]espite the tensions that invariably exist between nature and our narrative discourse, we cannot help but embrace storytelling if we hope to persuade readers of the importance of our subject."[31] Although not an environmental historian per se, Austin implies that the importance of the subject of nature demands similar sensibilities in whatever form of linguistic representation: awareness of the interaction between teller and hearer, of multiple narrative possibilities, of the limitations of language, and of the power and complexity of nature as subject. Through an embrace of all of these, stories of the land can potentially transport the audience as close to actual experience as possible; like Austin upon hearing stories of local history upon first arriving in California, her audience might be taken "there . . . wherever the story was."[32]

For when she first arrived in California, Austin felt that "the country failed to explain itself. If it had a history, nobody could recount it."[33] *The Land of Little Rain* posits that through a greater understanding of the relationships among language, power, and gender, we might more accurately begin to "recount" the story of the land through dialogue. Contextualizing this argument within the early twentieth-century nature movement, and in particular alongside the words of writers such as Muir, puts into relief Austin's sophisticated yet largely unknown vision. Such a context thus provides insight into the radicalness and foresight of Austin's critique. For in constructing a more self-aware and a more integrative (perhaps even "ecological") paradigm to represent nature, Austin anticipates both post-structuralist understandings of the ways in which nature is constructed by language and ecofeminist understandings of how such revisions might be liberating for both women and the land itself. She may not have had, as she suggests in my epigraph, "the gift of making it clear to others," but Austin's insights into the interrelationships among humans, nature, and language certainly have "become part of our social consciousness" today.

Part II

Expanding the Genre

Part II of this collection shows how a variety of American women writers adapted and expanded the traditional genre of nature writing within a literary genre in which we have been trained to overlook its presence.

Marcia Littenberg reveals the important role women writers play in unifying themes from the nonfiction nature essay and the regional sketch in late nineteenth-century writing. In publishing nature sketches, essays, and stories, popular publications such as the *Atlantic Monthly* promoted the blurring of these genres and created a popular market for nature writing, a need met largely by women writers. The regionalist focus that many women nature writers endorsed during this time period did not, as Littenberg points out, imply a narrowness of vision.

In the other essays in this section, close readings of popular works by authors as varied as Willa Cather, Gene Stratton-Porter, Zora Neale Hurston, Sarah Orne Jewett, and Florence Merriam demonstrate the keen ecological awareness mentioned by Thomas Lyon and its integration into a variety of literary formats.

Cheryl Birkelo argues that Gene Stratton-Porter, unlike her contemporaries John Muir and John Burroughs, was forced to submerge her serious environmental concerns in the guise of romantic literature. Stratton-Porter faced a criticism similar to that leveled at Susan Fenimore Cooper: Her fictional writing was characterized as overly sentimental. Her novel *The Harvester,* however, although overlooked by her reviewers, contained a carefully crafted ecological treatise that advocated a balance between preserving the land and allowing the land to remain economically productive. Stratton-Porter's work demonstrates the difficulties one encounters when attempting to move beyond or between unnecessarily rigid genre distinctions, and the inability of contemporary critics to appreciate the power of fiction within the nature writing tradition.

In Mary Ryder's close study of Willa Cather's Nebraska novels, Ryder makes a convincing case for a reconsideration of Cather as an ecofeminist writer. While many critics in the past have focused on the importance of Cather's characters, both male and female, to feminist criticism, Ryder outlines the fundamental importance of Cather's ecological concerns to her fiction. She demonstrates Cather's growing dismay with a world that "broke in two" as nature was swept aside in favor of human need and greed. Cather gives her male protagonists a female ethic that sounds the warning about the cost of our conquest of nature, and her work argues generally for a renewed attention to the role of fiction in the nature writing tradition in American literature.

In a similar way, Valerie Levy introduces us to the anthropological writings of Zora Neale Hurston and shows how Hurston elucidates the intimate connection between place and identity for Florida's black folk cultures. Drawing on Lawrence Buell's definition of an environmental text in *The Environmental Imagination* as one in which "the nonhuman environment is present not merely as a framing device but as a presence that begins to suggest that human history is implicated in natural history," Levy shows that Hurston's Florida writings for the Federal Writing Project occupy a central position in the nature writing continuum, where human history and natural history as so intertwined as to be inseparable. Levy makes the case in her essay that Hurston's FWP writing "brings nature to the forefront as a main character in the drama of American history and life."

Kelly Richardson adds another important layer to our understanding the complexity of the Maine writer Sarah Orne Jewett. Richardson shows how Jewett's *The Country of the Pointed Firs* and Dunnet Landing stories explore nature's benevolent, malevolent, and indifferent aspects as part of a multifaceted world in which humans do not necessarily dominate. Jewett's work, and perhaps, as an extension, much of what has been labeled regionalist writing, is significant in part because of the interconnectedness between human society and natural environment that dictates much of its action. As Richardson shows, the natural environment of Jewett's Dunnet Landing is neither simplistic nor overly sentimental. In noting a parallel to Dillard, Richardson emphasizes that Jewett's portrayal of the natural world is a complex and at times contradictory one, one that reflects a similar ambivalence in the society in which Jewett wrote.

In the concluding essay of this section, Jen Hill argues that ornithologist Florence Merriam proposed a new way of seeing nature, which invited women, children, and the uneducated to participate in the scientific discussions of the late nineteenth century. Contrasting Merriam's approach with that of scientists such as John Burroughs, Hill demonstrates how Merriam's writing marked a shift in ornithology toward the study of birds in their natural habitats. As with popular nature writing, Merriam's work built bridges between scientist and layperson, and between children and adults. In the process, she contributed significantly to expanding the genre of nature writing as well as its audience.

Marcia B. Littenberg

Gender and Genre

A New Perspective on Nineteenth-Century Women's Nature Writing

In the last decades of the nineteenth century, American popular culture embraced nature and nature study in a number of important ways that encouraged women writers. Women writers published in a wide variety of popular magazines, as well as in scholarly journals, ranging from the *Atlantic Monthly* and *Harper's New Monthly Magazine* to *Nature, Scientific Monthly, Audubon,* and the *American Naturalist.*[1] As this partial list indicates, a distinction was not always made between scholarly, scientific nature writing, and popular writing; the genre of nature writing during the late nineteenth century embraced not only appreciative personal essays and sketches but narratives of field work that combined science and sentiment. In addition to the narrative flexibility of nature writing during this period, women nature writers found a growing marketplace after 1860 in the popular journals and magazines that turned writing into a commercially viable career. As Richard Brodhead notes in *Cultures of Letters,* during the late nineteenth century, partly as a result of the growing number of popular literary and cultural journals and periodicals, "writing itself was being reestablished as a social activity in America, made the subject of a new scheme of institutional arrangements that stabilized the relations of authors to readers and solidified the writer's public support."[2] Women nature writers therefore had an unprecedented number of literary options and a wider, more diverse reading public available to them, depending on where they chose (or were able) to publish, and the particular form and style in which they chose to write. Recontextualization of nineteenth-century American women's nature writing thus requires reconsideration of the significance of gender and genre to its popular appeal.

Nature writing, by men or women, as science or as entertaining sketch, shares a common rhetorical purpose: to change and to deepen readers' awareness of the world around them and accessible to them, if they had but eyes to see. The nature writer provides those eyes as well as interpretive strategies that allow readers to appreciate and understand what is being described. This common rhetorical purpose links Henry David Thoreau to Rachel Carson, John Burroughs to Celia Thaxter and Annie Dillard. What differs is the historical and cultural context of these writers and the particular concerns and interests of their readers. During the last quarter of the nineteenth century, in spite of their expressed enthusiasm for nature study and nature writing, Americans demonstrated marked ambivalence toward nature. On the one hand, they expressed with great sentiment their appreciation for the "sublime" beauty of unspoiled nature, and, on the other, they recognized that without aggressive programs of preservation, "unspoiled nature" was threatened with extinction.[3] By the last decades of the century, nature appeared further removed from the everyday lives of most Americans, endangered by unchecked expansion, threatened by increasing industrialization and urbanization, and subsequently even more precious. Led by its most visible spokes*men,* John Audubon, John Muir, John Burroughs, and Theodore Roosevelt, a public environmental reawakening occurred in the last decades of the century comparable in its impassioned rhetoric and religious fervor to the Great Awakening.[4]

The public's appetite for nature study and nature writing was enormous. When they could not tramp off to the woods and mountains themselves, Americans enthusiastically and often indiscriminately devoured nature writing by amateur enthusiasts and professional naturalists alike. Nature writing was a form of retreat narrative.[5] Reading about nature provided vicarious experience, an excursion for urban readers, a day in the country, a summer among the pointed firs, a birding trip on a bronco, a morning in field or meadow with an informed guide. It was instructive, teaching city dwellers plant names, the characteristics of bird song, plumage, and habits, and introducing readers to the intricate relationships among plants, insects, animals, and birds in environments both exotic and near-at-hand. It awakened moral sensitivity to the importance of ecological relationships and thus encouraged the flowering of preservationist societies, such as the Audubon Society and later the Sierra Club, as well as the formation of garden clubs and bird-watching societies.

Although the most visible figures in this late nineteenth-century nature revival movement were male, women nature writers published extensively in the leading journals of the day and in a wide variety of popular magazines, including the *New Yorker, Harper's New Monthly Magazine, Atlantic Monthly, Lippincott's, National Geographic,* and *Nature.* In his study of nature writing in *The Environmental Imagination,* Lawrence Buell notes that "roughly half the nature essays contributed to the *Atlantic Monthly* during the late nineteenth century, the point at which the nature essay became a recognizable genre, were by female authors."[6] What helped to create this market for popular nature writing and to encourage women writers in this endeavor?

In part, it was the increased specialization of the natural sciences. The more specialized and academic the study of botany, zoology, or ornithology became, the less the general public felt connected by feeling and observation to nature. Combining first-hand observation with more accessible language and style, popular nature writing found an increasing market, much as the articles in such journals as *Popular Science, Natural History,* or the *Smithsonian* have for lay readers today. Because they served as guides and interpreters of the natural world, popular nature writers provided an important link between a public that felt itself cut off from nature, nostalgic for a simpler rural life, and eager to learn more about the world around them. In addition, women nature writers almost exclusively dominated the market for children's nature books used in schools and read with pleasure and enthusiasm by generations of young readers.[7] Their names are less familiar to us not only because the scientific study of nature was dominated by male institutions and the leading conservation movements were spearheaded by male naturalists, but because the focus of many women nature writers was often on more local or domestic habitats rather than wilderness studies.[8] They characteristically wrote a type of regional portrait that focused on local habitats and often demonstrated a special empathy for the natural world.[9]

The style and tone of women's popular nature writing was also marked by "sentiment," an effusion of feeling for their subject and an appeal to its human interest. Frequently, women nature writers employed metaphors from domestic life directed at their largely young or female readership.[10] These stylistic traits contributed to the later disregard of the scientific value of women's nature writing of this period, in spite of the wealth of specific observational details it characteristically contained. Professional jealousy was also a factor in obscuring the contributions of women naturalists and nature writers. In the 1870s and after, many educated women, encouraged by the introduction of botany and other natural sciences into women's academies and colleges, joined scientific organizations and sought work in museums and observatories. By the 1880s and 1890s, as the history of women in science suggests, there was a negative reaction to the "encroaching" of women into major and visible positions in professional organizations. Florence Merriam, for example, clearly a leading American ornithologist and nature writer, was finally accepted into the prestigious all-male American Ornithological Union as its first female member in 1885, but only in the capacity of an "associate member," not as a fellow.[11]

The dismissal of women's nature writing as more sentiment than science was also the result of a particularly heated public literary debate carried on in the nation's leading magazines and newspapers at the turn of the century. Led by John Burroughs's attack on Rev. William J. Long in the pages of the *Atlantic Monthly,* this debate focused on both the style and purpose of nature writing. Burroughs argued that unlike scientific study, the literary treatment of nature "aims to give us the truth in a way to touch our emotions . . . and to satisfy the enjoyment we have in the living reality."[12] While both the natural scientist and the literary nature writer convey natural facts, the nature writer "observes and admires" while the scientific naturalist *collects.* "The former would enlist your sympathies and arouse your enthusiasm; the latter would add to your store of exact knowledge," notes Burroughs, who adds that both on occasion may be guilty of coloring or falsifying facts, but for different reasons. Burroughs acknowledges that the nature writer gives more than facts, "he gives you impressions and analogies and, as far as possible shows you the live bird on the bough [not the laboratory specimen]." The problem, Burroughs asserts, is not the different goals or methods of the scientist and nature writer but a tendency toward the false anthropomorphizing of bird, insect, and animal behaviors in the latter in order to invest nature study with "human interest." The false attribution of human behavior to animals and birds, Burroughs agues, betrays nature and diminishes nature writing to a kind of Sunday-school lesson.

Burroughs's specific criticism of Long and other "sham naturalists" who engage in the misrepresentation of nature for public consumption thus rests on an essential critique of style as much as on method. While not specifically aimed at women nature writers, combined with the jealously guarded professionalism of the natural sciences, Burroughs's critique of literary sentimentalism did much to obscure the real contributions of women naturalists such as Florence Merriam, Olive Thorne Miller, and Mary Treat, and accounted in part for the disregard of women's nature writing of this period by later literary historians, who tended to dismiss it as sentimentalism.

The goal of the women nature writers did differ in part from that of male naturalists such as Burroughs, although like them he believed that "the most precious things of life are near at hand" and require only the ability to see and understand.[13] Although they often provided extremely valuable field study observations, the primary goal of popular nature writers was to transmit their observations and experience to their readers, to create an imaginative awareness of the natural environment, and to shape readers' moral response to nature. Olive Thorne Miller, for example, pointed out in her *First Book of Birds* (1899) that she was less concerned with "the science of ornithology" than with arousing "sympathy and interest in the living bird." Men who studied dead birds, she notes, "can tell how their bones are put together and how many feathers there are in their wings and tails. It is well to know these things. But to see how birds live is much more interesting than to look at dead ones. [The watcher will be] surprised to find out how much like people they act. After studying live birds, he will never want to kill them. It will seem like murder."[14] Florence Merriam, the best pure ornithologist among the women writers, wrote *Birds through an Opera Glass* (1889), *Birds of Village and Field* (1898), as well as numerous articles for *Audubon Magazine* primarily for novice bird-watchers. Although her writings are based on careful field observations in a variety of settings, her goal is to translate her enthusiasm for nature study to amateurs. Her valuable observations about bird song, markings, colorations, and habits in a highly readable style are marked by occasional unscientific metaphoric comparisons, directed at the amateur reader, attributing dubious sentiments to birds, such as her description of the robin in *Birds through an Opera Glass* as a type of "self-respecting American citizen, sitting on a branch, whispering a little song to himself . . . full of contented appreciation of the beautiful world he lives in."[15] Here, it is clear, her desire to draw a useful lesson from nature whimsically colors her description of this common American bird.

Partly because of the different emphasis and tone of women's nature writing in this period, later historians comfortably ignored this large body of popular nature writing, dismissing both its seriousness as science and significance as literature. This is the legacy that feminist scholarship has begun to reclaim. Vera Norwood's study *Made from this Earth* and the work of Marcia Myers Bonta on American women naturalists, *Women in the Field: America's Pioneering Women Naturalists* and *American Women Afield: Writings by Pioneering Women Naturalists,* as well as more comprehensive anthologies of American regionalism, have begun to unearth women's nature writing of this period, recontextualizing this writing, and claiming an important literary legacy by shedding new light on women's contributions to nature writing and the environmental movement.[16] Most of these studies focus primarily on nonfiction nature writing. Both Norwood and Bonta begin their study with Susan Fenimore Cooper's *Rural Hours* (1850) and include Florence Merriam Bailey and Rachel Carson; only studies of American regionalism discuss Celia Thaxter and Sarah Orne Jewett's descriptions of the natural environment in any detail, although a notable exception to this critical division between environmental nonfiction and literary bioregionalism occurs in Lawrence Buell's study *The Environmental Imagination.*

It is my contention that the study of nineteenth-century American women nature writers should be broadened to include not only the nonfiction naturalist essay or "sketch" but the regionalist sketch, often classified as regionalist *fiction,* because both share a parallel literary endeavor. There is clearly an overlap in the subjects, style, and intent of these related genres: to increase readers' awareness of the web of relations—whether ecological or sociological and familial—within a natural landscape. Although the regional sketch often employs dialogue and depicts characters within a community located in a rural setting, and nature writing is more frequently a narrative of solitude, both of these genres demonstrate the writer's interest in natural history, provide detailed accounts of the natural landscape and close, personal observations of the natural world and the rhythm of life in a particular locale.[17] In addition, both share a common preservationist theme and often a similar tone, perspective, and focus. In the regional sketch, as in nonfiction nature writing by women, the writer serves as a guide, interpreter, and enthusiast, mediating between the reader and the environment being described, creating sympathy and concern for a world threatened by time, change, and external disruption. By reconsidering the regional sketch as part of the larger endeavor of popular nature writing, and by including portraits of rural people and rural communities that live in harmony with the natural world as part of a preservationist impulse, one can also appreciate more fully the role of women writers of this period and begin to reconstruct a female literary tradition in which these genres are more clearly interrelated.

The regional sketch and the nonfiction nature essay both gained recognition in the nation's leading literary and cultural journals during the last third of the nineteenth century. While it has become common practice to distinguish between the sketch, which contains fictional elements such as dialogue and characterization, and the nature essay, which shares many of the characteristics of the personal journal, it appears that many of the leading periodicals often did not draw such genre distinctions, treating both the "local color" sketch and the nonfiction nature piece as mutually supportive parts of an important textual and cultural agenda, to inform their readers about the wonders and delights found in nature through realistic narrative. The featured article in each issue of the *Atlantic Monthly,* for example, was often a descriptive travel narrative, a short essay about the natural wonders of a particular place, a natural history sketch, or a profile of an individual species, particularly birds or forest animals, as well as a sketch or story, often by Rose Terry Cooke or Sarah Orne Jewett. Never before or since this period have women writers been as widely represented in these parallel genres or found such favorable conditions for publication.

In his discussion of the "variegated character" of nature writing in the appendix to *The Environmental Imagination,* Lawrence Buell suggests that nature writing draws on the techniques and focuses of many genres.[18] Then, as now, the parameters of nature writing were not fixed. The mark of popular nature writing from the mid-nineteenth century, Buell notes, is the creative freedom it allowed its writers, combining empirical observation, popularized science, natural history, legend, folklore, gossip, and the picturesque.[19] Buell's comments are supported by an advice column in the *Atlantic Monthly* to its "Contributor's Club," which published essays anonymously, suggesting that the nature writer might use a variety of methods as long as the writer thinks of "himself" as a lens through which images of nature are filtered.[20] Buell himself includes both Sarah Orne Jewett's regional writing and Celia Thaxter's *Among the Isles of Shoals* (1873), which utilizes fictional methods as well as natural history techniques, in his study of "environmental nonfiction," pointing out the overlap between travel narratives, literary bioregionalism, and genre sketches. Buell suggests that nonfictional literary bioregionalism and local color genre sketches frequently overlap in their methods, their "finely calibrated environmental sense," their attachment to locale, and in their attempt to render imaginatively the look and feel of a place.[21] Because literary regionalism was dominated by women writers and nature writing more clearly identified with male writers, Buell suggests that in writing about nature, "male" and "female" literary traditions overlap.[22] A more convincing argument can be made for the merging of two related female literary genres, two aspects of regional writing, that explore the natural world and affirm the connections between the environment and the human community.

Although regional realism has on occasion been faulted for the narrowness or "impoverishment" of its vision, reading women's regional sketches as a form of "bio-regionalism," as part of a larger effort to write about nature in a particular locale, helps us to recontextualize both their subject and method.[23] Limitation of geographic scope, for the nature writer or regionalist, does not infer a narrowness of vision, as Thoreau's *Walden* so effectively proves. Even while observing trout beneath the ice of a pond he is capable of demonstrating nature's profoundest mysteries. Nature can provide a wealth of detail and an abundance of material. Helping readers see this, is, in fact, the purpose of both genres. In her work *Home Studies in Nature* (1885), for example, Mary Treat notes of her study of insects and spiders in her one-acre backyard plot, "the more I limit myself to a small area, the more novelties and discoveries I make in natural history."[24] The women regionalists would also prove that a wealth of material could be found in the smallest and seemingly most desolate of villages or even in a front garden.

It is clear that the nature essay and the regional sketch share a number of aesthetic concerns and a common objective, to offer a sympathetic portrait of nature. Both have what Buell terms "dual accountability": an attempt to represent the natural world truthfully, and conceptual or imaginative fidelity, that which makes the natural world in all its complexity accessible and comprehensible to readers.[25] The first requires careful observation, the second, not only literary talent but a sympathetic identification with the subject as well. In the popular nature essay as in the regional sketch, narrative elements were often combined with naturalistic observations, encouraging readers to appreciate the natural world or to visit scenic regional locales and translating through representational fidelity the writer's feelings about a landscape. What mattered was the author's ability to convey a truthful sense of place, one that linked descriptions and explanations of the physical environment with the writer's subjective, identifiable response to it.

As Richard Brodhead reminds readers, nineteenth-century regional writing was written for a different, more literate and privileged culture than it portrayed.[26] Similarly, popular nature writing flourished in post-bellum America precisely because Americans were beginning to recognize that nature was endangered. The elegiac feel of women's nature writing, combined with close, detailed observation, echoed the concerns of the reading public.

By reading nineteenth-century women's nature writing as a form of regionalism, moreover, one can perceive certain stylistic affinities between these related genres. One of these is the metaphoric comparison between nature and human experience, as I have shown earlier in the quotations from the bird studies of Olive Thorne Miller and Florence Merriam. In the regional sketch, the metaphor often works the opposite way, so that human behavior can be understood and appreciated in terms of nature. In the sketch "Root-Bound," for example, Rose Terry Cooke recounts a dialogue in which Mrs. Rockwell explains to her two female visitors why her plants blossom so profusely. She explains modestly that it is because they are "root-bound" rather than wandering willfully all over the garden and adds: "It's good for folks and flowers to be root-bound, I think, sometimes; especially if we want to bring forth good fruit."[27] The women nature writers often explained the behavior of birds through homely domestic metaphors that would be appreciated by their largely female readership, such as the way the male king-bird jumped up politely when the female returned to the nest or how he held the dragonfly in such a way that the young could nibble off small bites that they could swallow, or how the chickadee defended its nest or a mother crow disciplined her disobedient chick when it would not come when she called. Imbuing nature with comparisons to domestic life not only made it interesting to readers, it helped to create concern for its protection.

One important difference between the style and method of women's regional sketches and women's popular nature writing requires some explanation, however, particularly for contemporary readers for whom the emotional style and domestic imagery of the women nature writers appears to be sentimentalizing. The women regionalists openly rejected the sentimentalism of early women's fiction in favor of a less rhetorical rendering of truthful experience, however homely or prosaic it might first appear. For the regionalists, sympathy was elicited by readers'identification with the rural locale and the characters' lives; it developed from pictorial realism, the faithful rendering of realistic details.[28] The nature essay also relied on its accurate rendering of observational details; however, the narrator/observer more directly intervened to explain and interpret these details and elicit her readers' sympathy. The difference in rhetorical style is greatest in those works directed to young readers. However, the nonfiction nature sketch has always utilized rhetorical interpretation to some extent, since its "characters" cannot speak for themselves and the actions of nature, however dramatic, require the naturalist's explanation as well as the artist's vision.

Even the designation "local color" for regional sketches suggests the analogy with the pictorial arts common in nineteenth-century reviews and essays. In an 1886 essay in the *Critic*, for example, James Lane Allen compares the painter's and regional writer's use of color, decreeing that "the writer must lay upon his canvas those colors that are true for the region he is describing and characteristic of it."[29] He adds: "From a scientific point of view, the aim of local color is to make the picture of life natural and *intelligible*, by portraying those picturable potencies in nature that made it what it was . . . " The growing popularity of genre painting and the new emphasis on *pleine aire* composition in the visual arts may both have encouraged regionalist nature writers and provided an important cultural context in which their work was read and appreciated. The key aesthetic term for this period is the "picturesque," a term that was employed favorably by John Ruskin in *Modern Painters*, and had been adopted, nearly as gospel, by American artists and art journals, particularly America's leading art journal of the period *The Crayon*. The humble naturalism of the luminist painters, such as William Sidney Mount, Fitz Hugh Lane, and Asher Durand, who depicted small, intimate landscapes and quiet scenes of rural life, provides an important pictorial analogue to the aim and method of both fictional and nonfictional regionalist nature writing. The small, picturesque luminist painting, suffused with a quiet light, suggests a world, a time, and a place free of the dividing conflicts of the age. Moreover, these genre paintings, like regionalist nature writing, reveal carefully detailed observation of the natural world, particularly of plants and animals, which often dominate the foreground. They also suggest the organic connection between nature and rural human experience central to the regionalist purpose. The pastoral idea held for a moment the balance between science, religion, philosophy, and art, offering a glimpse of the world as it might be or might have been, whole and luminescent, before it was dissected and analyzed by science or destroyed by the march of progress.

In France, the Barbizon school of painting, whose most famous member was Jean-Francois Millet, deserted the salons for the open air, and sought in paintings of nature and rural peasants the same sense of simplicity, truth, and organicism that prompted Thoreau's original "experiment in living." Millet's American student, William Morris Hunt, an intimate friend of Celia and Levi Thaxter's, influenced through his art and instruction a rising generation of Boston artists, particularly women artists.[30] Hunt spread Millet's influence throughout the Boston artistic and literary community. Jewett's biographer, Paula Blanchard, cites some of the most dramatic examples of Hunt's influence on Jewett and her friends, including a series of photographs in the Millet style, some of which were used as illustrations for the 1893 edition of *Deephaven,* and a collection of "picturesque" stories, *A Week Away from Time.* However, the deepest impression on Jewett and her contemporaries was created by the picturesque rendering of the rural world in the paintings themselves. The paintings of Hunt and others lent a romantic charm and a classic dignity to portraits of rural life. The renewed interest in landscape art and in painting from nature inspired by Hunt and his followers encouraged women writers and artists. It further invigorated the genre of nature writing by creating a wider audience for "picturesque" writing as well as by connecting it to its classic pastoral lineage. In such an intellectual and cultural climate, women's nature writing flourished.

Both the aesthetic affinities between women's popular nonfiction nature writing and the regional sketch, and the cultural conditions that encouraged women writers to develop these related genres, suggest the need not only for historical recontextualization of women's nature writing but a reconsideration of these parallel female genres. It is important to recognize that the cultural conditions that encouraged women writers by providing them with journals that published their writing and a large, influential middle-class readership also extended the genre of nature writing.

Women's nature writing flourished in a cultural climate that valued nature but demanded change, that looked nostalgically back at rural life even as it was transforming it from reality into myth. In a nearly unprecedented way, the perspective offered by popular women's nature writing both in fiction and nonfiction found its place in American letters in the last part of the nineteenth century.

Cheryl Birkelo

The Harvester and the Natural Bounty of Gene Stratton-Porter

Gene Stratton-Porter, an early-twentieth-century popular fiction writer and chronicler of natural history, is especially known for her novels of the Limberlost Swamp of east-central Indiana. These texts include her original photography as well as personal anecdotes and observations. Appreciated primarily for her fiction with a message of land stewardship, Stratton-Porter was often panned by critics for writing only sentimental romance.

Such criticism may have resulted in part from the dominance of male writers in the field at the turn of the century. John Burroughs and John Muir are representative of the male writers of the environmental movement that occurred then and continues in society today. Burroughs initially wrote of his local eastern birds and their rural natural history and "established nature study as a popular pursuit, set many of the standards for nature writing, and gave the natural history essay its definitive form."[1] Stratton-Porter often refers to Burroughs as a source of influence in her studies, "and greatly admired and avidly read him, although he refused to read her."[2] Burroughs was strongly influenced by his life-long friends, poet Walt Whitman and President Theodore Roosevelt. These men also greatly influenced Stratton-Porter. One early public-speaking engagement Stratton-Porter gave was a critique of Whitman's poetry.[3] Stratton-Porter openly admired the President in her essay "By the People."[4]

John Muir was as western and wild as Burroughs was eastern and rural. Muir, as the founder of the Sierra Club in 1892, "wanted to preserve wilderness not for its timber resources or game but for its very wildness, beauty, and spiritual significance."[5] Muir's writings and beliefs were known and also shared by Stratton-Porter.[6] Other prominent male nature storywriters of the times included Bradford Torrey, Ernest Thompson Seton, and Jack London.[7]

Upon the publication of *The Harvester* (1911), Gene Stratton-Porter joined her male contemporaries in relating nature stories to an increasingly environmentally sensitive public. *The Harvester* was a novel seemingly focused on the Harvester David Langston's search for the beautiful woman he envisions in the swamp and on his overcoming numerous obstacles to win her love. While reviewers almost universally acknowledged that this novel, more than Stratton-Porter's previous ones, allowed a glimpse into the author's own attitudes toward nature and human nature, what has long been overlooked is the importance of *The Harvester* as an ecological treatise that advocates the careful stewardship required for preserving the land without compromising its productivity.

The Harvester tells of David Langston's life in Medicine Woods, which he shares with his dog Belshazzar, his horse Betsy, and the flora and fauna he cultivates and protects. The choice of his solitary lifestyle is left to the yearly response of Belshazzar as to whether the two remain alone in Medicine Woods or open it up to include a possible mate. Upon envisioning the ideal woman, Langston sets out with the help of his friend Dr. Carey, from the nearby town of Onabasha, to find her. He finds Ruth Jameson, but he must overcome the trials of curing Ruth's illness and discovering her distant relatives. Ruth chooses to share the natural beauty and serenity of Medicine Woods rather than life in the city with her grandparents and Doctor Harmon, who admired her greatly and who attended her dying mother. By sharing life with her Harvester, she can earn her own way, drawing original patterns for architectural and clothing design firms based on the local flora and fauna. Transferring natural specimens into art is yet another way in which the two preserve and share Indiana's natural bounty with urban dwellers. The book happily concludes with their marriage and resumption of their idyllic life in Medicine Woods. As might be expected from this summary of the plot, reviewers paid virtually no attention to the ecological issues that Gene Stratton-Porter raises.

The October 1911 *A. L. A. Booklist* deemed *The Harvester* "sentimental to a degree and rather attenuated," while the *New York Times* more generously observed with its response, "In spite of one or two glaring faults, [*The Harvester*] is worthy of warm commendation because of its strong individual note [and] its sincerity . . ."[8]

Such reviews focus on plot and sub-plot structure rather than on the explicit and implied messages with which Stratton-Porter endowed her story. Stratton-Porter readily responded to her critics, regarding their interpretation of her characters as "unreal," saying "so as deeply as I see into nature I also see human nature. I know its failings, its tendencies, its weaknesses, its failures . . . [but *The Harvester*] reproduces a picture true to ideal life, to the best men and women can do . . . so I lighten each nature book with enough fiction to make it readable to those not interested in nature work—sugar coat their pill, so to speak . . . I knew the fate of any nature book better than most folk."[9] Stratton-Porter was as good a judge of human nature as she was of animal nature. She pragmatically realized that although she saw *The Harvester* as a nature book, it would only be profitable if she included her "sugar-pill" of romance. Her daughter Jeanette Meehan recounted that her mother "considered intimacy with all outdoors and nature one of the best pathways to God and his teachings. So she wrote this book about a young man [who stayed] in the woods and earn[ed] a good living by things he found there. She based her story on growing ginseng and other herbs which can be sold to druggists and chemical supply houses."[10]

The novel's protagonist, David Langston, is more than the conventional hero, however; he is a Thoreau-like figure (the novel was dedicated to Thoreau), who "harvests" the medicinal plants of the swamp to preserve them from extinction *and* to sell them as life-saving pharmaceuticals. The farmers of the region misunderstand Langston's non-conventional farming practices, not recognizing that his resistance to draining the swampland stems not from his rejection of cash-crop farming but from the ecological ethic that native plants must be propagated and preserved in their natural habitat. Only by doing so can humans and nature form a harmonious co-existence, to the benefit of both.

Langston, early in the novel, recounts how his mother instilled his land ethic as they gathered seeds, leaves, and barks to pay for his education, an ethic and knowledge supplemented by the few extension pamphlets available. He expanded his knowledge base to include textbooks and drug pamphlets and so learned "every medicinal plant, shrub, and tree of his vicinity, and for years roamed far afield through the woods collecting."[11] After the death of his mother, he continued his solitary ways ignoring the times "when his [farming] neighbors twitted him with being too lazy to plow and sow, of 'mooning'over books, or derisively sneer[ing] when they spoke of him as the Harvester of the Woods or the Medicine Man."[12] He allowed no trees to be felled and no hunting of ducks or deer upon his lands, leaving the land instead as a preserve necessary for the plants' and animals' essential habitat.[13] Indeed, whenever he harvested a tree, he apologized, as is the Native American practice, for taking the spirit and life away from the specimen harvested.[14] In this way, Stratton-Porter again has the Harvester stand out as an individual different from the typical homesteader or farmer.

Stratton-Porter further explores her land ethic, foresight, perseverance, and patience through the Harvester's conversation with Dr. Carey: "Ten years ago I began removing every tree, bush, vine, and plant of medicinal value from the woods around to my land; I set and sowed acres of ginseng, knowing I must nurse, tend, [and] cultivate seven years. If my neighbors had understood what I was attempting, what do you think they would have said? Cranky and lazy . . . Lunatic would have expressed it better. That's close [to] the general opinion, anyway. Because I will not fell my trees, and the woods hide the work I do."[15]

Equally important, however, are the conscious distinctions that Stratton-Porter has the Harvester make when collecting specimens, distinctions concerning the needs of humans compared to the needs of the plants and animals that he and she champion: "I vow I hate to touch you . . . I got to stop this or I won't be able to lift a root. I never would if the ten cents a pound I'll get out of it were the only consideration . . . you are indispensable to sick folks . . . humanity comes first, after all.

Seems as if animals, birds, flowers, trees, and insects as well, have their right to life also."[16]

The Harvester and Gene Stratton-Porter were friends to all living creatures because the animals and birds knew "who fed and everyday protected them and so were unafraid."[17] In conversation with Ruth, David observes that animals and birds in particular do know and interpret fear:

I am familiar with them, and that is not correct. They have more to fear than human beings. No one is going to kill you merely to see if he can shoot straight enough to hit. Your life is not in danger because you have magnificent hair that some woman would like for an ornament. You will not be stricken out in a flash because there are a few bits of meat on your frame some one wants to eat. No one will set a seductive trap for you, and, if you are tempted to enter it, shut you from freedom and natural diet, in a cage so small you can't turn around without touching bars . . . I also have observed that they know guns, many forms of traps, while all of them decide by the mere manner of a man's passing through the woods whether he is a friend or an enemy. Birds know more than many people realize. They do not always correctly estimate gun range, they are foolishly venturesome at times when they want food, but they know many more things than most people give them credit for understanding. The greatest trouble with the birds is they are too willing to trust us and be friendly, so they are often deceived.[18]

When collecting plant specimens to present them for future enjoyment by Ruth in the Wildflower garden, the Harvester takes special care to see that habitat requirements are met by placing plantings in the "soil of the same character . . . [so that] few knew they had been transplanted."[19] Stratton-Porter makes the Harvester's philosophy echo her own on the ethics of all beings' right to life when he relates to Ruth why he (and Stratton-Porter) won't kill a moth to have as a permanent model for their drawings and carvings:

I must have a mighty good reason before I kill . . . I cannot give life; I have no right to take it. I am afraid.
Of what please? [Questions Ruth.]
An indefinable something that follows me and makes me suffer if I am wantonly cruel.[20]

Further, Stratton-Porter makes note of the condition of the lands and waterways in *The Harvester* and proclaims her stance on the intrusion of careless human exploitation upon the environment. When out gathering ginseng, Stratton-Porter observes that "a wry grin twisted [his] lips as he clambered over the banks of the recently dredged river, standing an instant to look at its pitiful condition and straight muddy flow."[21] Stratton-Porter personifies the river, giving it an emotional response of its own. She writes, "The Harvester again walked down the embankment of the mourning river [emphasis mine]."[22] The river bemoans the interference of greedy, indifferent human hands that alter its natural flow for commercial gain and the destruction of the ecological balance of the native soils, plants, and animals. Stratton-Porter puts the blame not on farmers in general, but on those farmers who do not respect and properly care for their lands. In the same scene, while observing the river, the Harvester declares: "Appears to match the remainder of the Jameson property . . . I don't know who he is or where he came from, but he's no farmer."[23] What Jameson epitomizes for Stratton-Porter is the profit-making agriculturist who callously resists environmentally sound tilling and watershed practices. The condition of Jameson's land adjacent to the river demonstrates this lack of concern for both preservation of the land's fertility and purity of the waters. Were Gene Stratton-Porter alive today, she would continue to be an outspoken supporter of, and for, modern no-till soil and water conservation practices. In contrast to Jameson, Stratton-Porter has the Harvester gather from the land enough of a yield to exist, while not exploiting the natural bounty for unethical gain. The Harvester comments about conducting life in harmony with the natural world: "You not only discover miracles and marvels in [the woods], you not only trace evolution and origin of species, but you learn the greatest lessons taught in all the world, early and alone—courage, caution, and patience."[24]

Critics have charged Stratton-Porter with excessive sentimentalism because they find her habit of anthropomorphizing elements of nature as "unscientific" and lacking the clinical objectivity supposedly necessary for "true scientific discourse." The label of "nature faker" was applied to popular nature writers of Stratton-Porter's time who wrote for the general public, but who so heavily laced their animal characters with human traits that they became unbelievable. This epithet was especially applied to William J. Long by President Theodore Roosevelt and John Burroughs in December 1902 and endured in the popular press for almost ten years. Such a prevailing sentiment was undoubtedly a factor in the critical neglect of Stratton-Porter's writings.[25] Bertrand Richards notes that Stratton-Porter was "particularly incensed" by Frederic Taber Cooper's remark that she was a "'nature faker' . . . and that the scientific value of her labor is largely discounted by the open and unabashed self-satisfaction of her style."[26] Yet, her familiar tone is comforting and engaging to the ordinary and mostly female readers of her time who would be perhaps intimidated by too much scientific jargon. A critic of *Moths of the Limberlost* said of the book that which can also apply to *The Harvester:* "The nature lover rather than the naturalist in the strict scientific sense of the word has been kept in mind thruout [*sic*] the preparation of the work whose aim is to teach . . . Her observations are scientifically valuable, her narrative is entertaining, and her revelations stimulating."[27]

Along with this is the manner in which Stratton-Porter presents the Harvester, David Langston, as a self-trained environmental scientist/pharmacist. Throughout the text, Stratton-Porter catalogues medicinal plants and their value to humanity and the economy to teach her readers about them. She has David ponder the medicinal value of honey that has been infused with the healing properties of certain plant extracts, and she has him compare the purity of wild to cultivated ginseng.[28]

His reputation as a provider to humanity thus becomes most important to him. He recounts: "Gathering stuff for drugs is very serious business. You see, I've a reputation to sustain with some of the biggest laboratories in the country, not to mention . . . I sometimes try compounding a new remedy for some common complaint myself . . . I want the money, but I want an unbroken record for doing a job right and being square and careful, much more."[29] One of the subplots of the novel involves the successful results the Harvester derives from a "nerve remedy" he had Dr. Carey test on appropriate patients, including Ruth.[30] He presents these findings to a medical convention and Stratton-Porter stresses his professional status.[31]

However, while these subplots reinforce the Harvester's message and character to help move the novel along, the love match between Ruth and Langston meld with Stratton-Porter's ideas of ecological conservation and preservation, but garnered the most negative criticism from reviewers. Yet, a close look at Ruth's role indicates that she, too, serves to fulfill Stratton-Porter's purpose as a nature writer. Like the Harvester, Ruth shows what Bertrand Richards calls "the general outlook" that Stratton-Porter supported, that "[t]here must be maintained a reciprocal and working balance between agriculture and nature."[32]

Although Stratton-Porter shows Ruth as initially motivated to sketch and harvest the wild plants and ginseng from the Limberlost region to pay off her debts, Ruth comes to appreciate the native landscape: "Before an hour she realized that she was coming speedily into sympathy with the wild life around her; for instead of shivering and shrinking at unaccustomed sounds, she was listening especially for them . . . instead of the senseless roar of commerce, manufacture, and life of a city."[33] She, as well as David Langston, "earn their livings from the flora, but they also make deliberate efforts to harvest wisely and to save threatened species."[34] At first fearful of the wild creatures and atmosphere of the swamp, Ruth, over time, allows David to teach her "the wonders" of the natural world and allows a moth to land on her hand and then fly to freedom.[35] Upon returning from town after their wedding, she wants to "worship" the plants and the natural setting of Singing Water bridge, which she admires "in silence," whereas the Harvester pragmatically tells of each plant's common name, medicinal function, and dollar value.[36] After the Harvester restores Ruth to her family, she refuses to stay in the city amidst her grandparents' wealth and protected, artificial lifestyle. Instead, she chooses to return to the "call of the wild [that] had been verified in the Girl."[37] Ruth parallels the biblical Ruth, who, gleaning in Boaz's fields, becomes his wife and allies herself with him.

The Harvester, then, incorporates many of Stratton-Porter's ecological concerns that she would continue to develop later in her nonfiction *Moths of the Limberlost* and *Friends in Feathers* (1917). After Stratton-Porter's death, *McCall's Magazine* in 1927 compiled her essays to be published posthumously as *Let Us Highly Resolve.* In these essays, she urged emphatically that "we do all that lies in our power to save comfortable living conditions for ourselves and the spots of natural beauty that remain for our children."[38] As her essays indicate, she realized fully from personal observation and history that the limits of nature's bounty were being reached in her home region, as well as nationally and globally. The closing lines in her essay "Shall We Save Natural Beauty" read: "There may not be coal and iron, [and other natural resources] at the rate at which we are using it, to supply coming generations. Any thoughtful person realizes that there will not. Certainly to plant trees and to preserve trees, to preserve water, and to do all in our power to save every natural resource, both from the standpoint of *utility and beauty* [emphasis mine] is a work that every man and woman should give immediate and earnest attention.[39] Thus, the stewardship and agricultural practices that Stratton-Porter advocates in *The Harvester* are those of an individual highly motivated to teach the untrained citizens who made up her reading audience how to conserve the natural beauty of not only her home territory of Indiana but of all landscapes "for the good of suffering Humanity."[40]

Like Stratton-Porter's nonfiction nature studies, her novels have been seriously underrated as contributions to women's nature writing. *The Harvester* merits reconsideration as an early twentieth-century tract written by an American woman to explore and expound upon the necessity of preserving Nature's bounty and the dilemma of appropriating her resources for the advancement and care of the human community.

Mary R. Ryder

Willa Cather as Nature Writer
A Cry in the Wilderness

Willa Cather's evocation of a sense of place, enhanced by her particularized landscapes and taxonomical detail, has long been documented, but only recently has critical attention focused on what Susan Rosowski, in her seminal article "Willa Cather's Ecology of Place," has called Cather's "ecological dialectic."[1] Cather's connection to the land, like that of the female protagonists of her early novels, was one of "love and yearning."[2] Yet, Cather is rarely considered a "nature writer."[3] Her precise and sensitive descriptions of the natural world could alone qualify her as a leading woman nature writer, but her ecological concerns were more far-reaching than merely to record nature's beauties and explore her mysteries.

A careful study of her later so-called "Nebraska novels" reveals Cather's growing disaffection with a world that disregarded or disrupted nature's delicate balance, a world in which thoughtless and greedy men raped the land for profit. For Cather, the world "broke in two in 1922 or thereabouts," and her two subsequent novels—*One of Ours* (1922) and *A Lost Lady* (1923)—are crucial to understanding the ecological issues that she recognized as part of that breakage.[4] In both works, Cather uses a male protagonist to voice her concerns, but their cries in the wilderness go unanswered, leaving them and the land ripe for destruction.

Much of Cather's fiction is "ecofiction," what Nicholas O'Connell calls "a hybrid genre combining the concerns of nature writing with those of narrative fiction." As O'Connell points out, ecofiction calls for "a new relationship between people and place, a relationship that emphasizes the spiritual dimensions of landscape and the need for people to treat the land with respect and reverence."[5] While even the earliest of Cather's works illustrated her appreciation of and respect for the land, as well as her understanding of its spiritual dimensions, most critics considered this emphasis on nature fundamental to her realistic settings or evidence of her appropriation of the pastoral or romantic tradition.[6]

More recently, however, critics have begun to place Cather among those whose focus on nature falls under the aegis of ecofiction, if not ecofeminism. Glen Love, in his Past President's address at the 1989 Western American Literature Association meeting, lauded Cather as one who moved "beyond the pastoral conventions to confront the power of nature that rebuffs society's assumptions of control." Placing Cather among those who never ignore nature's "primal energy and stability," as well as its pastoral qualities, Love assigns her an "eco-consciousness" rather than an "ego-consciousness" that would subdue and control nature.[7] Kathleen Nigro, in her 1997 unpublished dissertation, carries this claim further, relating a "nurturing reverence for nature" in Cather's fiction and situating Cather in opposition to the "traditional controlling approach to nature which has prevailed throughout American history."[8] Susan Rosowski, too, makes a strong case for ecological principles shaping Cather's arts with *O Pioneers!* as pivotal in articulating her land ethic.[9] The 1997 International Willa Cather Seminar offered new approaches to understanding Cather's relationship to the land, with special focus on her formative years in northern Virginia, including her acquaintance with the rich and abundant species of plant life in the region. John and Cheryl Swift's conference presentation, "Natives and Transplants: Cultivating the Wild in Cather's Remembered Gardens," argued convincingly that her childhood exposure to nature, coupled with scientific training at the University of Nebraska, led Cather to portray her "most particularlized landscapes . . . ecologically," rather than merely taxonomically.[10]

While none of these more recent voices go so far as to label Cather an ecofeminist, the ecofeminists' emphasis on reclaiming and redefining women's nature writing prompts this reassessment of Cather's work.[11]

Unlike those ecofeminists whose agenda are often political, utopian, or even reformist, Cather's focus is on the interconnectedness of humans and their natural environment. Cather's early works *O Pioneers!* (1913), *The Song of the Lark* (1915), and *My Ántonia* (1918) offer positive portraits of women interconnected with the natural world and attentive to its caretaking. From Alexandra Bergson's raising alfalfa to rejuvenate the land—an experiment considered foolish by her brothers—to Ántonia Shimerda's devoted watering of young trees, Cather acknowledges that "the great fact was the land itself" and it was "rich and strong and glorious."[12] The tutelary Genius that broods over the prairie landscape demanded homage from human interlopers and responded only to those who would protect its integrity. Such a response requires accepting, as does Thea Kronborg in Panther Canyon, a connection to a primordial world of the Great Mother in which an "ethic of responsibilities or care" supersedes self-assertiveness and desire for gain.[13]

These values are those of an ecofeminist, and Cather clearly links representations of nature with gender in her works, rejecting "a patriarchal environmental ethic that has conceptualized land as 'woman'" to be subdued.[14] From childhood, Cather recognized the need for stewardship of rather than dominance over the land. While Cather characterized the prairie as the greatest happiness and curse of her life, she always fought to preserve its native beauty.[15] The little girl who gathered prairie flowers in "great armfuls" and cried over them because they were "so lovely, and no one seemed to care for them at all!" became the mature writer whose detailed descriptions of nature were more than nostalgic longings.[16] First attracted to scientific studies at the University of Nebraska, Cather found herself exposed to the ground-breaking work of Charles Bessey, who "revolutionized botany by directing attention away from people's taxonomies and toward the field where one might study nature."[17] Here she also befriended F. E. Clements who, along with his wife Edith Schwartz, co-authored the one book Cather said she would rather have written than all of her novels—a botany on the wild flowers of the West, *Rocky Mountain Flowers*.[18] In public lectures, she advocated maintaining and appreciating native flora and fauna and was adamant about protecting trees. Once, when speaking to the Hastings, Nebraska, Women's Club, Cather made an appeal "especially [for] the giant cottonwoods," which second-generation pioneers were chopping down. She urged the planting of more trees, even if they took up "available ground" and meant the "sacrifice [of] a few bushels of wheat."[19] Fully aware that the cottonwood was, as she put it, "the only tree of beautiful form that grows easily and naturally in this state without any care," Cather appealed to farmers "to let stand those great trees," arguing that in low-lying areas they could draw up moisture from potential sloughs. She went on to encourage farmers' greater attentiveness to the "groves of ash and native elms," thinning them as necessary for their survival.[20]

While never an activist for political or social issues, Cather "took an active part in agitating for forest conservation and observation of Arbor Day, adopted first in Nebraska in 1872."[21] That her hunger "for the hills of home" (ostensibly Virginia) might have prompted her strong interest in trees is credible, but her respect for the conservation movement is rooted in her life on the prairie.[22] Her visit as a reporter to Brownville, Nebraska, to record the fortieth anniversary of the town's founding, resulted in a column in which the young Cather valorized a local resident who shared her ecological concerns. She describes Robert W. Furnas, once the state's governor and now the owner of an orchard estate and a leader in horticulture, as "a sort of forestry missionary to the whole country around him," a "connoisseur of trees," saying also that, "His solid, patient methods are a pleasant contrast to all the fizbang and sham so common in the West. He has the correct judgment and bold unerring insight that men who are much and close to nature always have, and . . . he sees perhaps further into the future and real resources of the state than any man on this side [of] the river."[23]

Cather was particularly concerned with preservation of forests and wetlands, both of which were disappearing before progressive farming techniques. Her love of the region's waters figures prominently in her early fiction, both as nostalgic places of escape and as an important element for nourishing the land and its inhabitants. Again, her childhood rambles along Back Creek, Virginia, undoubtedly sparked her interest in water conservation, but of equal importance was her experience on the open plains, which, in a given season, could parch and blow away. Her 1901 story "El Dorado: A Kansas Recessional" opens with a discouraged description of "a muddy little stream" that fails to support a few "pathetic little trees," which are mysteriously akin to the area's few human inhabitants. The monotony of the plains and the unproductive fields lead Cather to conclude that "Nature always dispenses with superfluous appendages."[24] Acutely aware of the delicate balance in nature, though, Cather records the land's renewal, "washed fresh and clean" after a thunderstorm, and allows that "Nature seems sometimes to repent of her own pitilessness."[25] Nourishing waters so sorely needed on the prairies appear most often in these early stories as "brown and sluggish" streams that flood bottomlands in spring and create unexpected sandbars that sprout growth of willow seedlings.[26] Cather attributes to these waters two moods: "the one of sunny complaisance, the other of inconsolable, passionate regret."[27] The marshes and river banks, which serve as places of escape to childhood wonder for so many of her characters (such as Niel Herbert, Jim Burden, and Douglass Burnham), also serve as the focal point for Cather's ecological concerns. Crazy Ivar's pond in *O Pioneers!* is a haven for migrating birds; the banks of the river at Black Hawk with their elder bushes and wild grapevines are hosts for orderly colonies of bees. Like Jim Burden, Cather had a "friendly feeling" for these waters that were so vital to the region's survival.[28]

Such ecological principles inform Cather's works and set up a dialectic that embraces gender as well. But, by 1922, Cather had admittedly suffered from both personal and social crises that led her to indict modern culture in her "war" novel *One of Ours,* a work that also reflects her long-standing concern that modern Americans' disconnectedness with the land would result in ecological crisis. The threat was two-fold: first, the masculine ethic of rights over the land had displaced the feminine ethic of responsible caretaking; second, a social impetus for acquisition and standardization threatened the natural order of things. For Cather, World War I was the watershed, the point at which America's "agrarian dream was lost" and pre-war values disintegrated.[29] Cather decried the cultural changes, which were fast shaping American life, what she called in a 1921 interview "This rage for newness and conventionality," the signpost of "a superficial culture."[30] The eagerness of the American people to emulate foreign style and fashion, to replicate items by mass-production, and to separate themselves from "traditional modes of living associated with the soil" resulted in a standardization of life that Cather resisted.[31]

Urbanization and industrialization threatened to defile the fragile landscape, as well. The incipient Jazz Age, with its indulgence in sensual pleasure and flagrant materialism, put in imminent danger the harmonious relationship between humans and nature that the first-generation pioneers, like Alexandra Bergson, had established on the frontier. Randall notes "a marked change . . . both in subject and tone" in Cather's texts after World War I and a significant change in her treatment of nature.[32] Her nostalgic yearning for stability and wholeness emerges in a jaundiced view of the materialistic ideal of progress. "Times may change," she said, "inventions may alter a world, but birth, love, maternity, and death cannot be changed."[33] These essential elements of existence Cather knew were of all life, and in her later novels she would create characters responsive to the maintenance of that cycle. Anticipating contemporary ecofeminist concerns, Cather assigns to her protagonists what Lorraine Anderson calls "a feminine way of being in relation to nature": "This way is caring rather than controlling; it seeks harmony rather than mastery; it is characterized by humility rather than arrogance, by appreciation rather than acquisitiveness."[34] But, Cather does not discriminate by gender in choosing her spokespersons. Using male protagonists who are responsive and sympathetic to ecological issues, Cather finds a vehicle for criticizing modern callousness toward nature. Through these male voices Cather can condemn the "masculine" land ethic of dominance and control—an ethic shown by Jim Burden in his desire to possess *his* Antonia and by Alexandra's brothers and later Ivy Peters—without condemning all men.

One of Ours is the story of the Nebraskan youth Claude Wheeler, whose sensibilities and idealism place him in opposition to the modernist ethics espoused by the progressive citizens of Frankfort, and especially by his father, Nat Wheeler, and his brother Bayliss. Nat Wheeler views the settlement of the region as "his own enterprise" (*OO* 8) and ascribes to the notion that "the land was made for man," for reaping profits (*OO* 57).[35] In answer to his wife's query of why he bought more land when they already had more than they could farm, Mr. Wheeler replies, "Just like a woman . . . You might as well ask me why I want to make more money, when I haven't spent all I've got" (*OO* 56). Nat Wheeler epitomizes what Nancy Chodorow and Carol Gilligan describe as a sense of self separate from the land, a sense that justifies ownership and exploitation of it.[36] That he also dominates his wife Evangeline is, therefore, expected; the land and his wife are simply different aspects of vulnerable femaleness.

The spiritual dimensions of the land are lost on him, but not on his son Claude who, as the voice of Willa Cather, exhibits a closeness to nature more often associated with women. To spite his wife, who has complained that the cherries are beyond her reach, Mr. Wheeler ruthlessly chops down the tree, and the child Claude responds "like a little demon." He condemns his father as a "damn fool," and is convinced that "God would surely punish a man who could do that." Evangeline Wheeler cautions her son that the act was her husband's "perfect right" (*OO* 25) and rationalizes that maybe by thinning out the tree, the other cherry trees will benefit. Cather clearly knows the importance of pruning and thinning, just as she notes in her novel the importance of plowing in the spring to help the land hold its moisture (*OO* 25). But, in Claude's perception of the fallen tree as a "bleeding stump" (*OO* 25), Cather personifies and personalizes the attack. Here, in a revisioning of the country's patriarchal mythos, Cather describes betrayal and indifference and a wanton destruction of all that is beautiful.

Claude's brother Bayliss exemplifies the second of Cather's worries—the drive to acquire and standardize life, to blot out the natural. Having achieved "considerable financial success" in the farm implement business (*OO* 9), Bayliss would turn Frankfort into Gopher Prairie, modernize Main Street, and through his labor-saving inventions, further separate its citizens from the land that had nourished them physically and spiritually. Claude decries both his father's and Bayliss's disregard of the fragile ecosystem, which they would reshape. With his mother's sensitivity, he feels that is not right to "have so much land,—to farm, or to rent, or to leave idle, as they chose" (*OO* 68). He bitterly resents the changes this second generation of pioneers would impose on the region and compares the current state of things to that of his childhood years:

The farmers took time then to plant fine cottonwood groves on their places, and to set osage orange hedges along the borders of their fields. Now these trees were all being cut down and grubbed up. Just why, nobody knew; they impoverished the land . . . they made the snow drift . . . nobody had them any more. With prosperity came a kind of callousness; everybody wanted to destroy the old things they used to take pride in. The orchards, which had been nursed and tended so carefully twenty years ago, were now left to die of neglect. (*OO* 85)

Claude consequently allies himself with Ernest Havel, a Bohemian neighbor whose rational response to environmental issues complements Claude's own emotional response. The Frankfort community considers Ernest a freethinker, but Claude admires Ernest's genuine appreciation of a bittersweet vine along the banks of Lovely Creek and his respect for the small creatures that inhabit the Wheelers' timber claim. Still, Ernest is not plagued by the sense of disaffection with modern life that haunts Claude, nor does he seek out the natural world as a sanctuary. While Ernest is content to farm, to accept the dignity of his labor and things "as they are," Claude believes "there ought to be something—well, something splendid, about life, sometimes" (*OO* 46).

Claude seeks that splendidness in nature, relishing country walks, the painfully sweet songs of larks, and "the sticky cottonwood buds . . . on the point of bursting" (*OO* 101). His father's timber claim becomes "his refuge" (*OO* 174), and in describing it so, Cather follows the literary, and largely masculine, tradition of depicting "nature as a sacred place where only solitary, single, and chaste men go to cleanse their spirits and be one with God."[37] In the timber claim, Claude feels "unmarried" and free from the passionless alliance he has with his missionary-minded wife Enid (*OO* 174). The natural world serves in Emersonian fashion as a conduit for the imagination, a means of reclaiming the essential self. Cather writes that Claude "would have died defending" that timber claim (*OO* 174). But, his defense is not just self-protectionism; he would defend the harmonious ecosystem it represents, as well.

On this claim, Claude discovers a family of quail, the presence of which he reveals only to his wife and to Ernest Havel. The quail are the only ones left in the region, as Claude himself is perhaps the only one left to fight for their preservation. Claude is sure that the birds have "learned enough about the world to stay hidden in the timber lot" (*OO* 149); still, he scatters corn in the grass and later plants peas in the ditch to attract the quail and keep them from wandering into neighboring fields where they would likely be shot. Claude would preserve his quail family and all it represents from what Ernest Havel calls "the harvest of all that has been planted" (*OO* 136)—that is, the aggressive violence that led to a world war. Unable to find a soul mate in a world dominated by destructive impulses, Claude, filled with romantic visions of making the world safe, escapes to that war in Europe.

Claude, like his creator, at once becomes a Francophile, finding in the French culture the perfect whole, a people living in harmony with their world. Whereas the farmers in Frankfort were cutting down cottonwoods because they were ordinary, these trees grew in abundance in France "as if they had been there for ever and would be there for ever more" (*OO* 275). French farmers bordered their fields with trees and hedges, providing shelter for birds and small creatures, and raised patches of mustard alongside their crops. "Didn't they know that mustard got into wheat fields and strangled the grain?" the American doughboys ask. "Didn't they know that trees took strength from the soil?" (*OO* 275). Here, profit motive, even in the face of war, does not displace ecological stability. Claude revels in the clumps of blue cornflowers growing everywhere, the wild morning-glories and Queen Anne's lace by the roadsides, and the ethereal beauty of forests and glades. Significantly, his first meeting with Madame Joubert, at whose home he billets, occurs under a cherry tree, still alive and productive in the midst of man-induced chaos. Old M. Joubert tends the plants and vines, "clipping off dead leaves and withered flowers" (*OO* 283) in his efforts to preserve the world for his grandchildren.

When Madame Joubert excitedly inquires about Claude's walks in the woods and asks if the heather is in bloom, Claude concludes that she "cared a great deal more about what was blooming in the wood than about what the Americans were doing on the Garonne" (*OO* 287). Even in the most heavily damaged villages, survivors did not complain about their lost goods—"their linen, their china, and their beds." "This war has taught us all how little the made things matter. Only the feeling matters," remarks Mlle. de Courcy (*OO* 312). The French people's sadness, she notes, is only in the loss of their trees. When a one-armed French veteran describes how he will tend four blasted locust trees to restore them, Claude comments, "How much it must mean to a man to love his country like this; . . . to love its trees and flowers; to nurse it when it was sick, and tend its hurts with one arm" (*OO* 313).

Claude does not survive to return to Nebraska, and readers might roundly condemn Cather for conveniently excusing Claude from bringing his ecological message to his home region and its people. But, Claude is neither savior nor martyr. He is instead a hapless victim to cultural values that typecast men like him as less masculine and less American than their counterparts. His condemnation of America's irresponsible stewardship in the name of profit making is Cather's. In a male protagonist, Cather captures the female ethic of ecological responsibility but believes that such an ethic no longer has a place in her world.

In the last of her novels with an exclusively western setting, *A Lost Lady*, Cather continued to explore the passing of the heroic age of the pioneers and, with it, the decline of an environmental ethos. Again she uses a sensitive youth as her observer, but unlike Claude Wheeler, whose romantic idealism limits his response to impending ecological disaster, Niel Herbert holds an androcentric view that bars him from protecting the very things he wishes to save—the natural beauty of the land and its resident goddess, Marian Forrester.

The central symbol of Niel's conflict is the Forrester marsh. Captain Forrester, a retired expansionist who brought the railroad and "civilization" to the plains, preserves this wetland for its natural beauty and its wildlife. As Nina Schwartz notes, "like the foresters of old, he acts as a warden keeping others out of the nobility's playground, not only leaving the marsh uncultivated but also refusing to allow hunting in his meadow."[38] In his youth, Niel and the other town boys gravitate to the marsh for picnics, frolics, and fishing. With innocent abandon, they spend long summer days in harmony with this unspoiled landscape, harming nothing and accepting Mrs. Forrester's cookies as manna from heaven. Niel notes that "any one but Captain Forrester would have drained the bottomland and made it into highly productive fields" (*LL* 9). Niel, early on, identifies himself with the Captain as one who shares the same values and admires the beautiful, but what Niel fails to recognize is that just as the Captain's former occupation involved subduing and deflowering the plains, his own attitudes harbor a latent paternalism toward both the land and Marian Forrester.

When the rude Ivy Peters invades the town boys' holiday, intending to "bag a few" ducks at sundown in the Forrester marsh (*LL* 19), Niel is repelled by Ivy's arrogance and by his reptilian looks. Ivy poisons their paradise, living up to the nickname "poison Ivy" that he earned after doing away with several of the town's dogs. His inherent cruelty, however, is, as Demaree Peck notes, evidence of an "aggressive and violent sexuality" which would subjugate the female to his control.[39] Capturing a woodpecker, Ivy at first wrongly identifies it as male. Then remarking "All right, Miss Female" (*LL* 21), he slits both its eyes with a tiny blade from his taxidermy kit (ordered, by the way, from *The Youth's Companion*, probably the most influential magazine of the time in defining what it was to be a true American boy). The bird flies about blindly and desperately, bumping into branches, and finally finding its own hole. Niel's failed efforts to climb the tree and to put the bird out of its misery is the precursor of his later failed efforts to rescue Marian Forrester from the "cold, vivisecting cruelty" of Ivy's possessiveness.[40]

Niel ascribes to the pastoral fantasy of what Annette Kolodny describes as "a daily reality of harmony between man and nature based on an experience of the land as essentially feminine—that is, not simply as land as mother, but the land as woman, the total female principle of gratification."[41] In Niel's view, Ivy's attack on the female woodpecker constitutes a violation of both the land and Marian Forrester. What Niel does not acknowledge, though, is that his dedication to keeping and preserving this pristine world is a projection of his own paternalism. Not unlike Ivy, he would own Mrs. Forrester and the marsh, not with Ivy's intent of degrading her and profiting from the land, but with the intent of making both conform to his ideal.

As a young man Niel believes himself driven by an "impulse of affection and guardianship" for Mrs. Forrester and the natural world with which he associates her (*LL* 80). He thus justifies as protectionism his hatred of Frank Ellinger, Marian Forrester's lover. Making a devotee's early morning journey to place flowers on the altar of Mrs. Forrester's window sill, Niel marvels that he had never before seen the day "before men and their activities had spoiled it, while the morning was still unsullied, like a gift handed down from the heroic ages" (*LL* 81). The "almost religious purity" of the morning air inspires "in all living things something limpid and joyous" (*LL* 80), Cather writes, but she allows her protagonist only a temporary idealism.

As Morton D. Zabel notes, *A Lost Lady* reflects Cather's "feeling that the inspiring landscape of the prairies, deserts, and mountains . . . had been obliterated by a vulgar and cheapening modernity."[42] After the Captain suffers financial failure, Ivy Peters rents the Forresters' meadow land, drains the swamp, plants wheat, turns a neat profit, and spites the Captain by hunting along Sweet Water Creek. Ivy, as his name implies, is the vine that will choke off and eventually kill the beautiful Spirit of Place. Niel, like Cather, perceives of this change "as a mark of historical degradation."[43] In describing this historical change, Cather uses language that connotes demise of the natural world, as well:

[Men like Ivy Peters] would *drink up* the mirage, dispel the morning freshness, *root out* the great brooding spirit of freedom . . . The space, the colour, the princely carelessness of the pioneer they would destroy and *cut up* into profitable bits, as the match factory *splinters the primeval forest*. All the way from the Missouri to the mountains this generation of shrewd young men . . . would do exactly what Ivy Peters had done when he drained the Forrester marsh. (*LL* 102, emphasis mine)

The subsequent "air of proprietorship" that Ivy displays about the Forrester place is more than pride in having dispossessed the great men of the pioneer era (LL 112). When Niel asks Marian if she doesn't miss the marsh, she is evasive and claims that she never had time to go there anyway and that she needs the money it brings in as productive land. Mrs. Forrester has reluctantly capitulated to the patriarchal ethic that land is woman to be owned and used. As an ecofeminist, Cather resists such capitulation but knows that it does happen. As Peck aptly puts it, "By having Ivy exert an 'air of proprietorship' over both Marian's land and her body, Cather suggests that these two kinds of claims are equivalent."[44]

Niel, unable to assert his own proprietary claim and unable to preserve his Edenic, childhood world, like Claude Wheeler before him, finds an avenue of escape. He makes a "final break with everything that had been dear to him" (*LL* 160), retreats to New York to take up a career as an architect, and leaves Marian Forrester and the land to the inevitability of change, and not change for the better. Like Cather, Niel sees that "The people, the very country itself, were changing so fast that there would be nothing to come back to" (*LL* 160). Indeed, after 1923, Cather never came back to Nebraska as the sole setting for a novel. *A Lost Lady* thus becomes a lament for the loss of fragile beauty—natural and female—before the forces of modernity, patriarchy, and profit-making. The lady is lost and so is the land.

Cather's sympathies were clearly with her "right-minded" male protagonists who espoused an environmental ethic that would not reconstruct nature but would reconstruct human relationships with the natural world. She was not, however, optimistic that such an ethic would prevail in the post-war era. "We come and go, but the land is always here," Alexandra Bergson had remarked in Cather's 1913 novel *O Pioneers!*[45] But by 1922, when Cather said "the world broke in two," her female protagonists, responsive to and supportive of the natural cycle of things, have been displaced by a male-dominated society, and the Genius of the land has fled. Those few individuals who respond sensitively to the land, wishing to preserve and embrace its pristine femininity in all its forms, human and non-human, are as ones crying in the wilderness. As Claude Wheeler notes, the "wise, unobtrusive voice [of nature], murmuring night and day, [was] continually telling the truth to people who could not understand it" (*OO* 127). Cather, like her character Niel Herbert, surely had "weary contempt" (*LL* 161) in her heart for a world gone astray, a world in which the giant cottonwoods were reduced to splinters, the wetlands drained, and the birds blinded, all in the name of progress.

Valerie Levy

"That Florida Flavor"
Nature and Culture in Zora Neale Hurston's Work for the Federal Writers' Project

> [T]he world is a great big old serving platter, and all the local places are like eating plates. Whatever is on the plate comes out of the platter, but each plate has a flavor of its own. . . .
>
> This local flavor is what is known as originality. So when we speak of Florida folklore, we are talking about that Florida flavor that the story and song makers have given to the great mass of material that has accumulated in this sort of culture delta.
>
> —Zora Neale Hurston, "Go Gator and Muddy the Water"[1]

I.

In her essay "How It Feels To Be Colored Me" (1928), Zora Neale Hurston exhibits an indefatigable love of life and love of place and regards her youthful self as "the first 'welcome-to-our-state' Floridian."[2] Known for her anthropological studies and taste for local color, the self-proclaimed "everybody's Zora" proudly associates herself with her home state of Florida and even more closely with the state's first all-black incorporated community, Eatonville. Throughout her life, Hurston felt a connectedness with her home and often chose it as a site and resource for a substantial amount of her anthropological research and literature. In 1938, she returned to Florida to supplement her income by working for one year as a relief editor and writer for *The Florida Guide* and *The Florida Negro,* two publications issued during the Great Depression by the Work Projects Administration (WPA) and the Federal Writers' Project (FWP).[3] Called upon by the FWP's folklore director Benjamin Botkin to write about the black folk culture and history of Florida, Hurston followed the dictate of her memorable character Janie Crawford, who tells Pheoby Watson, "you got tuh *go* there tuh *know* there."[4] In embarking upon her field work for the Jacksonville "Negro Unit" of the Florida FWP, Hurston was able to make the most of her anthropological experience gained studying under the renowned Franz Boas and of her own personal knowledge of the state and the people. She added to this familiarity by traveling extensively throughout Florida in order to "provide a 'fresh first-hand viewpoint' of not only the highway vistas one would expect along the routes, but of regional variations, local history, and lore."[5] What Hurston discovered during her explorations of the indigenous black folk culture of the state was that Florida, in her words, has "the most tempting, the most highly flavored Negro plate around the American platter."[6]

It is easy to recognize the elements in Hurston's FWP writing that highlight her expertise as a cultural anthropologist and folklorist, but what is equally significant in these essays and sketches is Hurston's contribution to the literary canon as a nature writer—that is, as Lawrence Buell clarifies in *The Environmental Imagination,* one who writes "environmental texts" in which "the nonhuman environment is present not merely as a framing device but as a presence that begins to suggest that human history is implicated in natural history."[7] Just as the black Southern literary tradition arises out of slave narratives and the oral tradition, Florida's distinct "flavor" and black culture arise out of the landscape itself. Hurston brings to the Florida FWP an African-American point of view that shows how the people's lives have been informed by the environment and also how, conversely, the people have affected the character of the land through their close contact with it. As a nature writer, Hurston's descriptions of Florida's multifarious regions, industries, resources, and peoples blur the lines between anthropology, folklore, realism, and fiction. Language use becomes the key to understanding Hurston's view of nature and culture, and her rhetoric, which incorporates naturalistic references, analogies, and imagery, mirrors the intricate attachment of humans to the land. Indeed, in Hurston's Florida, the fluid relationship between nature and black folk culture is as organic as the state's oranges and alligators.

II.

The wide variety of writing styles and topics that characterize Hurston's FWP work can be ascribed to her investigation of the many intersections of place and culture that contribute to Florida's inimitable flavor. While some essays and sketches provide basic records of her discoveries, others combine what she calls "business" and "between-story conversation."[8] Some pieces capture the folklore and linguistic nuances of the African-American community; others describe the physical and historical contexts out of which such cultural traditions grew. In essence, Hurston writes a series of pieces as varied as the milieu she illuminates. At the core of such a marked mixture of genres and themes lies an inherent paradox upon which Hurston bases her subsequent observations: "Florida is still a frontier with its varying elements still unassimilated," she writes. At the same time, the African-American culture there is seasoned by the state's various cultural groups: "The drums throb: Africa by way of Cuba; Africa by way of the British West Indies; Africa by way of Haiti and Martinique; Africa by way of Central and South America. Old Spain speaks through many interpreters. Old England speaks through black, white, and intermediate lips. Florida, the inner melting pot of the great melting pot America."[9]

The seeming incongruity of Florida as an unassimilated place and Florida as a melting pot sustains its force in two ways. Because Florida is both a crude frontier of sorts and a cultural melting pot, it offers a unique mix of peoples and elements. Because it has such an extraordinary blend of peoples, landscapes, and elements, Florida attracts both foreigners and pioneers alike. Hurston treasures Florida's diverse population and untamed landscape. She does not deem nature the "converse of human civilization"; instead, nature is an integral part of human civilization.[10] In Hurston's view, "No country is so primitive that it has no lore, and no country has yet become so civilized that no folklore is being made within its boundaries."[11]

In an attempt to organize a plan for recording and describing the black folk art in Florida, Hurston clarifies the paradox of Florida's diverse elements and attributes by dividing the state into four geographical areas: West Florida, Northeast Florida, Central Florida, and South Florida. Her preliminary research sketch, "Proposed Recording Expedition into the Floridas," indicates how the farming and resources of each region shaped the history and the lives of those who lived and worked there. In this way, Hurston's role as nature writer is to augment the connection she sees between land and lore.

West Florida, or what Hurston refers to as Area I, is regarded mainly as the plantation region whose chief agricultural commodities are cotton, corn, and tobacco. Because the land's natural resources have called for the extensive farming undertaken by slaves in the antebellum days, the racism of the Jim Crow South is more prominent in this plantation region. As Hurston puts it, "people live under the patriarchal agrarian system," and "[t]he old rules of life hold here."[12] As both a defense mechanism and a way of signifying or gaining power, African-American slaves developed finely-tuned senses of humor, which Hurston attributes directly to their plight as slaves in America. She argues that "what has always been thought of as native Negro humor is in fact something native to American soil." Her case in point is a comical tale about why African Americans are "black": When God was handing out color to his clay creations and needed some physical space from one particularly anxious group of people, he yelled "Get back," which the group mistook as "Get black," so, as the storyteller explains, "they just got black, and we been keeping the thing up ever since."[13] Other lore originating from or indirectly out of slavery characterizes what Hurston calls "the wish-fulfillment projection." In stories of the Florida prison camps, Big John de Conquer, the hero/trickster figure of slavery days who could "outsmart Ole Massa, God, and the Devil," evolves into Daddy Mention, the new folk hero capable of hairpin escapes and superhuman activity. Thus, as Hurston observes, Big John de Conquer and Daddy Mention by extension represent "the story that all weak people create to compensate for their weakness."[14] From the history of American slavery to the lingering condition of racial subjugation in the plantation region, the black folk culture of West Florida evolved out of the close contact the African slaves had with the American soil and with the white landowners. This contact weakened the slaves' ties to their African roots more so than their Bahamian and West Indian counterparts, but it also helped them to develop new traditions specific to their situations and sense of place.[15]

On many levels, Hurston acknowledges, accepts, and even relishes West Florida's "melting pot" qualities. Area I, in fact, is not merely a plantation region consisting of white and black; it is also a coastal one featuring maritime settlers of quite a different ilk. Hurston notes, "The shipyards and the like are the culture beds of other maritime folk creations." Because West Florida has such a rich mixture of peoples and natural resources, it is dubbed home of the "Spanish-French-English-Indian fighting tradition."[16] Negro folk tales and Creole songs reflect the resultant struggles of four different groups of people trying to control the area. The effect of West Florida's various offerings is that the richness of the land has directly influenced the richness of the culture and vice versa. But this can be said of all of Florida. For this reason, Hurston looks upon Florida as qualitatively and quantitatively superior to every other state in the nation: "There is not a state in the Union with as much to record in a musical, folklore, social-ethnic way as Florida has."[17] Although she concedes that California is also culturally rich with its Chinese, Japanese, and Filipino population, she does not see the same amalgamation of tradition, society, and landscape with the Asiatic cultures of California as she does with the European and African cultures of Florida.

To Hurston, part of what makes Florida's cultural appeal thrive is that its natural resources and landscape determine its industries, which, in turn, impact the type of art created in the region. Hurston makes this association when she avers that Northeast Florida and Central Florida (Areas II and III) are "so full of varied industries that [they are] full of songs and story."[18] Like West Florida, Northeast Florida is also "a conglomerate of many cultures," from the "Georgia-Alabama Cracker" and the English, French, and Spanish descendants to the turpentine workers, river men, and maritime folk. In this region, the culture arises out of the myth-evoking seascape in the form of such tales as "Pap" Drummond's account of pirates and hidden treasures.[19] So intrigued was Hurston by Florida's maritime culture that she revisited it as source material for her 1948 novel, *Seraph on the Suwanee,* in which Jim Meserve enjoys life in New Smyrna with the commercial fishermen, exotic sea creatures, and culture of people who "spend freely and enjoy life vigorously when they can."[20]

In Northeast Florida and Central Florida, the great pine trees that peppered the landscape gave rise to the lumber and turpentine industries, which flourished in the earlier part of the twentieth century and which opened up not only jobs but an entire way of life for the African Americans who worked as laborers in this portion of the state. For part of her field research, Hurston visited the "wild, lawless" Cross City in 1939 and interviewed John McFarlin, a turpentine foreman and woods rider.[21] Almost obsolete by the 1930s, the turpentine industry, in McFarlin's words, "'[t]aint like sawmills and such like that. Turpentine woods is kind of lonesome."[22] In her FWP essay, "Turpentine," which, Pamela Bordelon points out, was intended as a life history of the black turpentiners, Hurston observes that the "lonesome" life of the turpentine workers inspired a special type of community and culture. The workers' days began at 6:00 A.M. and ended at 5:00 P.M. All day they would chip, chop, and dip for the sap or "gum" of pines. It was tedious and often dangerous work, and the art that arose out of it reflects this repetition and the need for comic relief. Songs became spoken legends and Everymen became heroes. "Uncle Bud," as Hurston explains in a recorded conversation with folklorist Herbert Halpert, is not a work song but "sort of a social song for amusement . . . so widely distributed it is growing all the time by incremental repetition . . . [and is] known all over the South." These "social songs" were not sung at all but "yelled down" by all the workers, who would contribute verses when their turns came. Hurston clarifies: "Everybody puts in his verse when he gets ready. And 'Uncle Bud' goes and goes and goes."[23] The repetition of the verses echoes the tedium of draining one pine tree after another. At the same time, hearing the humorous and sassy lyrics about Uncle Bud's farming dilemmas and how his girlfriends "rock their hips like a cannon ball" lightened the burden of the turpentine workers.

In less dire need for comic relief, Central Florida's citrus workers fashioned a more melodic art form than the social songs of the turpentine woods. In the groves, the workers'days were not unbearably tedious as were the turpentine workers'. Much of their time was divided between picking, loading, and packing the various citrus fruits—from the most common of oranges, grapefruits, and tangerines, to the less familiar satsuma, tangelos, and tango range. The grove owners, meanwhile, tended to the planting, cultivation, pruning, watering, and fertilization of the crops. In *Seraph on the Suwanee,* Jim regards the citrus workers' relatively easy lives with awe and thinks the workers are "obviously shirking their duty."[24] But the workers in the groves had their own share of dangers, as they came directly into contact on a daily basis with some of nature's most alarming creatures: black snakes in the trees, rattlesnakes on the ground, and the grampus (a kind of scorpion) in the rotten wood in the groves.[25] The art form that expressed for the workers this bitter-sweet crop and bitter-sweet life was the quintessential bitter-sweet African-American art form, the blues. While clipping and loading the various citrus fruits, workers in the groves would sing melodies about "women and likker."[26] The blues ethos Hurston observes expresses what Cheryl Wall says is "a central premise of Hurston's work: material poverty is not tantamount to spiritual poverty or experiential deprivation."[27] As a sign of spiritual wealth, the blues music of the citrus workers reflects the lives they led and the art that was harvested along with Central Florida's most famous crops.

Despite the ubiquitous nature of citrus fruits in Florida, the most dominant feature in Northeast Florida in the 1930s was the railroad centers. Unlike the turpentine and citrus industries, the railroad industry was not determined by the landscape; rather, the industry directly impacted the environment. Hurston viewed the Talahand railroad camp and other such construction sites as places abundant in songs, chants, and stories. As the workers and the hammer gangs would build the railroad structure by shaping and spiking lining bars, each bar weighing nine hundred pounds, they would sing chants and pull back the rails as leverage to go forward. As Hurston explains, "they use the rhythm to work [the rails] into place."[28] These work songs boast of "the Florida boys" who built the railroads and whose chants have been distributed all over the state. Some songs are short, some contain thirty and forty verses; some have a fast rhythm, others are slower. Such factors depend upon "the mood of the liner," who starts the singing and without whom "the men would not work." As the railroad men worked and sang, they literally changed the face of landscape and the means of travel and communication across the American continent. Thus, the formation of the railroad occurred contemporaneously with the formation of the work song, a specific and rather advanced African-American art form.

In comparison to West, Northeast, and Central Florida, South Florida, or Area IV, exemplifies a most vibrant spirit and culture because it has a "cultural preserver" in its natural landmarks, Lake Okeechobee and the Everglades. Much of South Florida's local color derives from myriad sources: from the Greeks of Tarpon Springs, the Latin colonies of Tampa, the Caribbean and South American cultures of Miami, and the Bahamian and Cuban cultures of the Keys and Palm Beach. This foreign culture, says Hurston, "has not yet [been] absorbed into the general pattern of the locality, or [is] just beginning to make its influence felt in American culture."[29] Here, too, African-American traditions thrive without the constant threat of being diluted by other cultural influences. In Hurston's FWP writing, she introduces her readers to the nightly dances and celebrations that originated in West Africa and that were held every night in the Everglades around the bean fields and sugar mills. The Fire Dance, for instance, a West African New Year's celebration, conjoins singing, dance, and ritual and marks the cultural phenomena of man's intimate communion with nature. In one portion of the dance, the Crow Song signals the appearance of "the Crow," who dances in "perfect rhythmic imitation of a flying buzzard seeking food." Hurston continues, "He enters, finds food, takes some in his beak, and flies off."[30] In documenting native dance cycles, Hurston endorses the sanctity of the black folk tradition, the human reverence for nature's creatures, and Florida's unique union of the old and the new.

III.

Florida's relative isolation makes it an ideal place for black folk expression, and Hurston's use of a rhetoric that illuminates the hybridization of nature and culture becomes evocative of the entire body of her FWP work because part of her mission is to show how language grows out of the landscape and ecosystem. In her autobiography, *Dust Tracks on a Road* (1942), Hurston acknowledges the role of nature in her own discourse and says that Southerners' metaphoric speech comes "right out of the barnyard and the woods."[31] In many of Hurston's literary works, her characters' names reflect their particular place in the natural world. Tea Cake of *Their Eyes Were Watching God* (1937), for example, is alternatively known as Vergible Woods, a name that echoes nature in its peculiar fusion of "vegetable" and "verdure" and that consequently reinforces his associations with nature and with the close relationship to the earth that he and Janie Crawford share.[32] In her FWP writing, Hurston's language is likewise colored and seasoned with the flavors of culture and nature. As if to mimic the fluidity of Florida's nature-culture relationship, Hurston incorporates in her nonfiction work natural elements in the very language she uses to describe such phenomena. To Hurston, African-Bahamian folk art is not simply *part of* America; rather, it "seeps into the soil of America."[33] The songs of Northeast and Central Florida are not simply abundant; rather, they "sprout in this area like corn in April."[34] Such expressions show not only how the culture is a part of the terrain but how the terrain becomes fertile with this particular influence and will proffer a special "crop," indigenous simultaneously to Florida's black people and to Florida's land.

In "The Sanctified Church," a piece written for the religion chapter of *The Florida Negro,* Hurston draws upon natural imagery when she regards the inclusion and exclusion of African religious characteristics into Christianity as something of an organic process gone awry: "the sanctified church, is a revitalizing element in Negro music and religion. It is putting back into Negro religion those elements which were brought over from Africa and grafted onto Christianity as soon as the Negro came in contact with it, but which are being rooted out as the American Negro approaches white concepts."[35] Hurston speaks of African religion as if it were a plant being "grafted onto" or "rooted out" of American forms of Christianity. An avid gardener herself, Hurston's rhetorical choices are apropos. Further, her use of the passive voice in this portion of her discussion seems to suggest that, like many other African heritages, traditions, and values, African religion is exploited or tossed aside as the mood of "white concepts" dictates. This precarious reverence and disregard for African religion is not unlike the relationship European pioneers had with the American land, a relationship in which the American frontier was alternately viewed as a savage wilderness or as a welcoming Indian maiden/virgin.[36] For Hurston, the great loss lies in watching

African-American religious expression relinquish "its vibrancy and feeling" as "blacks became more prosperous and joined mainstream Protestant denominations."[37] In "The Sanctified Church," however, Hurston is able to uphold the survival of African music and religion; her active predicate "putting back" demonstrates the potency of revivalist efforts and underscores the endurance of pure African religious forms.

The folklore itself also reproduces and reflects this synthesis of nature, culture, and language. "Go Gator and Muddy the Water" and "Other Florida Guidebook Folktales," for example, feature vibrant naturalist-oriented stories told to Hurston by the people she met on her expeditions through the state. In the African-American version of the popular "Jack and the Beanstalk," Florida's storytellers give the tale a Floridian slant and revolve the plot around the tremendous fertility of the rich muck lands of the Lake Okeechobee region. In this tale, the corn grows so quickly that one brother "rises magically with the corn . . . and begins selling his roasting ears to the angels."[38] The various stories and quips Hurston records unite Florida's African-American people in their common bonds: their storytelling heritage and their ties to Florida. Living with the Florida heat or hurricanes, for example, becomes a terrific source of fodder for the development of stories, myths, and language in general. In one tale, the heat in Tampa is so hot that men literally melt out of their suits. In another, the wind on the Florida west coast "blowed so hard till it blowed a crooked road straight. Another time it blowed and blowed and scattered the days of the week so bad till Sunday didn't come until late Tuesday evening."[39] Hurston cites one dialogue that illustrates the tradition of dramatization as a way of telling a story. In this competition of wit, two helpers and a storyteller try to "out-do" the others' descriptions of "ugly." The storyteller begins by asking the other two, "What is the ugliest man you ever seen?" The first helper responds, "Oh, I seen a man so ugly that he could set up behind a jimson weed and hatch monkeys." The second helper says, "Oh, that man wasn't so ugly! I knowed a man that could set up behind a tombstone and hatch hants." The storyteller, who has set himself up to "win" the competition, replies, "Aw, them wasn't no ugly men you all is talking about! Fact is, them is pretty mens. I knowed a man and he was so ugly that you throw him in the Mississippi River and skim ugly for six months."[40] The reference in the punch line draws upon a familiarity with one of South's most famous features, the Mississippi River, and, in all of these examples, the hyperbolic quips match the hyperbolic spirit of Mother Nature.[41] It is as if the hotter the sun beats down or the stronger the wind blows, the taller the tales grow.

Even Hurston's descriptions of African-American music reflect its ties to nature. The Negro blues, for example, originate out of the human propensity to create art, which Hurston defines as "the setting up of monuments to the ordinary things about us, in moment and time." The blues, or "feelings set to strings," developed out of a desire to chronicle man acting in and as part of nature. Hurston says, "The singing grew like this . . . First a singing word or syllable repeated over and over like frogs in a pond; then followed sung phrases and chanted sentences as more and more words were needed to portray the action of the battle, the chase, or the dance. Then man began to sing of his feelings or moods, as well as his actions."[42] In agrarian or hunting communities, humans may be said to be closer to the earth, relying upon it more directly for food and clothing. The art and culture grow out of this experience. In Hurston's explanation of the evolution of Negro blues, it is interesting to note how she herself, a modern woman, draws upon nature for her own expression. Just as the pre-industrial peoples look to the actions of man in nature to create art, Hurston looks to nature (in this case, to frogs singing in a pond) in order to sculpt the language she uses to describe the development of the blues. Hers is a meta-language that utilizes the duality of the nature-culture bond in order to underscore the complex relationship between the two.

At times, the presence of nature in Hurston's writing functions not just as a mirroring device but as an indispensable force, without which the story, novel, or essay could not exist. Henry David Thoreau could not have written *Walden* without Walden Pond. In *Their Eyes Were Watching God,* the hurricane proves to be an even more formidable presence in Janie's life than Nanny, Logan Killicks, and Joe Starks combined.[43] JoAnne Cornwell comments in her discussion of *Their Eyes,* "nature is not only filled with that same vital force that animates humans, it is equally rambunctious. Each new day springs out from hiding on to the town of Eatonville: 'Every morning the world flung itself over and exposed the town to the sun.'"[44] Likewise, in her FWP writing, Hurston often personifies nature (or shows how black folklore personifies it) and brings nature from its invisible backdrop status to the forefront of both plot and structure. The land's vital force both competes against and works with the people's vital force, functioning alternately as foil and hero. In one quip based upon the hurricane of 1925, the hurricane and the storm become friends who sit down, eat breakfast together, and work as accomplices conspiring to "go down to Miami and shake that thing!"[45] To some degree, the hurricane and the storm are complex in their ability to be the "bad guys" we love to hate. And the wonder of this particular instance is that Hurston is able to achieve this relative complexity in just two sentences. In her more developed tales, as Morris and Dunn remark, Hurston distinguishes her characters by drawing upon the plants and animals they resemble.[46] "Uncle Monday," for instance, utilizes Florida's most prominent symbol, the alligator, for the creation of Uncle Monday, the regal medicine man from Africa who turns himself into a gator. The FWP version of the tale is an adaptation especially expanded upon by Hurston for the Florida guidebook, as it provides readers with an account of Uncle Monday's slave heritage and participation in historical events specific to the state of Florida, such as the Seminole War. As a multifaceted character, Uncle Monday exemplifies the perseverance of Africans in America, the powerful symbolism of the alligator in Florida, and the "consequences of pride and arrogance." He serves as the foil to Old Judy Bronson by revealing her weaknesses and teaching her the error of allowing petty, professional jealousies dictate one's life. This alligator-king also plays the role of protagonist-hero, his great truth being, "The foolishness of tongues is higher than mountains"[47] In the tale, Hurston transforms an already mythic creature indigenous to the state of Florida into human form, demonstrating how the awesomeness of the alligator's power to frighten can be fortified by the human propensity for wisdom and insight.

In Hurston's world, then, people and nature share a certain communion that contributes to the redolent folklore and organic language of Florida's African-American people. Hurston's discerning definition of folklore points to the reason why the two forces (nature and culture) are so closely related: "Thinking of the beginnings of things in a general way, it could be said that folklore is the first thing that man makes out of the natural laws that he finds around him—beyond the necessity of making a living. After all, culture and discovery are forced marches on the near and the obvious. The group mind uses up a great part of its life span t[r]ying to ask infinity some questions about what is going on around its doorstep."[48] Accordingly, among Florida's African-American population, many of the questions being asked of infinity pertain to the state's unique weather patterns, landscape, and creatures. As Hurston explains, "Folklore is the art of the people before they find out that there is any such thing as art, and they make it out of whatever they find at hand."[49] What the people find at hand is nature itself.

Because Hurston understood so well the reciprocity of the relationship between people and the earth they work, she deserves to be recognized for her tremendous acts of discovery and recovery in the realm of nature writing. Like Henry David Thoreau, Hurston recognized the power humans exercise over nature. In "The Bean-Field" chapter of *Walden,* Thoreau views his daily work as "making the earth say beans instead of grass."[50] Hurston's language and observations in her FWP work similarly indicate her awareness of both the influence humans have over nature and of nature's power to "speak" to humanity. Her FWP writing unveils the ways in which African Americans function(ed) in and with the land around them; it celebrates Florida's unique landscape and the natural world at large; it represents nature through historical and social vantage points; it brings nature to the forefront as a main character in the drama of American history and life; and it embraces the reverberation of nature, art, and imagination. In many ways, Hurston's FWP writing expands not only the genre of nature writing, but her distinction for having a divided or double voice, as Henry Louis Gates called it.[51] In *Their Eyes Were Watching God* and many of Hurston's other literary works, she utilizes both the dialect of her people and the more "standardized" voice of an omniscient narrator in order to create a "speakerly text." In her essays and sketches for *The Florida Guide* and *The Florida Negro,* Hurston generates yet another dual voicing, one that shows that she is aware of the nature-culture relationship she observes around her and that, more importantly, she is an active part of that wonderful connection. As Janie felt secure in knowing "mah tongue is in mah friend's mouf," so too might Florida's African-American people be assured in having Hurston speak for them, elucidating "that Florida flavor" as only Hurston could.[52]

Kelly L. Richardson

"A Happy, Rural Seat of Various Views"

The Ecological Spirit in Sarah Orne Jewett's

The Country of the Pointed Firs and the

Dunnet Landing Stories

> "I think," said Kate, "that the more one lives out of doors the more personality there seems to be in what we call inanimate things. The strength of the hills and the voice of the waves are no longer only grand poetical sentences, but an expression of something real, and more and more one finds God himself in the world, and believes that we may read the thoughts that He writes for us in the book of Nature."
>
> —Sarah Orne Jewett, *Deephaven*[1]

> But Nature repossesses herself surely of what we boldly claim.
>
> —Sarah Orne Jewett, "An October Ride"[2]

Along with themes of community, gender, and rural and city locations, discussions of nature play a significant role in Sarah Orne Jewett's writing, as these quotations from her 1877 novel *Deephaven* and her 1881 sketch "An October Ride" suggest.[3] Many critics have explored this element of her work, often implicitly pointing to Jewett's ecological leanings. In his early 1962 study, for example, Richard Cary argues that "through consistent use of analogy and symbol, Miss Jewett reasserts the reality of this primeval interdependence between man and nature."[4] Josephine Donovan likewise examines Jewett's environmental concern, stating that "in her most important fiction, she mourned the passing of a world in which nature was still a sanctuary, in which community was still a possibility, and in which women of strength were the prime sustainers of the communal experience."[5] Donovan also discusses Jewett's concerns about technological progress because of the potential environmental consequences.[6] Most recently, Marcia Littenberg has argued that Jewett, along with Celia Thaxter, "sowed the seeds of an early ecofeminism in the rocky New England coastline," asserting that "the power of Jewett's and Thaxter's writing lies in its expression of female sensibility, its sympathetic portrait of women's lives and women's perspectives, and its 'loving' relationship to the natural world."[7] As these critical reviews suggest, Jewett understood nature to contain both spiritual and natural lessons.

In this essay, I will explore how Jewett's work has affinities with nature writing because of her ecological focus and assertions that the human and nonhuman worlds are intimately intertwined, assertions best illustrated in her 1896 work *The Country of the Pointed Firs*.[8] I begin by briefly discussing ecocriticism as a critical approach. Next, I examine Jewett's use of nature in her famous "A White Heron." I then turn to her *The Country of the Pointed Firs* to illustrate her ecological concerns through a discussion of key characters who have a personal connection with their surrounding environment. Because religious ideologies often define attitudes toward nature, I will end by analyzing the different representations of spirituality in the work. Ultimately, while we may not be able to classify Jewett's *The Country of the Pointed Firs* specifically as "nature writing" as it is most frequently defined in contemporary criticism, I will suggest that this work represents a significant contribution to expanding our understanding of the environment.[9]

In recent years, fields such as ecocriticism have developed that can assist in reconsidering Jewett in this role of "nature writer." Ecocriticism is still characterized by much disagreement over definition and purpose; however, Cheryll Glotfelty provides a useful, general explanation to the term in her introduction to *The Ecocriticism Reader*.[10] She defines ecocriticism as "the study of the relationship between literature and the physical environment." Conceding the variety of interpretations of that idea, she points to the common belief that humans and the physical world interact in connective ways.[11] In other words, ecocriticism does not elevate humans over nature but instead emphasizes their place as part of an overall scheme of dynamic, natural activity.

For many readers, Jewett asserts her clearest ecological message in one of her most famous and frequently anthologized works, "A White Heron," from 1886. Here, she represents nature in a predominantly harmonious and almost sacred way, as Sylvia saves the heron from the intrusive and destructive presence of the hunter. The clear theme of preservation and protection resonates with turn-of-the-century movements toward wilderness protection and environmental awareness such as the 1892 establishment of the Sierra Club.[12] However, what distinguishes this earlier work from *The Country of The Pointed Firs* is the sharp distinction made between nature and society that somewhat undermines the preservationist message because it forces an either/or decision for Sylvia, the "lonely, country child."[13] Such dualistic thinking simplifies the human's relationship with the environment, suggesting that nature and society must remain separate. Sylvia remains somewhat detached from her surroundings. As Sarah Way Sherman notes,

"Sylvia's communion with nature does not mean that she herself is embedded in that instinctual ground but that she protects it, oversees it, loves it from above."[14] Sylvia's role as overseer asserts a dominance over nature, even if it is a benevolent, protective one.

To investigate the more complex connections that exist between people and nature in Jewett's writing, we must turn to her later, more mature work, *The Country of the Pointed Firs* and the other stories associated with Dunnet Landing: "A Dunnet Shepherdess," "The Queen's Twin," "William's Wedding," and "The Foreigner." Here, Jewett's representations of nature call for a renovation of attitudes about humans and the environment, a call that illustrates Jewett as a proponent of ecology. By asserting that nature exists as a live animate force that can be malevolent, benevolent, or indifferent in its interaction with the human world, we see Jewett's respect for the environment. Sharp, divisive dualism does not appear in Jewett's Dunnet Landing works because nature consistently impacts on the lives of the village inhabitants while community relationships remain, reminding readers of the intersections between nature and society.

With the Dunnet Landing stories, *The Country of the Pointed Firs* reflects Jewett's concern to educate an urban audience about the complexity of rural life by detailing the visit of an unnamed female narrator to the rural community of Dunnet Landing one summer.[15] Presuming that this town will provide her with a quiet retreat, she instead comes to see the complexity of this environment through interacting with the landscape and townspeople, especially the town's wisewoman Mrs. Todd, whose interaction with nature provides the strongest model of interconnection; former sea captain Littlepage, whose unsettling experiences with nature have haunted him; Mrs. Todd's brother, William, and mother, Mrs. Blackett, who live on the symbolic and pastoral Green Island; Joanna Todd, whose story of isolation shows how nature serves as a companion; and Esther, the Dunnet Shepherdess, who provides a female counterpart to William. Peppered among the narrator's interactions with these characters are descriptions of Dunnet Landing and the surrounding islands, which show how nature nurtures as well as isolates, develops as well as deprives, gives as well as takes away. As Jewett asks her audience to move beyond stereotypical readings of character, she likewise asks us to do the same with nature.

Jewett begins the novel by stripping from the narrator any illusions or assumptions she has about life in a small, rural town. Nature does not exist simply to provide her with a pleasant backdrop. No escape exists from nature in this community, not even when she rents the schoolhouse to concentrate on her writing, only to find that bees are visitors. At the same time, recording the natural environment becomes difficult, as the narrator describes how "the sentences failed to catch these lovely summer cadences" (*CPF* 18). This failure points to another example of nature's power; artistic representation pales in comparison to interacting with the physical surroundings in an active and integrated way. A later episode, described in the Dunnet Landing story "A Dunnet Shepherdess," confirms these initial impressions. Instead of writing, the narrator abandons her work to go trout-fishing. As she leaves, she notices "a flutter of white go past the window as [she] left the schoolhouse and [her] morning 's work to their neglected fate" (*CPF* 219). Instead of bemoaning the disorganization or the lost time, she leaves without guilt having reflected on the value of trout-fishing:

If there is one way above another of getting so close to nature that one simply is a piece of nature, following a primeval instinct with perfect self-forgetfulness and forgetting every-thing except the dreamy consciousness of pleasant freedom, it is to take the course of a shady trout brook. The dark pools and the sunny shallows beckon one on; the wedge of sky between the trees on either bank, the speaking, companioning noise of the water, the amaz-ing importance of what one is doing, and the constant sense of life and beauty make a strange transformation of the quick hours. (*CPF* 218)

The narrator has learned that to "catch the cadences" of nature, one must actively participate. The wind, swirling the pages of her work, establishes its dominance over the written word, challenging ideas that nature can simply be recorded.

The models that she observes in the community confirm the narrator's more connected relationship to nature. The most influential figure is Mrs. Todd, from whom the narrator rents a room for the summer. Herbalist, healer, and community leader, Mrs. Todd closely interacts with nature. Her power is detailed not only in her sympathetic treatments of others, but also in how that sympathy is developed from her close ties to the land. On one of their walks together, the narrator ob-serves: "I glanced at the resolute, confident face of my companion. Life was very strong in her, as if some force of Nature were personified in this simple-hearted woman and gave her cousinship to the ancient deities. She might have walked the primeval fields of Sicily; her strong gingham skirts might at that very moment bend the slender stalks of asphodel and be fragrant with trodden thyme, instead of the brown wind-brushed grass of New England and frost-bitten goldenrod. She was a great soul" (*CPF* 266). Not only does this passage clearly align Mrs. Todd with nature, but it also imbues her with a timeless, eternal quality, connecting her with the land in a permanent way. This association has led Sherman to suggest that "the setting [is] . . . personified as a woman" and Marilyn Mobley to note that "Mrs. Todd embodies the spirit of the land": Both conclusions point to how place and person intersect in this novel.[16]

Mrs. Todd clearly shows her connection with nature in her care of her herb gar-den. Gwen Nagel discusses how Jewett's gardens often show "a landscape of life and continuance in a mutable world" and that many characters "find expression in gardens which become a landscape of the self and an emblem of human connected-ness."[17] Such a comment aptly explains how Mrs. Todd creates a kind of middle ground between nature and society in her well-tended garden. Caring for her herbs, Mrs. Todd demonstrates her awareness of nature and her interest in main-taining a relationship with it. She, however, also uses the medicinal herbs and her role as resident wise woman as a way to continue relationships with the human community. The narrator is particularly curious about those herbs that she cannot identify and Mrs. Todd neglects to explain those that she senses "might once have belonged to sacred and mystic rites, and have had some occult knowledge handed with them down the centuries; . . . it seemed sometimes as if love and hate and jealousy and adverse winds at sea might also find their proper remedies among the curious wild-looking plants in Mrs. Todd's garden" (*CPF* 4–5). Jewett uses the language here to demonstrate how Mrs. Todd has learned how to merge two worlds into one. Had Jewett simply described the herbs as being representations of "love and hate," a basic dualism would have been suggested. Instead, she adds a third element, jealousy, and a fourth, "adverse winds," integrating the emotional realm with the natural one. This use of language becomes another subtle reminder that nature cannot be reduced to a simple dualism. It is under the influence of one of these mysterious herbs "that used sometimes to send out a penetrating odor late in the evening" that Mrs. Todd reveals to the narrator the story of her lost love (*CPF* 9). Laurie Crumpacker points to what has been Mrs. Todd's ability to combine her natural understanding with her desire to perceive and to participate in her community, arguing that "her garden, with its rows and fences, lets us know that some of her wildness has been tamed and that she represents nature integrated into the community."[18] While Mrs. Todd clearly understands the natural world, she definitely recognizes the importance of bringing that understanding into contact with society.

If Mrs. Todd is emblematic of Dunnet Landing and of nature to some extent, then noting how Jewett complicates our perceptions by portraying her negative characteristics is important to understanding how Jewett also broadens our understanding of nature. Despite her self-sufficiency and the way she can sympathetically connect with characters such as Abby Martin, a woman who believes she is Queen Victoria's twin, Mrs. Todd does not always emerge as an ideal spiritual creature. She certainly possesses a great deal of knowledge, but while she sells her herbs for medicinal cures, she does not educate others enough to make herself indispensable. The narrator observes, for example, that when she dispenses her Indian remedy and other "certain vials," Mrs. Todd provides "long chapters of directions," maintaining "an air of secrecy and importance to the last" (*CPF* 5).[19] Later in the story, another example occurs when she demonstrates the regard she feels for the narrator when she surprises her with a certain tea, telling her "I don't give that to everybody" (*CPF* 47). The narrator accepts the drink but wonders about a certain herb that she cannot identify. As a healer, a possessor of secrets about health and matters of life and death, Mrs. Todd remains powerful. Like the nature she embodies, her knowledge reminds others of her dominant position in the community and their dependence on her.

Not only does Mrs. Todd establish her position by her herbal work, she also feels secure enough to make certain judgments about several of the other characters. For example, at the Bowden family reunion, she employs stereotypes when she compares Mari' Harris to a "Chinee" (*CPF* 169).[20] Also, when she observes her husband's cousin at the reunion, she tells the narrator that "I hate her just the same as I always did; but she's got on a real pretty dress" (*CPF* 171). Another example occurs in her statements about Elijah Tilley, a local fisherman who is most notable for his almost obsessive honoring of his dead wife's memory. She concedes that "'Lijah's worthy enough; I do esteem 'Lijah, but he's a ploddin' man" (*CPF* 206). Elijah, in contrast, thinks of Almira in better terms, telling the narrator "there ain't a better hearted woman in the State o' Maine" (*CPF* 204). In these pronouncements, Mrs. Todd indicates her judgmental attitude about these individuals. In fact, it is easy to forget, given her prominence in the work, that Mrs. Todd journeys to Green Island only to see her mother, to the Bowden family reunion, and to Mrs. Abby Martin's, the "Queen's twin." Captain Littlepage and Elijah, like William, seek out the narrator. As the "foreigner," it is she who can see with fresh eyes the need for inclusion of these individuals. Overall, then, in her characterization of Mrs. Todd, as with her depictions of nature, Jewett challenges readers to look beyond a single view of subjects to understand their multifaceted and sometimes unpleasant aspects.[21]

The next significant character that the narrator meets is Captain Littlepage, who is a retired sea captain. His travels have led him to believe in certain problematic ideas that have isolated him from the larger community. If we were to dismiss him or to see him as a stereotypical character, he would simply seem a harmless but confused man who has lost his place in the world. However, delving into his character and listening to his story provide us with some important lessons about nature in its more disturbing aspects. The narrator first observes Captain Littlepage during a funeral procession, another reminder of how the Dunnet Landing community ultimately includes its misfit figures in its rituals. Later, Littlepage comes to the schoolhouse, and, after introducing himself, recounts to the narrator a story about a shipwreck in which he was involved and his subsequent refuge at a missionary site. While waiting for a supply ship, he meets Gaffett, a man who tells Captain Littlepage of "the waiting place." An uncharted area with "blowing gray figures" (*CPF* 37), this "waiting place" becomes the representation of "the next world to this" (*CPF* 35). Having learned that nature is unpredictable, Captain Littlepage does not dismiss the tale; rather, he listens charitably and comes to believe the story, conveying it to the narrator with an "alert, determined look" (*CPF* 41). Though age has made it impossible for him to test the theory of this place, his belief in it is obvious as the narrator notices with what "bewilderment" he observes the map in the schoolroom (*CPF* 41).

This episode is important for many reasons. While some have seen this tale as a sad indication of Captain Littlepage's inability to move into the present, it also shows how his experience with nature has taught him not to doubt, that nature can still challenge the boundaries humans have set for it on charted, neat maps.[22] The tale also continues the association between spirituality and nature. Unlike a Transcendental illumination, however, this experience fails to elevate Captain Littlepage; it, instead, increases his alienation, since no one believes his story. As the narrator observes later in the novel, "There was a patient look on the old man's face, as if the world were a great mistake and he had nobody with whom to speak his own language or find companionship" (*CPF* 143). Also, no certainty exists in his story, only questions that the earth could contain a spiritual crossroads of sorts. Dismissing the idea that a spiritual domain could be housed on earth reinforces the belief in a separation between nature and people. Even if the tale represents a fanciful creation, it nevertheless demonstrates that the boundaries we establish for the world's geography may not always be solid constructions, reminding us that humans can still be taught and that nature can still surprise.

In contrast to Littlepage's frightening vision, Jewett portrays one site where nature and spirituality interact in a more positive way: Green Island. Sherman refers to this place as a "pastoral center on the periphery of the civilized word."[23] As the narrator and Mrs. Todd observe the distant island from the mainland, the narrator notices that "a gleam of golden sunshine struck the outer islands, and one of them shone out clear in the light, and revealed itself in a compelling way to our eyes . . . The sunburst upon that outermost island made it seem like a sudden revelation of the world beyond this which some believe to be so near" (*CPF* 45). This harmonious picture continues as she meets Mrs. Todd's mother, Mrs. Blackett, in whom she observes an amazing tact, and her brother William, whose "spirit of youth" impresses the narrator (*CPF* 69).[24] He offers to show her around the land and eventually takes her to the island's highest point, where the view gives her "a sudden sense of space, for nothing stopped the eye or hedged one in,—that sense of liberty in space and time which great prospects always give" (*CPF* 71). William's pride in his "native heath" provides another positive connection between a person and the land.

Despite Mrs. Todd's criticism that he should be more social and more ambitious, William emerges as a thoughtful character who represents the best aspects of living in a natural environment. His patience, gentleness, and quiet manner have suggested to some critics that his character exists as a vehicle for Jewett to complicate our understandings of gender.[26] While debate exists over this interpretation, his existence has been decidedly marked by interacting with the land in a consistently active and personal way. One reason this normally shy person connects with the narrator at their first meeting is that he observes her digging for potatoes, which demonstrates that the narrator is not just a vacationing voyeur, but open to making some kind of connection with the land (*CPF* 67). William senses her authentic interest, proving himself a good judge of character despite his isolation. Observing the natural landscape has taught him also how to read human nature. Jewett, however, does not relegate William or Mrs. Blackett to stereotypical figures; she presents us with multiple views of these characters. Mrs. Blackett engages in social activity so much during the Bowden Reunion that Mrs. Todd describes her as a "queen" of the festivities (*CPF* 161). William, too, demonstrates his interest in others by courting and eventually marrying Esther, the local shepherdess. By presenting these characters in different moments, Jewett continues to remind readers to avoid one-dimensional readings of character and, in turn, nature.

In addition to these characterizations, the visit to Green Island contains a key example of how the landscape cannot only instruct us but also become a site for people to isolate their memories and pain. Specific locations, then, become places deeply intertwined with human feeling, a connection that argues for the importance of overcoming dualistic assumptions about society and nature. Mrs. Todd, for example, takes the narrator to a part of Green Island "where pennyroyal grew." She explains that the spot "'tis kind of sainted to me," as it is the place where she and her husband would visit when "courtin'" (*CPF* 76–77). She can also observe from this perspective the place where his ship was lost, another memorial of nature's powerful destructive force. Her memories are not simply of her husband; the pennyroyal also reminds her of another whom she loved that she could not marry because of his family's disapproval. No cold cathedral invokes the kind of emotion from Mrs. Todd that the landscape does. So connected is she with this spot that the narrator equates her with a timeless representation of authentic, quiet feeling: "She looked away from me, and presently rose and went on by herself. There was something lonely and solitary about her great determined shape. She might have been Antigone alone on the Theban plain. It is not often given in a noisy world to come to the places of great grief and silence. An absolute, archaic grief possessed this country-woman; she seemed like a renewal of some historic soul, with her sorrows and the remoteness of a daily life busied with rustic simplicities and the scents of primeval herbs" (*CPF* 78).

Here, Jewett reemphasizes how Green Island stands out as an area that nourishes its inhabitants while also serving as a location for connecting to moments of personal significance, both emotional and spiritual.

The land not only acts as a repository for memory, but can also demonstrate benevolence, as illustrated by the story of Joanna Todd, a woman whom the narrator never meets but who powerfully affects her thinking about people and nature. Perceiving herself to be damned and unfit for society when her fiancé runs off with another woman, Joanna retreats to and creates a world on Shell-heap Island. The island itself, though small, remains a site of local mystery, with stories about Indians, rumored to have been cannibals, and one chief, in particular, who was said "ruled the winds" (*CPF* 100). While Mrs. Todd rejects ideas about cannibals, she knows that some one has lived on the island because of the presence of "wormwood," which is a "planted herb" (*CPF* 109). This mysterious history echoes Joanna's life.[26] She requests no one visit her, and one would have to have the cooperation of natural forces even to attempt a trip; Mrs. Fosdick remarks: "'Twas a bad place to get to, unless the wind an' tide were just right; 'twas hard work to make a landing" (*CPF* 103). Despite the isolated location, the island nourishes its resident recluse physically and spiritually and enables her to live alone, without consistent societal intervention.

To illustrate the separation that Joanna feels from the community of Dunnet Landing, Jewett includes a story about an early visit that Parson Dimmick and Mrs. Todd made to Joanna. In sharing the story with the narrator and Mrs. Fosdick, Mrs.

Todd describes how, on the day of the journey, she sensed that Joanna "saw us coming" (*CPF* 114). Instead of hiding herself, Joanna demonstrates that she is still aware of social rituals by greeting her guests. Despite the lack of visitors, Joanna has also maintained a home that Mrs. Todd describes as very comfortable. While Mrs. Todd's eye catches sight of a Bible, the parson lacks such observational skills and treats Joanna poorly. Believing her to have completely abandoned the faith, he presents to her a negative judgmental portrait of God. Mrs. Todd wants a more "fatherly" treatment, but she describes how "'twas all about hearin' the voice o' God out o' the whirlwind; and I thought while he was goin' on that anybody that had spent the long cold winter all alone out on Shell-heap Island knew a good deal more about those things than he did. I got so provoked I opened my eyes and stared right at him" (*CPF* 119). In some ways, the "voice of God out o' the whirlwind" represents the societal judgment of Joanna's action; however, Mrs. Todd's interpretation locates God in the expression of natural occurrences. The scene also emerges as an example of how Jewett underscores her point with humor; Mrs. Todd's reaction to the Parson shows another side to her character and argues against strict judgment of such characters as Joanna.

The minister's inability to influence Joanna demonstrates her alienation from traditional expectations; however, her decorative rugs, the comfortable feel she manages to create in her home, and the shoes that she makes out of the island's rushes demonstrate that this retreat should not necessarily be seen as negative, especially if we consider her relationship with nature.[27] If one sees nature and society as necessarily distinct things, then it would follow that Joanna's choice reflects a negative choice of isolation; however, by viewing nature as a living, guiding force, it becomes more of Joanna's companion. Joanna has created a life for herself on the island that harmonizes with nature. As Mrs. Todd remarks, "she seemed so above everything common" (*CPF* 121). In contrast, Mrs. Todd criticizes the minister, saying that "he seemed to know no remedies, but he had a great use of words" (*CPF* 123). Viewing nature as a living force that serves as Joanna's companion assists us in understanding her as something other than a bitter spinster figure. Nature becomes a friend to this person who feels so alienated from certain elements of her culture.

Joanna's separation could suggest the kind of dualistic thinking about nature and culture against which I have been arguing. However, what marks her situation as resistant to dualism is her ability to immerse herself in a natural environment, while retaining a place in the community. In her story, the two spheres often merge. The community remembers her; fishermen bring her gifts; she asks for Mrs. Blackett to come when she dies; and she can signal to others if she experiences any trouble. Her funeral testifies to this sense of community. Mrs. Todd recalls:

'Twas a pretty day, and there wa'n't hardly a boat on the coast within twenty miles that didn't head for Shell-heap cram-full o' folks, an' all real respectful, same's if she'd always

stayed ashore and held her friends. Some went out o' mere curiosity, I don't doubt,—
there's always such to every funeral; but most had real feelin', and went purpose to show it.
She'd got most o' the wild sparrows as tame as could be, livin' out there so long among
'em, and one flew right in and lit on the coffin an' begun to sing while Mr. Dimmick was
speakin'. He was put out by it, an'acted as if he didn't know whether to stop or go on. I
may have been prejudiced, but I wa'n't the only one thought the poor little bird done the
best of the two. (*CPF* 124–25)

Her death brings together both nature and culture in this scene, and the sparrow's
song reminds us not only of nature's care for Joanna but also of her harmonious
relationship with it. Jewett, who clearly sympathizes with Joanna, presents this
story as a way to demonstrate that coexistence with nature becomes more important than attempting to elevate ourselves above it.

This story of isolation, of this "saint in the desert," generates a sense of curiosity and interest from the narrator. Traveling to Shell-heap Island, she observes that
other "pilgrims" had also traveled there. The visit makes her conclude: "In the life
of each of us, I said to myself, there is a place remote and islanded, and given to
endless regret or secret happiness; we are each the uncompanioned hermit and
recluse of an hour or a day; we understand our fellows of the cell to whatever age
of history they may belong" (*CPF* 132). By equating Joanna's situation to a timeless, universal, and spiritual one, Jewett's narrator reminds readers that we can be
provided with the same comfort from nature if we learn to interact with it in a personal and interdependent way.

In "A Dunnet Shepherdess" Jewett presents us with another female character
whom nature has refined although she has been isolated: the shepherdess Esther.
As with William, the natural landscape has developed in Esther a youthfulness
and strength that immediately impress the narrator. As with Mrs. Todd, the narrator speaks of her in larger than life terms. When she first sees Esther, she remarks
that she "stood far away in the hill pasture with her great flock, like a figure of
Millet's, high against the sky" (*CPF* 224). Elsewhere, she compares her with Ata-
lanta and Jeanne d'Arc (*CPF* 228–30). These heroic associations contrast with her
life of caretaker, not only of her sheep, but also of her demanding, invalid mother,
appropriately equated with a "warlike Roman emperor" (*CPF* 225). Esther, however, does not appear lonely. She takes her role of shepherdess seriously, as the
narrator learns that she will spend the night out in the moonlight if necessary to
care for her sheep (*CPF* 227). Her isolation has neither crippled nor damaged her
psyche. Indeed, her strength of character becomes even more remarkable with
Jewett's description of the landscape being "the wildest, most Titanic sort of pasture
country up there; there was a sort of daring in putting a frail wooden house before
it, though it might have the homely field and honest woods to front against. You
thought of the elements and even of possible volcanoes as you looked up the stony
heights" (*CPF* 223). When Esther smiles at the narrator, she describes it as "a
smile of noble patience, of uncomprehended sacrifice, which I can never forget.

There was all the remembrance of disappointed hopes, the hardships of winter, the loneliness of single-handedness in her look, but I understood, and I love to re- member her worn face and her young blue eyes"(*CPF* 231). Her experience mir- rors William's in Jewett's portrayal of the potential for nature to influence people in positive ways, by developing patience and character. Esther's marriage to William also shows her need for some kind of societal inclusion. As a bridge between nature and society, Esther becomes a culminating instance of Jewett's presentation of a range of characters who are often isolated in some way but who benefit from their time in nature.

Throughout each of the previous characterizations, Jewett has presented gener- ally positive experiences.However, she does not simply create an idyllic pastoral representation. As with contemporary nature writers such as Annie Dillard, who include the violent and unpredictable in their discussions of the environment, so too does Jewett remind her audience of the sense of danger that nature contains. Noting where Jewett focuses on nature's "vigor" interferes with human society is important in demonstrating her argument that while nature can be indifferent to us, we remain dependent on it. In *The Country of the Pointed Firs,* Jewett begins showing her awareness of nature's threats with the title image. The pointed firs de- marcate the region in a way that maps do not. While this image appears at first to contain nature, Jewett shows how such containment is problematic because of nature's potential threat. For instance, before the trip to Green Island, Mrs. Todd and the narrator look out at the view, and the quaint, idyllic image of the book's title becomes threatening. The narrator observes: "We were standing where there was a fine view of the harbor and its long stretches of shore all covered by the great army of the pointed firs, darkly cloaked and standing as if they waited to em- bark" (*CPF* 44–45). The word "army" suggests a certain order in nature, the com- parison reminds readers of nature's violence. The "army" could also be viewed as a protective force for the community; still, this picture of uniformity becomes threatening in its stillness, making the reader realize what power remains hidden from human eyes.

Not only do the trees sometimes take on dark characteristics, but the sea also invokes awe because of its potentially destructive and uncaring force. With its lo- cation, Dunnet Landing's population has first-hand knowledge of not only the profit that whaling and seafaring can bring but also the destruction. Paula Blanchard's observations of nature in *Deephaven* apply here as well; she notes how "the darker side of nature," the men claimed by the sea, lurks behind any idyl- lic representations.[28] For instance, the narrator learns at the ostensibly utopian Bowden family reunion that the graveyard is dominated by women, missing the graves of the men lost at sea, out west, or in combat. At the same time, Jewett bal- ances this threat by underscoring how sea travel can often positively influence a person, as it has for many of Dunnet Landing's population. One reason that Cap- tain Littlepage bemoans the decline in seafarers is that sailors and their families "were some acquainted with foreign lands an'their laws, an' could see outside the battle for town clerk here in Dunnet; they got some sense o' proportion" (*CPF* 28). Embedded in these remarks, we observe another example of Jewett's consistent double-sided commentary on nature. Though it can discourage provincial attitudes, it possesses the ability to destroy.

Jewett additionally reinforces negative natural images in many of the narrator's observations about characters and events, a discussion that emphasizes the link between the physical and spiritual. One important instance of this method occurs during the Bowden family reunion, when this idyllic small town begins to look confining to the narrator. Observing Mrs. Todd's active and vivacious exterior, she realizes that small town life, despite its close relationship with nature, does not promise happiness. Instead, by equating herself to a botanist, she attempts to take on a more objective stance of the very thing affecting her. By concentrating on "the thousand seeds that die," the narrator reminds us how, despite the calmness and relative pastoral quality to the land, isolation in nature does not always nourish (*CPF* 174).

Another occasion occurs as the narrator and Mrs. Todd walk to meet Mrs. Abby Martin, the "Queen's twin." On the way, Jewett employs a relatively common technique of using the landscape as a way to reflect an individual's state of mind. A sense of inaccessibility characterizes the environment surrounding Abby's house and property, a reflection of Abby's own mental inaccessibility. As Robin Magowan notes, "Pastoral is fundamentally an art in perspective; if only, because in it nature does not exist as other than a transparent medium, a psychic mirror reflecting a one-to-one correspondence between the person and the setting in which he is imaged."[29] As they struggle through the landscape, Mrs. Todd tells the narrator how difficult the land they are crossing has been to farm. She notes how the land has defeated three families; perhaps the defeat would be "good" in teaching more respect for the land, but Jewett stresses that the failures come about because of the land's attitude. While the view from a hill they travel on "is as beautiful as anything in this world" (*CPF* 261), Nature becomes greed personified in Mrs. Todd's description:

"I've known three good hard-workin' families that come here full o' hope an' pride and tried to make something o' this farm, but it beat 'em all. There's one small field that's excellent for potatoes if you let half of it rest every year; but the land's always hungry. Now, you see them little peakéd-topped spruces an' fir balsams comin' up over the hill all green an' hearty; they've got it all their own way! Seems sometimes as if wild Natur' got jealous over a certain spot, and wanted to do just as she'd a mind to. You'll see here; she'll do her own ploughin' an' harrowin' with frost an' wet, an' plant just what she wants and wait for her own crops. Man can't do nothin' with it, try as he may. I tell you those little trees mean business!"

I looked down the slope, and felt as if we ourselves were likely to be surrounded and overcome if we lingered too long. There was a vigor of growth, a persistence and savagery about the sturdy little trees that put weak human nature at complete defiance. One felt a sudden pity for the men and women who had been worsted after a long fight in that lonely place; one felt a sudden fear of the unconquerable, immediate forces of Nature, as in the irresistible moment of a thunderstorm. (*CPF* 265)

These "unconquerable, immediate forces of Nature" in many ways haunt the backdrop of this small community. Nature's intentionally destructive spirit is obviously something that Mrs. Todd interprets; nature remains ultimately indifferent to those families it usurps. Nevertheless, the personification employed here reminds readers that nature reflects a living spirit of some kind—another reminder to the reader that nature should not be viewed as something designed solely for humans' use but instead exists on its own terms.

One of the final images of the novel reinforces the potentially predatory quality of this natural spirit. Departing from Dunnet Landing at the end of the novel, the narrator observes from her boat how "the tide was setting in, and plenty of small fish were coming with it, unconscious of the silver flashing of the great birds overhead and the quickness of their fierce beaks" (*CPF* 212). In this powerful closing image, the narrator again points out the destructive quality of nature, an observation particularly appropriate because she herself is going out into an unfriendly world. This final representation also suggests that natural images may also be applied to a societal setting, an application that discourages bifurcating nature and society into two different and warring camps.

Not only does Jewett's work ultimately present a holistic community and balanced representation of nature, it also addresses the role of spirituality and religion in this ecological awareness. As discussions in *The Ecocriticism Reader* illustrate, religious views are central to discussions about nature writing. Ecocritics often criticize conventional patriarchal Christianity, arguing that it encourages an exploitation of natural resources by setting humans apart from nature and by advocating the idea that the earth is here for use rather than worship.[30] Mark Stoll explains the unfairness of much of this criticism, asserting that many Christians see this "higher" view as requiring a responsible attitude toward nature; still, the negative view remains.[31] Paganism, in contrast, generally asserts the interconnectedness of humans with the natural world and the belief that God/spirituality does not transcend nature in some otherworldly realm but exists here on earth, housed in specific locations. Many have seen paganism as an approach more likely to result in a kinder treatment of natural resources than Christianity, and paganism often becomes included in ecological discussions for its animistic view of nature.

It would be misleading to say that Jewett is completely Christian or completely pagan in her approach; she certainly presents both spiritual approaches.[32] Christian thought clearly exists in the novel: individuals attend church; the minister remains a religious authority, despite negative comments from Mrs. Todd; Mrs. Blackett and Joanna both own Bibles. Also, Captain Littlepage introduces powerful associations between this small community and the prelapsarian Eden when he joins the narrator in her schoolhouse workplace and, observing the landscape around them, quotes a description of Eden from *Paradise Lost*—"a happy, rural seat of various views" (*CPF* 21).[33]

Pagan associations counterbalance such views, particularly in Jewett's inclusion of a pantheistic attitude. Mrs. Todd emerges as the clearest symbol of a pantheistic or pagan stance, an association explored by many feminist critics. The narrator observes signs that could be evidence of a witch-like power: the telepathic bond she and Mrs. Blackett seem to share and her ability to read the weather ((*CPF* 54–55, 134). In the Dunnet Landing story "The Foreigner," Mrs. Todd also sees a ghost of Mrs. Tolland's mother as the young woman lies on her deathbed (*CPF* 255–56). Elizabeth Ammons describes her as one of "Jewett's witches," while Josephine Donovan associates her with a kind of "matriarchal Christianity."[34] Sarah Way Sherman, likewise, has explored the use of goddess imagery in Mrs. Todd.[35]

Rituals take on pagan-like significance, particularly in one of the work's most famous passages of the Bowden reunion procession:

The plash of the water could be heard faintly, yet still be heard; we might have been a company of ancient Greeks going to celebrate a victory, or to worship the god of harvests in the grove above. It was strangely moving to see this and to make part of it. The sky, the sea, have watched poor humanity at its rites so long; we were no more a New England family celebrating its own existence and simple progress; we carried the tokens and inheritance of all such households from which this had descended, and were only the latest of our line. We possessed the instincts of a far, forgotten childhood; I found myself thinking that we ought to be carrying green branches and singing as we went. So we came to the thick shaded grove still silent, and were set in our places by the straight trees that swayed together and let sunshine through here and there like a single golden leaf that flickered down, vanishing in the cool shade. (*CPF* 163–64)

Jewett's consistent portrayal of nature in this way illustrates how pervasive "pagan" associations are in the world of Dunnet Landing. In general, Jewett depicts spirituality being *contained* in nature rather than somehow being symbolic. Josephine Donovan explains this quality as part of Jewett's Swedenborgian influence, noting that Jewett favored the belief "that the spiritual interpenetrates the material. From this she developed an animistic view, one that appears to go beyond the Swedenborgian doctrine of correspondence . . . toward a theory that erases the divide between spiritual and material, seeing the transcendent as incarnate in the physical."[36] By presenting both Christian and pagan approaches, Jewett shows her interest in presenting multifaceted views of religion as well as character and landscape. Her inclusion of both also shows that whatever one's spiritual background, a connection exists between people and the land. Thus, she implies ecological responsibility. As Christopher Manes notes, citing Hans Peter Duerr, "people do not exploit a nature that speaks to them."[37]

Considering Jewett through this ecological lens serves as another example of Marjorie Pryse's label of her as a border-crosser.[38] From the title image that focuses clearly on the environment to the holistic community observed in Dunnet Landing, Jewett presents a complex and ever-present natural world that impacts humans in ways that point to our relationship to but not elevation above nature, a key to ecological inquiry. Jewett's work reflects the complex societal attitudes toward nature; her writing also resonates with nineteenth-century nature women writers such as Susan Fenimore Cooper.[39] In their introduction to *Jewett and Her Contemporaries,* Karen Kilcup and Thomas Edwards explain that viewing Jewett in this role of border-crosser "recuperates the writer as a powerful model for cultural and theoretical transformation."[40] Jewett's promotion of a holistic approach to the natural and the cultural world demonstrates her continued relevance to contemporary readers. Moreover, the ecological spirit that emerges in *The Country of the Pointed Firs* and the Dunnet Landing stories suggests that Jewett's message may lead to another "cultural and theoretical transformation," one that contributes to an increasing awareness of human-nature relations and to the importance of investigating the environmental messages that fictional works contain and create. As Willa Cather notes in her famous 1925 preface, Jewett's "'Pointed Fir' sketches are living things caught in the open, with light and freedom and air-spaces about them. They melt into the land and the life of the land until they are not stories at all, but life itself."[41]

Jen Hill

The Best Way Is the Simplest
Florence Merriam and Popular Ornithology

Florence Merriam's 1889 book *Birds through an Opera-Glass,* one of the first popular guides to bird watching, establishes Merriam's concerns with observation and classification in its opening lines:

We are so in the habit of focusing our spyglasses on our human neighbors that it seems an easy matter to label them and their affairs, but when it comes to birds,—alas! not only are there legions of kinds, but, to our bewildered fancy, they look and sing and act exactly alike. Yet though our task seems hopeless at the outset, before we recognize the conjurer a new world of interest and beauty has opened before us.

The best way is the simplest. Begin with the commonest birds, and train your ears and eyes by pigeon-holing every bird you see and every song you hear. Classify roughly at first,—the finer distinctions will easily be made later.[1]

The first lines set out what, for Merriam, is a recurring strategy throughout the book, to compare bird watching to people watching, bird behavior to people's behavior. It is perhaps her easy comparisons of the human and the natural worlds that lead to the critical appraisal of Merriam's early work as "soft science," and its subsequent dismissal from the history of academic ornithology. The classification of Merriam's popular writing as "children's literature" excuses her readers from any serious interrogation of her narrative and scientific methodologies. In the pages that follow, I will suggest reading Merriam's prose not only for what it has to tell us about birds, but also about the then newly emergent science of ornithology and the creation of scientific observers. Are Merriam's seemingly simplistic anthropomorphizing descriptions more than a "mere" displacement of the social onto nature? Might "the best way" to read Merriam be "the simplest"? Or, to put it another way, what do we find when we apply the same skills of close and careful observation Merriam advocates in the field to her *texts*? Pursuing these questions reveals that Florence Merriam uses her role as an early popularizer of ornithology to propose a way of seeing, which newly enfranchised women, children, and the uneducated in the scientific projects of the late nineteenth century. In turn, her advocacy of these "new" scientists reshaped the discipline of ornithology.

By the publication of Merriam's works in the late nineteenth century, the field of ornithology already was shifting its focus from the taxonomic study of dead specimens to the behavioral study of birds in the field. This shift in field method marked the beginnings of a concern with habitat—the organism's place in the world. If we recognize that Merriam locates bird watching as a conscious human practice in a clearly articulated natural habitat, we can interpret her anthropomorphic descriptions as attempts to track both the activities of birds and those of the watching human eye. Thus, the metaphorics of the "opera-glass" are not naive. Rather, they acknowledge the centrality of the human perspective in the project of ornithology.

In *Birds through an Opera-Glass* and the later *Birds of Village and Field,* the human experience of nature is linked to the Transcendental tradition by Merriam's self-conscious emulation of her one-time mentor, the inheritor of Thoreau's mantle and so-called "father" of the nature essay, John Burroughs. But though they borrow subject and scope from Burroughs, the texts ultimately discard both his rhetorical and field techniques. The solitary male communing with nature who inhabits the work of Thoreau and Burroughs is, in Merriam's texts, replaced by a more inclusive category of naturalists. Necessarily, for Merriam, the popularization of natural sciences changes both the means and the ends of the pursuit of knowledge. In addition to communicating her vast knowledge of bird species and habitats, Merriam reminds her readers of the continuity of the social and the natural, and of the mutually constitutive act of seeing: Close observation changes both the seer and the seen, revealing the complexity of nature and of the scientific subject.

The few biographers and critics familiar with Florence Merriam tend to group her early work with that of her contemporaries, Mabel Osgood Wright, Neltje Blanchan Doubleday, and her good friend Olive Thorne Miller. But Merriam, twenty years younger than Miller, belonged to the next generation of women nature writers, and her early work, although superficially similar to Miller's, already reveals much of the sophisticated field scientist Merriam was later to become.

Born in upstate New York to an upper middle-class family in 1863, Merriam was already an avid amateur bird watcher when she attended recently founded Smith College between 1882 and 1886.[2] Her first book, *Birds through an Opera-Glass,* was issued by Houghton Mifflin's Riverside Press in their series "The Riverside Library For Young People," in order to help "not only young observers but also laymen to know the common birds they see around them."[3] More publications followed, the better-known of which include *A-Birding on a Bronco* (1896), *Birds of Village and Field* (1898), and, in 1902, her classic piece of ornithological research, *Birds of Western America,* which for more than thirty years was the standard guide to western birds. She was the first woman admitted to the

American Ornithological Union in 1885, was made its first female fellow in 1929, and in 1931 won the AOU's Brewster Award for her contributions to ornithology. Merriam's embrace by the scientific community is in part responsible for the relative obscurity of her early popular writing: Those in ornithology prefer to remember her more-scientific contributions, and the literary community remembers Merriam as a scientist. As neither fish nor fowl, Merriam is difficult to place.

The recent rediscovery of Merriam's work by critics falls short of exploring its complexity. Vera Norwood zeroes in on Merriam's use of "female metaphors of domesticity," by focusing on the most "domestic" of Merriam's works, *A-Birding on a Bronco,* which despite its western locale is very much a narrative about eastern geography and interior domestic space projected on a western landscape.[4] Norwood describes the narrator of *Birds through an Opera-Glass* as "patiently" watching her subjects, "with no more threatening weapons than a parasol and a consuming interest in courtship and family behavior."[5] Certainly this voice is present in Merriam's text in passages such as the following about blackbirds: "When the nest-building commenced, our gay chevalier complacently permitted his meek little wife to perform the main part of the labor, while he would perch himself on a limb as near . . . as possible and taunt or ridicule his opponents."[6] But Merriam and other female writers were not alone in the practice of anthropomorphism. Ernest Thompson Seton, the writer, artist, and founder of the North American Boy Scouts—whom Merriam met during her Smith years—had his work dismissed by "serious" readers (rightly so, no doubt) due to his humanizing of animals. "Serious" nature writers such as John Burroughs also anthropomorphized, as did the famous biologists and authors of the theory of natural selection, Charles Darwin and Alfred Russell Wallace. The criticism has always been that the use of such imagery allowed societal prejudices and social constructions to contaminate what would otherwise be seemingly "objective" scientific conclusions. Critical work in the history of science has revealed the impossibility of scientific objectivity, but even without this perspective, singling out women nature writers of the late nineteenth century for their participation in such rhetorical strategies is unfair and at best, unhelpful. The issue is not one of whether or not Merriam anthropomorphized, but what these incidences reveal to us about Merriam's views both of nature and her society.

Other sections of Merriam's text reveal a more complex function of Merriam's anthropomorphism, even while her descriptions of bird families do at times borrow from recognizable nineteenth-century domestic ideology. Nesting goldfinches resemble a late-Victorian middle-class couple expecting their first baby, with an attentive husband lovingly attending to his wife:

When the blue eggs are laid upon their thistle-down bed in the compact round nest in the apple-tree, the father bird watches us anxiously till he knows that he can trust us near his mate, but when once sure of our good faith, will feed her in our presence. How tenderly he calls out as he comes to her! The quality of his note has changed entirely since spring. Instead of the per-chic-oree that told only of his delight in his free life in the air, his call is now a rich, tender "dear, dear, dear-ie, and a gentle, homelike, dear, dear, dear."[7]

While passages like these are clearly informed by popular conceptions of gender and romance, they co-exist with descriptions that reveal an author just as likely to interrogate as to embrace the status quo. Take the following passage concerning female birds carrying names based on male plumage, including the black-throated blue warbler:

Like other ladies, the little feathered brides have to bear their husbands' names, however inappropriate. What injustice! Here an innocent creature with an olive-green back and yellowish breast has to go about all her days known as the black-throated blue warbler, just because that happens to describe the dress of her spouse! The most she has in common with him is a white spot on her wings, and that does not come into the name at all. Talk about woman's wrongs! And the poor little things cannot even apply to the legislature for a change of name![8]

Earlier in the text, Merriam writes of bobolink behavior, "Shades of short hair and bloomers, what an innovation!"[9] Strangely, the bird she describes using the discourse of the early women's movement is *male*. Her projection of human gender roles onto the natural is soundly rejected again when, investigating vireo families in *Birds of Village and Field,* Merriam mistakes a male vireo tending the nest for a female. "Afterward I heard the vireo song from her, and concluded that *she* was the *father* of the family, left on guard while the mother was taking her rest."[10] By bringing her (incorrect) assumptions about gender roles to the attention of the reader, Merriam reveals the limits both of applying human social categories to the natural world and of the categories themselves. Passages like these indicate Merriam's knowing employment of human analogies and her gendering of the animal world as she interrogates both animal and human behavior.

The complexity of Merriam's imagery makes the inclusion of her writing in a category of late nineteenth-century women writers known for their "reticence and gentle quality" seem oversimple.[11] Not surprisingly, Merriam narrates from her sensibility at the edge of the culture, as an educated social activist who in 1891 worked at Jane Addams' Hull House and spent several winters working in Grace Dodge's working girls' clubs in New York City.[12] Merriam's out-of-the-box descriptions force her readers to recognize the shaping sensibility of the narrator/observer, and in so doing, to question the seemingly unassailable, male facade of scientific culture in the nineteenth century. The result is that the complicated taxonomy of academic ornithology, with its formal Latin names and distinct species families, is replaced by legible information accessible to the attentive, conscious human eye. In her words: "The best way is the simplest."

By opening up science to *any* attentive eye, Merriam enlarges the category of scientist to include those who lack formal training, a move that enables the under- and uneducated to imagine themselves as serious interrogators of the natural world. Merriam addressed children in her writing, but she knew that her books would also be read by those responsible for children's education, a category largely composed of middle-class women. These women, Evelyn Morantz-Sanchez notes in her article on women's participation in ornithology, were already being invited into the science in the late nineteenth century by ornithology's shift away from laboratory study towards fieldwork.[13] Obviously, one way Merriam models scientific achievement by a woman is by her own successful publication, but she also establishes the possibility of female authority by frequently citing female sources. The prefatory note of *Birds of Village and Field* singles out, as managing to observe birds "in the narrow margins of our busy lives," Mrs. Sara Hubbard who "in a shrubby back yard in Chicago, close to one of the main thoroughfares . . . has seen fifty-seven species in a year, and her record for ten years was a hundred species," as well as a Mrs. E. B. Davenport from Brattleboro, Vermont, who recorded seventy-nine species.[14] At points in the body of her text, Merriam even surrenders a portion of her narrative to female correspondents such as Helen M. Bagg, who contributes a "racy description" of a blackbird "wedding" to *Birds through an Opera-Glass*.[15]

In addition to modeling active scientific behavior in her own writing, and by citing female observers, Merriam employs a not-so-subtle narrative technique to naturalize and popularize scientific subjecthood. Using the first person, she moves from the centrality of the observer's *eye* to the centrality of the observer's "I." That eye/I, in much of her writing, records Merriam's own experience and observations, but, in her early guidebooks, she moves fluently from descriptions of her own experiences to situations that feature a collective "we" and then on to employ "you." This shift to the second person, interestingly, does not generalize or collectivize the experience described, but instead acts to personalize the activity by increasing its immediacy. The author-scientist disappears, to be replaced by the amateur enthusiast reader in her place. In her description of the habits of the ruffed grouse, for example, Merriam swiftly shifts from her experience in the first sentence of a paragraph—"Here, one of his favorite covers is in a quiet spot where I go to gather ferns"—to the second person and, by the end, a gentle imperative: "But pass by this hiding place, and a sudden *whirr* through the bushes, first from one startled bird and then another, tells you they have flown before you. Approach the drumming-log when the air has been resounding with exultant blows—the noise stops, not a bird is to be seen."[16] The substitution of "you" for "I" lends Merriam's own experience and authority to the reader, while the imperative gently commands the reader to go forth as an observer herself.

Thus, it is nearly impossible for readers to passively experience Merriam's texts; they are exhorted and cajoled to action. Observation, while the central activity of bird watching, is but one part of the active scientific inquiry she urges. Again and again, Merriam advocates formalized participation in ornithology by urging her readers to record what they see. Early in *Birds through an Opera-Glass*, while discussing shortcuts to visual categorization, Merriam states, "The first law of field work is exact observation, but not only are you more likely to observe accurately if what you see is put in black and white, but you will find it much easier to identify the birds from your notes than from memory."[17] By insisting that her readers discipline their observations, she completes their recruitment to scientific subjecthood. Merriam's own observations—anthropomorphic and other—lead women and children to science, but it is *their* observations, in their words, that will make them scientists.

While recording and representing may cultivate a vocabulary of scientific discourse in these new scientific subjects, Merriam is equally aware of the changes these "new" voices (and eyes/"I"'s) will have on science. The opening of the category of scientist, however, to people of differing genders and classes, redefines that category. In Merriam's work, the scientific study of birds necessarily leads to a new relationship between human and natural. For instance, she advocates camouflage for effective scientific observation. "But the best way is to keep perfectly still and let the birds show me just where the nest is, though of course it is only a matter of a few minutes more or less," she writes in *Birds of Village and Field*. "I sit down in the grass, pull the timothy stems over my dress, make myself look as much as possible like a meadow, and keep one eye on the bobolinks, while appearing to be absorbed with an object on the other side."[18] Once the scientist erases differences between self and surroundings, she is open to a new understanding not only of the factual "science" of bird relations, but of a possibility of cross-species communion between bird and human, illustrated in the following passage about the "least flycatcher" or "chebec":

The Least [flycatcher] is a most friendly little bird who quickly responds to kindness . . . [one] was the pet of a lady whose shrubby yard had many nesting birds. Almost every day through the summer, when she would go out to water her garden at five o'clock, the Chebec would come flying in to have her give him a shower-bath. While waiting for her to get out the hose, he would 'fly down on the fence and begin his talk'; then she would come up within five or six feet of him and turn the hose upon him gently . . . After telling about all the attractive ways of the friendly bird, the little lady concluded: "Now you do not wonder that I called him the darling little fellow, for I really have an affection for him." And then she went on to say that, although she lived by herself in her cottage, she found so much companionship in her birds and flowers and trees that she could never be lonely.[19]

Here, Merriam's description of human/animal relations is more complex than it first appears. The "lady" is "scientific" only in the loosest definition: She observes the birds and, as evidenced by her writing to Merriam, records her observations. Her relationship with the chebec in its domestic intimacy, however, transcends the traditional scientific split between observer and observed. In addition to understanding the behavior of the chebec *species,* the woman understands the uniqueness of the single "friendly bird" who provides her with "companionship." She then generalizes from her relationship with the bird to her greater relationship with the natural world at large. She recognizes herself not as being alone ("by herself in her cottage") in terms based on conceptions of human society, but as being part of the larger fabric of nature. The bird becomes a "fellow," in all of the word's meaning; that is, she and he are "fellow" creatures who share characteristics.

The unnamed woman's use of the word "fellow" in this way and the possibility that imaginative communion across species exists for Merriam, returns us once again to the author's anthropomorphizing descriptions of birds. By describing bird behaviors through their human analogs, Merriam erases at least some of the differences between human and avian species, introducing the idea of a cross-species sympathy that is the basis for an environmentally sensitive science. It would be incorrect to claim for Florence Merriam responsibility for ornithology's changed field techniques and methodology. Indeed, a historically specific moment in the popularization of bird study made Merriam's own entry into the science possible. Indicating the vast groundswell in the popular study of nature, Merriam phrased the new environmental movement in terms of disease, claiming that at Smith, "we all caught the contagion of the woods."[20] Her own scientific and literary development was linked to the male ornithological mainstream not only by her brother, Collins Merriam, who became an important governmental ornithologist, but when, as founder of the Smith College Audubon Club in the spring of 1886, she invited the venerable nature writer and bird watcher John Burroughs to visit. Burroughs, a white-bearded friend of Wallace Stevens, was then America's best-known living nature writer, and Merriam wrote to her brother that Burroughs opened "a new world . . . to [the students] with a gain in spirituality, which realized my highest hopes for the influence of the work."[21] Already, for Merriam, nature study and ornithology performed serious ideological "work," and we sense, in her use of "spirituality" that she already understood the connections between human and nature that would so profoundly influence her work. Burroughs, who would later note in his diary a re-meeting of "Mrs. Bailey" (Merriam's married name) in 1906, was a little less "spiritual" about his visit to Smith, writing to a friend, "my walks and talks with the Smith College girls would set every old bachelor like you fairly wild. Twice a day I walked with fifty of them, and every night I was surrounded by them; never was a wolf so overwhelmed by the lambs before."[22]

Later, Florence Merriam and the venerable Burroughs would also disagree on more than their perceptions of his Smith visit. His first book, the 1871 *Wake-Robin,* was one of Bailey's "favorites," and there is no doubt the young Merriam modeled her early endeavors on his very successful publications.[23] *Birds through an Opera-Glass,* like *Wake-Robin,* is a collection of previously published periodical writing, so there are structural similarities as well as many rhetorical ones.[24] But Merriam outgrew her mentor, and certain passages in *Birds through an*

Opera-Glass suggest that she consciously wrote against the predominantly male ornithological establishment, critiquing both its methods and its ends. Passages by Burroughs and Merriam that discuss the Blackburnian warbler render their different approaches and focuses in sharp relief. First, Burroughs:

From those tall hemlocks proceeds a very fine insect-like warble, and occasionally I see a spray tremble, or catch the flit of a wing. I watch and watch till my head grows dizzy and my neck is in danger of permanent displacement, and still do not get a good view. Presently the bird darts, or, as it seems, falls down a few feet in pursuit of a fly or a moth, and I see the whole of it, but in the dim light am undecided. It is for such emergencies that I have brought my gun. A bird in the hand is worth half a dozen in the bush, even for ornithological purposes; and no sure and rapid progress can be made in the study without taking life, without procuring specimens. This bird is a warbler, plain enough, from his habits and manner; but what kind of warbler? Look on him and name him: a deep orange of flame-colored throat and breast; the same color showing also in a line over the eye and in his crown; back variegated black and white. The female is less marked and brilliant. The orange-throated warbler would seem to be his right name, his characteristic cognomen; but no, he is doomed to wear the name of some discoverer, perhaps the first who robbed his nest or rifled him of his mate,—Blackburn; hence Blackburnian warbler. The burn seems appropriate enough, for in these dark evergreens his throat and breast show like flame. He has a very fine warble, suggesting that of the redstart, but not especially musical.[25]

What shocks the modern reader—Burroughs' unapologetic killing of the bird simply to resolve his indecision—was commonplace in nineteenth-century ornithology. (Audubon himself was described by Robert Penn Warren as "the greatest slayer of birds that ever lived").[26] Certain knowledge, for Burroughs, requires possession—the bird in hand—as a necessary link between looking and naming. By shifting from the first person subject who discovers and kills to a disembodied second person —"look on him and name him"—Burroughs implicates the reader in his act. Even more discomfiting is his implication of the bird in its own death by linking its name origin to an original "Blackburn" who "rifled him of his mate." Thus, not only is Burroughs' reader implicated in the entire bloody history of ornithology, but so, by their very existence—their looks, their names—are birds themselves. By the end of the passage, though, Burroughs' bird lives again: his throat and breast showing "like flame" against the evergreens and his "very fine warble" the last we hear of him. This rhetorical sleight of hand, from live bird to dead bird and back to the live, acts to erase the violence of the death, enabling the dead specimen to represent the many who still live. We are left with the certainty of named, living birds and a fading—imagined, even—memory of a dead one.

Merriam writes of the Blackburnian warbler in a strikingly similar breathless tone, seeking, as Burroughs does at the outset, to capture the bird's elusiveness in the tempo of her prose. The outcome, however, is entirely different:

Now and then you are fortunate enough to get a near view of this exquisite bird, but he has an exasperating fondness for the highest branches of the tallest trees. You can see there is something up there, but as you throw your head back and strain through your opera-glass, you fancy it is some phantom bird flitting about darkening the leaves. The seconds wear into minutes, but you dare not move. Your glasses don't help you to see through the leaves, but you feel sure that something will appear in a moment, over the edge of that spray or on the end of that bare twig, and it won't do to miss it. So when your neckache becomes intolerable you fix your eyes immovably on the most promising spot, and step cautiously backward till you can lean against a tree. The support disappoints you, your hand trembles as much as ever, and your neck is growing stiff. You make a final effort, take your glass in both hands, and change your focus, when suddenly a low, fine trill that you recognize from being accented on the end like a redstart's comes from a branch several feet higher than before over your head. Your neck refuses to bend an inch more. You despair. But all at once your tormentor comes tumbling though the leaves after an insect that has gotten away from him, and you catch one fleeting glimpse of orange that more than repays you for all your cramps.[27]

Birds through an Opera-Glass was the "first bird book that did not presuppose shooting" as necessary to ornithological study, and the most obvious difference in these passages is the status of the bird at the end of each.[28] But also notable is Merriam's use of the second person, which places her reader in the scene with the bird at all times. She acknowledges the importance of imagination to scientific vision—the "phantom bird"—but is careful to balance it with the centrality of the physical eye and its technological aid, the opera-glass. Together, imagination and close observation enable recognition of the bird. Naming the object involves the scientific subject as much as it does that which is to be named. The "hunt" in this case is as much for subject position as it is for the bird; a battle against discouragement and not against the bird itself. In this is an echo of the transcendentalists: seeing involves disregarding personal discomfort, erasing self to enable vision even as that self is central to the project of seeing. The "reward" is "one fleeting glimpse," but that is enough, because in that glimpse is an affirmation of the observer as well as the identification of that which is observed. The result is similar to what Randall Roorda names in his work on Thoreau, "retreat"; that is, that the "climax to the movement to lose the human comes in the recognition that the human has indeed been lost, that a place/condition marked as pure nature has been attained."[29] Paradoxically, it is in that moment of retreat—the erasure of the subject's discomfort by the vision of the bird—that the reassertion of the centrality of the scientific subject occurs.

That scientific subject is, in the words of Merriam, a *naturalist,* as she states with emphasis in her bird watching guide, *Birds of Village and Field.* Her concern with scientific specificity, sympathy with the natural world, and non-intrusive observation merge in the figure of the new scientist in the following passage:

A most perplexing moment once came to a Lark. He found himself on a fence between a hawk and a collector! To which should he expose his brilliant breast? His brothers in the locality, at sight of this same collector, had promptly turned their backs to him, looking back at him only over their shoulders, but this bird kept his back to the Hawk and stood facing the man. As the collector was a Naturalist, the bird's trust was not misplaced, and he lived to again sing his joy to his mate.[30]

Both her redefinition of the scientist's role and the sophistication of self-consciously transcendental passages like Merriam's account of the Blackburnian warbler can inform a reading of other "simpler" passages in Merriam's text, including those that anthropomorphize her avian "friends." If one effect of her comparisons of human and bird behavior, seen earlier, is to establish the links between species, she also employs anthropomorphism to comment negatively on her fellow humans. She likens the catbird to "some people, [because] he seems to give up his time to the pleasure of hearing himself talk. With lazy self-indulgence he sits by the hour with relaxed muscles, and listlessly drooping wings and tail. If he were a man you feel confident that he would sit in shirt sleeves at home and go on the street without a collar."[31] Not only does Merriam's anthropomorphic imagery act as a mnemonic to enable her readers to recognize the male catbird the next time they see him, but it also allows them to remark on the difference between them—the newly formed, methodical, *sympathetic* people of science—and their undisciplined, and in this case gendered male, human counterparts. The natural world instructs, not only about nature, but about the human world as well.

To conclude, let us return once again to John Burroughs and his visit to Smith College. "Never," he wrote his friend, in a kind of reverse-anthropomorphism that was prescient in a way he could not have imagined, "was a wolf so overwhelmed by the lambs before." The non-violence of Merriam's approach to nature and the subtlety of her rhetorical strategy is misleadingly lamb-like, and, if the "wolf" is the male scientific establishment of the late nineteenth century, certainly her work changed the make-up of that world by enabling new scientific subjects and new science.

Part III

Nature Writing in the Twentieth Century

Part III of this collection explores the rich variety and complexity of work by modern American women nature writers, demonstrating the breadth and the scope of the evolution of the genre.

Michael Branch examines the work of Marjory Stoneman Douglas, who wrote tirelessly on behalf of the Everglades. Eloquently combining natural and cultural history, Douglas brought to public consciousness the idea of the Everglades as a river of grass, and worthy of preservation. Branch outlines the importance of a writer such as Douglas who, in the same way as Jewett or Hurston, is closely identified with a specific bioregion, and who emphasizes the importance and value of a carefully nurtured sense of place. With her sensitive and insightful combination of the natural, political, and cultural history of the Everglades, Douglas makes the argument for the Everglades as an integral component of the history of Florida rather than the backdrop setting in which an exclusively human, political development evolved.

Suzanne Ross introduces the work of Sally Carrighar, an author who urges that we reorient our perspective on the natural world. In crediting the animal's position in nature, Carrighar places humans and nature as mutual participants in a dialogue. Her stance extends the anthropological positions of authors such as Hurston even farther than before. Uniting close field observation with an appreciation for the complexity of the social and biological world of the creatures she studied, Carrighar as nature writer translated for her subjects and advocated for a more thoughtful awareness of the total natural world that surrounds us.

In Tamara Fritze's examination of the ephemeral quality of gardened land, women literally write with and on nature, as their gardens of domesticity, dominion, subversion, abundance, or abandonment provide a wide middle ground where culture blends with nature. In these various gardens, each author-gardener works out her own sense of place.

Karen Cole's examination of Elizabeth Lawrence's history of southern market bulletins demonstrates how writing forges a connection between gardeners and gardens. Serving as a site for an extended conversation, the market bulletins recorded southern women's voices and their relationship with the garden landscape around them. Lawrence's study preserves the vernacular traditions of southern gardens and the voices of those who gardened them and spoke of them. Her essay underscores an important function of women nature writers, their ability to build and maintain communities that linked the trained with the untrained, the expert and the amateur.

By drawing attention to the farm women who speak in market bulletins and the abandoned gardens of gardeners past, Cole and Fritze both broaden the scope of the participants in nature writing as well as the type of "texts" they write. They provide an important reminder in this discussion of women writers that the dialogue of our relationship with the land, especially for some women, took place literally outside their own door.

In Rena Sanderson's essay, the writings of rancher Linda Hasselstrom remind us that nature, with or without a human presence, is a constant and harsh struggle between life and death. As a counter to the more romantic strains of nature writing discussed in the earlier essays in this book, Hasselstrom's realism, born from her real-life experience in the west, records the hardships and the joys of working an oftentimes unforgiving land. Here, as with Carrighar, humans and nature can share a conversation, but in Hasselstrom's writings the mutual dialogue is one of survival against harsh realities, where pain and loss are constant factors. As with authors from Cooper to Jewett to Cather and Douglas, Hasselstrom demonstrates a close and comprehensive understanding of her region as well as of a human society developed in response to, and in conjunction with, the environment around her.

Karen Waldron's essay on Leslie Marmon Silko returns us, in some ways, to Emerson. The "I"—or "Eye" of Emerson's essay "Nature"—cannot for Silko be substituted for the centrality of the land itself. As Waldron shows, for Silko, one cannot have consciousness *of* the landscape, only *within* the landscape. In the process, Silko deconstructs the traditional locus of the (Western) mind meeting nature; where the two are inseparable, there can be no "meeting." As Waldron points out in her essay here, Silko's writing asks not only the traditional questions posed by most nature writers, "What do you see/feel/know of the land?" Silko concludes, rather, that "everything depends on *how* humans are in the landscape." Writers such as Silko, who operate outside the European tradition, challenge the simple notion that we need simply more information about our natural world. We need rather to begin by adopting a whole new way of seeing what is already before us.

Tiffany Ana López and Phillip Serrato look at the ways that noted Chicana writer Gloria Anzaldúa creates a dialogue focused on the interrelationship of issues of identity and the environment in the important, yet often neglected, field of children's literature. In two children's books, *Friends from the Other Side* and *Prietita and the Ghost Woman,* Anzaldúa writes across genres, translating her philosophy of Chicana feminism and environmentalism for a juvenile readership. Anzaldúa's children's books are perhaps most notable for their deliberate expansion of her project of 'conscientizacion' of readers for the purpose of encouraging social change. Often ending her children's books with the theme of unfinished work, in the books discussed here, Anzaldúa utilizes the trope of healing to advance a politics of community building and the reconciliation of differences in ways that empower the child and emphasize each individual's responsibility for the healing of social divisions and preserving the environment.

Jenny Davidson's essay explores Annie Dillard's own "appropriation" of Thoreau in her *Pilgrim at Tinker Creek*. Davidson demonstrates how Dillard has achieved a reconciliation of two seemingly opposed strands of nature writing, the normally male-dominated hunting myth and a more domestic, feminine trope, that of the garden or the house. Dillard, like Austin, has "appropriated" Thoreau, and by assuming the intertwined and interdependent roles of the hunter and the shaman in her quest to "see" nature, she gains a dual vision, and imagines a cyclical power that encompasses paradoxical images of both beauty and violence. The power of her text lies in its ability to reject an overly simplistic dualism, to build bridges and fill gaps in the genre.

In the concluding essay of this collection, Annie Ingram traces the grassroots responses to cases of environmental injustice and follows that response into literature. With a close reading of three late twentieth-century novels—Barbara Kingsolver's *Animal Dreams,* Ana Castillo's *So Far From God,* and Linda Hogan's *Solar Storms*—Ingram outlines these authors' activist perspective and illustrates how the environment for them is nothing less than the entire arena in which we live and work and play. In many ways, the literature of environmental justice endorses the dismantling of the nature/culture dichotomy that Silko's work rejects, and poses the same question: *How* are we in the landscape?

Michael P. Branch

Writing the Swamp
Marjory Stoneman Douglas and
The Everglades: River of Grass

After meeting environmental preservationist and writer John Muir in the Sierra in
1871, Ralph Waldo Emerson remarked in a letter that it gave him great joy to find
Muir "the right man in the right place."[1] Emerson's observation applies to many
important figures in America's rich tradition of literary natural history: those authors for whom graceful, vital, and forceful writing is, like the fragrance of a
flower or the agility of a bird, a direct product of the richness and beauty of the
natural environment that engenders it. One thinks not only of John Muir in
California's high Sierra, but also of Henry Thoreau in the fecund forests of New
England, Aldo Leopold on his modest sand county farm in Wisconsin, Edward
Abbey in the glowing slickrock desert in Utah, Wendell Berry on the home
ground of his farm in Kentucky, and Rick Bass in the dripping forests of
Montana's Yaak Valley.

Of course "the right man in the right place" has very often been a woman, as
the work of Susan Fenimore Cooper, Mabel Osgood Wright, Mary Austin, Rachel
Carson, Ann Zwinger, Annie Dillard, Linda Hogan, Terry Tempest Williams, and
many other gifted American women nature writers so clearly demonstrates. Writing by American women has often been deeply concerned with issues of landscape, place, and the ethical relationship between the human and nonhuman
worlds. From the letters and diaries of the seventeenth century, to the garden
books and travel narratives of the eighteenth century, to the literary rambles and
place-based fiction of the mid-nineteenth century and the more political environmental tracts of the late nineteenth century and early twentieth centuries, through
the ecologically informed and artistically ambitious environmental literature
produced since the mid-twentieth century, American women have offered important literary representations of their home places and their home land. Indeed, the

125

sustained regional and bioregional emphasis that distinguishes many works by women nature writers has helped turn ecocritical attention toward literary descriptions and celebrations of local landscapes, thereby helping readers recognize the immense value of attachment to one particular place.

In the case of the Everglades—the unique wetland wilderness that has from time immemorial occupied the southern part of the place we now call Florida—Marjory Stoneman Douglas was clearly the right person in the right place. Originally from Minneapolis, Marjory Stoneman lived in Rhode Island and Massachusetts before graduating from Wellesley College with a degree in English in 1908. After a failed marriage to Kenneth Douglas, she moved to Miami in 1915 (then a new city with fewer than four thousand inhabitants), where she lived with her father and worked as a staff journalist at the *Miami Herald*. During World War I, she served in the U.S. Naval Reserves and worked with the American Red Cross in Europe from 1918 to 1920. During the 1920s, she lived in Miami and was an active social reformer who championed causes including civil rights, prison reform, family nutrition and counseling, legal aid, affordable housing, job training, literacy, and women's rights.

It was also during the 1920s that Douglas launched her literary career by writing short stories, approximately fifty of which she published between 1925 and 1943, mostly in the *Saturday Evening Post*. While the plots of these stories usually center on love or adventure, the landscape of the Florida Everglades figures prominently in Douglas's short fiction. She later wrote three novels and a half-dozen books of nonfiction, many of which are set in the Glades, and nearly all of which engage regional landscape, weather, flora, and fauna. Even Douglas's journalistic pieces, nonfiction essays, brochures, government reports, and autobiography are essentially rooted in the landscape and creatures of South Florida, and her literary and activist career might be described as a progressive movement into greater intimacy with and celebration of the little understood majesty and vulnerability of the Everglades.[2]

From the 1940s through her death in 1998 at the age of 108, Douglas's was the most eloquent and influential literary and activist voice raised on behalf of the Everglades, and, like the nationally prominent nature writers in whose company she belongs, Douglas's life and work were intimately braided with her place. Using literature as her political and aesthetic vehicle, she was among the first to awaken Americans to the beauty and importance of the Everglades, and she remained vigilant at the front lines of the long battle to protect the Glades from urban encroachment, agricultural pollution, and the alternating desiccation and inundation caused by excessive human interference with the natural hydrological cycle of this unique wetland ecosystem. Douglas was one of Florida's most esteemed citizens and one of America's finest literary exponents of the ecological, aesthetic, and spiritual value of wilderness. She participated in a long tradition of women's advocacy of social and environmental causes, and she was the first American woman nature writer to forcefully combine preservationist and literary goals in service of America's greatest wetland ecosystem. As one former Florida governor remarked, Douglas was "the poet, the sledgehammer advocate, the constant conscience of the Everglades for half a century."[3]

The year before Marjory Stoneman Douglas died, 1997, marked the fiftieth anniversary of the publication of *The Everglades: River of Grass* (1947), her first and most important book, and the book that earned the Glades a place of honor on America's imaginative, literary, and preservationist map. Thoroughly researched and gracefully written, *River of Grass* has a rhetorical control and topical scope that make it as impressive today as when it was first published as part of Rinehart and Company's acclaimed Rivers of America series. Indeed, the now common understanding of the Everglades *as a river*—albeit wide, shallow, and heavily vegetated—is attributable to Douglas's own insights. Asked by Rinehart to write a book on the Miami River, she insisted that the Miami was ultimately inextricable from the Everglades, and she proposed instead that the entire South Florida wetlands be discussed as an immense "river." Arguing that the Miami River "is only valuable as part of a system," Douglas began a five-year writing project on what she referred to as the "river of grass."[4] It has since been widely noted that this appealing concept of the Glades as a river "educated the world as to what the Everglades meant."[5] Working without the benefit of scientific ecology, which would develop during the decades following the publication of her book, Douglas nevertheless understood the vital importance of the hydrological, geological, climatological, and biological interconnections that comprise what would later come to be called the "Everglades ecosystem."

Precisely because it is crafted so as to account for the complex matrix of forces that have historically influenced the Glades, *The Everglades: River of Grass* is a difficult book to categorize. While it is written in nonfiction prose, the book's narrative dramatization of historical and natural historical information gives it the feel of a novel, and the charm of a chronological procession of short stories such as those Douglas once regularly published in the *Saturday Evening Post*. While it is replete with precise scientific details, *River of Grass* filters those details through a poetic sensibility, rendering the landscape with a lyrical intensity characteristic of such classics of American nature writing as Henry Thoreau's *Walden* (1854), John Muir's *My First Summer in the Sierra* (1911), or Aldo Leopold's *A Sand County Almanac* (1949), and reminding us of the graceful combination of natural historical and literary aims that characterizes Susan Fenimore Cooper's *Rural Hours* (1850), Rachel Carson's *Silent Spring* (1962), and Terry Tempest Williams's *Refuge* (1991). Just as the book defies the constraints of a particular genre, so too does it resist any single disciplinary discourse, instead invoking insights from history, geology, hydrology, anthropology, zoology, botany, politics, economics, journalism, and literature to offer a rich, multidimensional view of South Florida's cultural and natural history.

It is perhaps best to think of *River of Grass* as a work of epic literary environmental history written in the tradition of Gilbert White's *The Natural History of Selborne* (1789) or William Bartram's *Travels* (1791): "epic," because Douglas's scope moves from the prehistoric geological formation of the Florida peninsula through current issues in ecosystem degradation and restoration; "literary," because her imaginative use of language and her gift for narrative are essential to the success of the story she tells; "environmental," because the attention she gives to landscape, place, and nonhuman beings constitutes an important, early attempt to describe the Glades in ecosystemic terms; and "history," because Douglas's accounts of Native American inhabitants, European explorers, mercenary pirates, early settlers, escaped slaves, avaricious land developers, devout conservationists, and concerned citizens offer an unusually inclusive and accurate picture of the nature-culture interactions that have shaped the South Florida landscape.

Although *River of Grass* made a number of important contributions to the literature of Florida and the American literature of nature, three distinctive aspects of Douglas's accomplishment deserve particular attention: the imaginative presentation of prehistoric South Florida; the sensitive treatment of Native American culture; and the relentless demand for environmental protection of the endangered Everglades. In combining environmental history with Native American history and preservationist politics, Douglas helped identify issues that would be of special importance and value to later American women nature writers including Carson, Zwinger, Hogan, and Williams.

"There are no other Everglades in the world." So begins "The Nature of the Everglades," the opening chapter of *River of Grass*.[6] This first chapter, which after more than a half century remains the best literary essay on the natural history of the Glades, is remarkable for its powerful imaginative reconstruction of the ancient, subtropical wilderness that exists beneath and before what is now Florida. By starting her book at a point before the beginning—that is, before the advent of human history—Douglas accomplishes a number of things that contribute to the overall success of *River of Grass*. First, she compels readers to engage in the salutary exercise of envisioning a natural world devoid of the human presence, thereby establishing an ecological baseline against which the environmental degradation caused by future human activities might be measured. Her approach asks us to consider the importance and beauty of a river of grass that exists outside the flow of human history, a place so essentially inhuman in its purposes that we can begin to comprehend it only if we meet it on its own terms. Douglas's imagination takes us back up the "river of time," beneath the surface of the land, and into the ancient, labyrinthine oölitic limestone foundation of the Floridian plateau:[7]

If all the saw grass and the peat was burned away there would be exposed to the sun glare the weirdest country in the world, thousands and hundreds of thousands of acres of fantastic rockwork, whity gray and yellow, streaked and blackened, pinnacles and domes and warped pyramids and crumbling columns and stalagmites, ridgy arches and half-exposed horizontal caverns, long downward cracks, and a million extraordinary chimney pots.

Under the sun glare or the moonlight it would look stranger than a blasted volcano crater, or a landscape of the dead and eroded moon.[8]

By showing us an aboriginal Florida that is perfectly complete despite the absence of *Homo sapiens,* Douglas insists upon the essential facts too often obscured by the fantastic and touristic rhetoric used to describe the Sunshine State: The earth was here before us; it was not created to serve our interests alone; and we must condition ourselves to abide by its limitations or perish in our ignorance.

Douglas's vision of prehistoric south Florida is also effective in its dynamic use of natural historical details. "The Nature of the Glades" offers an impressionistic literary collage in which a wood stork's–eye view of the ancient wetland is created through patient descriptions of resident plants and animals: saw grass, scorpion, pine, palmetto, deer, quail, rattlesnake, woodpecker, dragonfly, ant, grasshopper, warbler, lizard, Resurrection fern, live oak, caracara, Banyan, ilex, eugenia, mastic, manchineel, poisonwood, coral snake, orchid, tree snail, moth, butterfly, palm, cypress, panther, bear, otter, heron, ibis, bald eagle, osprey, man-o'-war bird, white pelican, raccoon, alligator, crocodile, mangrove. "[A]s if she were using a wide-angle lens," writes Melissa Walker, Douglas's artistic eye "takes a sweeping view of southern Florida and focuses on one spot after another until her panorama is rich with detail."[9] The use of such panoramic description allows Douglas to create the literary correlative of biodiversity, for hers is a style of environmental writing calculated to reflect and express the nearly incomprehensible plenitude and diversity of the aboriginal Everglades ecosystem.

Douglas's imaginative vision of prehistoric Florida also includes a fantastic menagerie of ancient beasts, because the Florida peninsula, which was never reached by the encroaching Pleistocene ice sheets that repeatedly engulfed most of the North American continent, nurtured a number of impressive megafauna, including:

tapirs and huge peccaries and two kinds of capybaras and a giant tortoise-armadillo seven feet long with flexible armor, and a great strange mammal called a glyptodont, with head and tail and feet sticking out of his heavy immovable shell. These were the peaceful beasts, the leaf eaters. The flesh eaters followed them into Florida. There came an Asiatic lion, yellow as sunlight, and cougars and true wildcats, and a long-legged cat more than half lion . . . There was a small wild dog and a big early hyena with crushing jaws. Wolves ran in ferocious packs, dire wolves, larger than any wolf man has seen, howling across the open swamps to the bellowing of alligators and crocodiles, their slitted eyes gleaming back to the same moon.[10]

The descendants of many venerable, prehistoric animals—such as the dragonflies, snakes, and alligators so often associated with the Glades—have survived into our own Cenozoic era with relatively few additional evolutionary adaptations, and serve as a reminder of our direct connection with the deep past of the land. Indeed, the distinctive saw grass of the region is a giant sedge so ancient as to be among the oldest green plants on earth. By describing such life forms, Douglas suggests that a full appreciation for the Everglades must begin with an understanding of its ancient natural history. "Shapes of growing things which this land still bears in the Everglades are so much older than the region itself that thought grows dizzy contemplating them," she writes.[11]

Another of Douglas's most important accomplishments in *River of Grass* is her detailed and sympathetic account of Native American inhabitation of the Everglades. Ten of the book's fifteen chapters describe aspects of native culture or discuss incidents in the troubled history of Indian-white relations in South Florida. Douglas was ahead of her time in comprehending the vital importance of nature-culture interactions to a full understanding of place. Whereas most histories of the region published in the early- and mid-twentieth century have little to say about Native Americans or else reveal a superficial or mistaken understanding of indigenous culture, *River of Grass* offers a thorough and informed account of the various native peoples who have inhabited the Glades. Furthermore, Douglas's brand of historical anthropology anticipates the work of current environmental historians in its recognition that stories about landscapes are ultimately inextricable from stories about human dwellers in the land—the exigencies of place mold and shape patterns of human culture, and the ambitions and technologies of human communities cause them to alter the landscapes they occupy.

In the second chapter of *River of Grass,* "The People of the Glades," Douglas explores the mysterious world of South Florida's earliest human inhabitants, the Calusa, Tekesta, and Mayaimi peoples. Most engaging is her account of how a culture soundly adapted to the beneficence and limitations of nature and thereby flourished for thousands of years in a place the first Florida legislature would later flatly describe as "wholly valueless."[12] Calusa Indians crafted tools from shells, constructed palmetto-thatched platform dwellings in the shelter of breezy, inland bays, and, over hundreds of years, created hurricane-proof high ground by piling into immense middens the shells of the oysters that formed a staple of their diet. The Mayaimis, who lived near the giant lake later named Okeechobee, fired rope-clay pottery, constructed elaborate fish weirs, and repelled insects by applying fish oil to their skin or by burning smudge fires. The Tekestas harvested cocoplumbs and sea grapes and cabbage palm hearts, made bone awls and sharkskin rasps, and cut inland canoe paths through the estuarine forests of red mangrove. Although little is known of the early Glades Indians' religious beliefs and ceremonies, *River of Grass* demonstrates that the relation of culture to nature in South Florida during the earliest periods of human habitation was distinguished by a sustainable use of natural resources and by a deep respect for the nurturing power of the wetlands environment.

In chapters covering more-recent history, Douglas tells the story of Indian enslavement, war, and removal in South Florida—a compelling story of the resilience and courage of native peoples and the sheltering power of the nearly impenetrable depths of the Everglades. From the time of Ponce de León forward, early European explorers and fortune seekers decimated the native Indian populations of La Florida. "[N]ever before in history was a whole people destroyed, and by so few," writes Douglas. "A year, two years, three, after Columbus had opened up a New World, island after island of Indian people, by slavery, torture, hard work, homesickness and disease, thousands after thousands of men, women, chiefs, children, priests, fishermen, warriors, had been blotted out."[13] In the wake of several centuries of such genocidal destruction, the Muskogee-speaking Seminoles and Hitchiti-speaking Mikasukis of the inner Glades became unusually vigilant, thus allowing them to elude slavers by seeking refuge deep in the watery heart of the swamp. To them, the expansive tangle of the Glades wilderness represented freedom from war, relocation, disease, enslavement, and possible extinction.

The Glades also meant freedom to the many black slaves who escaped from bondage in the American South and fled into Spanish Florida in the decades preceding the acquisition of that territory by the United States in 1818. Over the course of generations, escaped slaves assimilated into Seminole, Mikasuki, and Creek culture so thoroughly that most blacks in the Everglades were, in fact, Indians—people who had been born and raised in tribal communities and who spoke no English. Once in American hands, however, Florida was opened to opportunistic slavers who often earned handsome rewards for capturing blacks and selling them "back" into bondage. Of course, most of these blacks were not escaped property but free Indians whose tribal sovereignty and individual rights had never been recognized. Douglas argues persuasively that the ferocity with which various Glades tribes—particularly those allied under the leadership of the famous Lower Creek leader, Osceola—resisted removal to the reservation lands of Oklahoma was inspired by their awareness that the blacks among them would be enslaved rather than allowed to emigrate west with their people. *River of Grass* not only tells this untold story of the persecuted Glades Indians, it does so in a way that celebrates their courage, their defiance, and the righteousness of their legal and moral claims to their land and way of life. Furthermore, Douglas's refusal to elide or soften the history of the Indian wars and their devastating effects upon tribal communities also forces readers to consider the ways in which destruction of the Everglades environment and disappearance of its native peoples have historically gone hand in hand. Anticipating later American women nature writers whose work would be informed by the insights of ecofeminism and environmental justice, Douglas was keenly aware that the subjugation of people and the subjugation of land are often cognate endeavors, and that the sustainability of local communities and local ecosystems need likewise be viewed as complementary ambitions.[14]

Of course, *The Everglades: River of Grass* is most widely recognized as a contribution to environmental literature, for it is the first book to offer detailed discussion of the ecological and aesthetic value of the Glades and to argue forcefully for its preservation. Douglas's popularization of the concept of the Everglades as a river helped reverse the long-held and ecologically uninformed view of wetlands as little more than miasmal swamps in need of redemption by human reclamation and use. Douglas was also prescient in her grasp of the systemic nature of the Glades, "one vast unified harmonious whole" to which she accurately referred as "the Kissimmee—Lake Okeechobee—Everglades watershed."[15] Douglas understood the vitality and complexity of hydrological cycles in the region, and she was among the first to recognize that the floods and droughts Floridians so fear are often caused by human interference with the natural system. Finally, *River of Grass* is remarkable for its heterodox assertion of an ecocentric environmental ethic. Published two years before Aldo Leopold's widely influential "The Land Ethic," which appeared in *A Sand County Almanac* (1949), Douglas's *River of Grass* argues for a valuation of the Glades made not in terms of human interests, but in terms of its importance as "the last refuge for the roseate spoonbill and the vast flocks of other Everglades birds, of the manatees, the crocodiles and the alligators, of the deer, the raccoon and the otter."[16] Thus, Douglas has in common with many other American women nature writers—one thinks, especially, of Susan Cooper and Mabel Osgood Wright before her, and of Terry Tempest Williams and Pattiann Rogers after her—a desire to help readers conceive of the "family" as an expansive ethical community that rightfully includes both human and nonhuman beings.

As a work of blistering environmental advocacy that prefigures Rachel Carson's *Silent Spring* and Sandra Steingraber's *Living Downstream* (1997), Douglas's *River of Grass* is particularly powerful as an exposé of the selfishness and ignorance that inspired various plans to drain the Everglades. Although minor opposition to wholesale drainage of the Glades dates from the early twentieth century, Douglas's absolute contempt for those who envisioned an agricultural "Empire of the Everglades" is unprecedented in its clarity and force:

[I]n all those years of talk and excitement about drainage, the only argument was a school-boy's logic. The drainage of the Everglades would be a Great Thing. Americans did Great Things. Therefore Americans would drain the Everglades. Beyond that—to the intricate and subtle relation of soil, of fresh water and evaporation, and of runoff and salt intrusion, and all the consequences of disturbing the fine balance nature had set up in the past four thousand years—no one knew enough to look. They saw the Everglades no longer as a vast expanse of saw grass and water, but as a dream, a mirage of riches that many men would follow to their ruin.[17]

Indeed, the tone here is particularly reminiscent of Carson's famous assertion that "[t]he 'control of nature' is a phrase conceived in arrogance, born of the Neanderthal age of biology and philosophy, when it was conceived that nature exists for the convenience of man."[18] To support her critique, Douglas patiently catalogs "the results of all these reckless ideas of drainage," explaining in detail how dug canals and ditches sucked nearly two million surface acres of life-giving water from the heart of the Glades.[19] In the wake of this improvident desiccation, populations of wading birds plummeted to ten percent of their historical numbers. The once peaty muck of the region dried up and blew away at the rate of a foot per year. The rains that once sustained the plants and animals of the region ceased to fall, and the consequent depletion of the aquifer resulted in salination of the wells settlers had begun to sink in new agricultural areas that had once been swampland.

Despite its subtext of tenacious optimism, *River of Grass* is a poignant literary reckoning of loss. It is an elegy delivered without sentimentalism, a dirge sung without self-indulgence. It is an unflinching look at the remarkable magnitude of human error and the exorbitant price exacted of every living thing when humans alter their environment without sufficient knowledge, compassion, and restraint. And Douglas's is a story not only of human folly and shortsightedness, but also of human avarice and cruelty. One thinks of the men who fire-hunted alligators for their skins, a single hunter sometimes taking ten thousand animals per month; or the plume hunters who devastated rookeries, driving the ibis and egret to near-extinction by scalping many thousands of birds in order to satisfy the demand of the fashion industry for aigrettes; or the poacher who murdered Audubon Society warden Guy Bradley in cold blood for his determination to protect endangered Everglades bird species from extermination.

As she counts such losses, Douglas also moves *River of Grass* dramatically toward an apocalyptic conclusion that is more reminiscent of John Milton than William Wordsworth. In her final chapter, aptly titled "The Eleventh Hour," she offers a terrifying vision of ecological holocaust in the Glades. Rather than offering a lyrical paean to South Florida's irreplaceable river of life, much of her final chapter describes an unnatural, desiccated landscape where fire sweeps uncontrollably and salt is drawn up through springs that once were fresh. "The whole Everglades were burning," she writes. "What had been a river of grass and sweet water that had given meaning and life and uniqueness to this whole enormous geography through centuries in which man had no place here was made, in one chaotic gesture of greed and ignorance and folly, a river of fire."[20] Like Rachel Carson, who asked readers to imagine a "silent spring" in which the devastating effects of pesticide use results in the extinction of songbirds, Marjory Stoneman Douglas compels us to imagine an "eleventh hour of fire and of salt" in which "[the only] Everglades in the world" vanishes forever in a glowing pillar of acrid smoke.

Although we are led to the brink of the abyss in this final chapter, Douglas concludes *River of Grass* just as she concluded her own long and rich life, with the abiding hope that the people of South Florida may at last "do something intelligent for themselves."[21] In the final sentence of the book she writes that "[p]erhaps even in this last hour, in a new relation of usefulness and beauty, the vast, magnificent, subtle and unique region of the Everglades may not be utterly lost."[22] In this spirit of demanding and qualified optimism, Douglas devoted the last half-century of her life to the preservation and restoration of the Everglades. From the publication of *River of Grass* until her death, she remained a tireless guardian of the Glades at every turn: helping to win the establishment of Everglades National Park during the 1940s, arguing against unchecked urban development during the 1950s, forming and leading the grassroots environmental group Friends of the Everglades in the 1960s, helping to successfully halt the construction of a jetport in the east Everglades in the 1970s, organizing citizens in favor of acquisition and reflooding of reclaimed areas of the historical Everglades during the 1980s, and endorsing restoration of the Kissimmee River and cleanup of agricultural pollution during the 1990s.[23] Douglas's leadership influenced the formation of the South Florida Ecosystem Restoration Task Force and the Governor's Commission for a Sustainable South Florida, and helped inspire the Everglades Forever Act and other important federal and state legislation designed to orchestrate and fund what is now widely recognized as "*the* precedent-setting [ecosystem] restoration project in the Western Hemisphere."[24]

Since its release in November 1947, *The Everglades: River of Grass* has continued to reach a wide audience. The book has been published by five presses, has appeared in six editions, has enjoyed well over twenty printings, and has often sold more than ten thousand copies annually. Douglas's various revised editions of the book have also helped *River of Grass* to grow organically, with added material that has provided useful updates on the current state of the Glades.[25] Indeed, Douglas's epic literary environmental history of the Glades continues to have an important impact upon policy makers at the highest levels. In awarding her the Medal of Freedom in 1993, President Bill Clinton called Douglas "a mentor for all who desire to preserve what we . . . affectionately call a sense of place" and during a bill signing ceremony at the White House several years later, Vice President Al Gore hoped that the legislation being enacted would "help us realize the dream of restoration of this 'River of Grass' so beautifully articulated by Marjory Stoneman Douglas nearly fifty years ago."[26]

The Everglades: River of Grass, now just over a half-century old, deserves a more prominent place in the canon of American literary natural history, and its author deserves wider recognition as an eloquent literary representative of—and tireless advocate for—our nation's only subtropical wilderness. Indeed, Douglas was among the most accomplished literary environmentalists of the mid-twentieth century, and she has earned a distinguished place among American women nature writers. In her best work, Douglas asked readers to imagine the deep history of Florida's natural environments and to acknowledge the ethical, aesthetic, and spiritual value of the nonhuman creation. She also wrote natural history with a keen appreciation for cultural history, and with a genuine sensitivity to the important role of native peoples in the story of the Everglades. Furthermore, Douglas was among the first American women nature writers to combine the early insights of scientific ecology and the rhetorical techniques of a place-based style of nature writing with the fiery conviction of an environmental advocate. This potent combination, which was to become so important to American women's nature writing of the late twentieth century, is the core accomplishment of *The Everglades: River of Grass.*

Marjory Stoneman Douglas lived well over a century and devoted much of her long, productive life to the literary celebration and environmental protection of one of the most lovely, biodiverse, and poorly understood landscapes in North America. In writing the swamp with such accuracy and power, Douglas clearly demonstrated that—like Henry Thoreau at Walden Pond or John Muir in the Sierra—she was indeed the right person in the right place.

Suzanne Ross

The Animal Anthropology of Sally Carrighar

Sally Carrighar was in her late thirties when she realized that she could "take back the stifled hope that [she] might devote [her] life to some kind of significant writing."[1] She was ill and despairing of the path her life had so far taken, which included work on the periphery of the Hollywood film industry and writing advertising copy in San Francisco. At this time, the late 1930s, while she was still living in California, a series of unusual and intense experiences with animals coalesced in an epiphany that her life work could be nature writing. "It would be a subject of inexhaustible interest," she later explained in her 1973 autobiography, *Home to the Wilderness,* "a supreme joy to be learning to tell it all straight and truthfully."[2]

In subsequent years, she wrote critically and popularly successful fictional natural histories, narratives chronicling the lives of individual animals within their natural communities in such books as *One Day on Beetle Rock* (1944), *One Day at Teton Marsh* (1947), *Icebound Summer* (1953), *Wild Voice of the North* (1959) and *The Twilight Seas* (1975).[3] In 1948, following the success of her first two books, Carrighar received the first of two Guggenheim fellowships to support her own study of arctic wildlife in Alaska. Through the 1940s, 1950s, and 1960s, she was a frequent contributor to periodicals such as *The Saturday Evening Post, Harper's* and *The Saturday Review.* Her books, which received strongly favorable reviews in publications such as *The New York Times,* are now unfortunately all out of print.

In writing her fictional natural histories, Carrighar set herself several tasks: to shift animals out of their object status in human-dominated stories and into the position of subjects of their own lives,[4] to give literary representation to a diversity of animal forms of consciousness,[5] to portray the interdependencies of individual animals in wilderness communities,[6] and, by these means, to suggest the extent of our common wild heritage.[7] In order to accomplish these goals, Carrighar resolved to ground her narratives in fact, undertaking at age thirty-nine an intense program of self-education in the young science of ethology, the study of the behavior and interactions of animals in their natural settings.[8] She required of herself that she make no "statements about what the animals feel beyond what can be safely assumed from the way that they act."[9] In doing so, she committed herself to a life of study and field observation.

In seeking to portray what the animals themselves perceive and find meaningful, Carrighar challenges herself and her readers to engage in a shift of perspectives. She takes as given that understanding does not stop at the species boundaries. Like an anthropologist engaged in field research as a participant observer, she grounds her animal narratives in the conviction that contact can be made. For this reason, Sally Carrighar's fictional natural histories remain relevant today as early texts in what we might term the fledgling discipline of animal anthropology.[10] Growing out of the interpersonal and holistic approach of participant observation, Carrighar's narratives offer powerful, ethologically sound portraits of animal selfhood that urge us to imagine more fully the implications of our common wild heritage.

Sally Carrighar's earliest publications appeared at a time when general interest in animal stories had dwindled. Extraordinary scientific strides in our efforts to understand and explain the world had fostered on the one hand an increasing sense of human distance from the natural world, and on the other, an impatience on the part of the reading public with the sentimentality that characterized much earlier work in this genre.[11] Along with the writing of Rachel Carson, however, Sally Carrighar's work heralds "the transformation and rebirth of the genre in the 1940s," according to Ralph Lutts.[12] In being animal-centered and empathetic to the struggles and joys of animal life, Carrighar's fictional narratives reflect what many critics see as a characteristically female ethic of care and responsibility. Her animal stories are early and influential efforts to move reports of wildlife study away from what Vera Norwood describes as the "dominant male narrative of contact with the wilderness . . . the search for and capture of the trophy animal."[13] Finally, Carrighar's fictional natural histories foreshadow by more than two decades approaches and concerns that have become the hallmarks of the work of female ethologists such as Dian Fossey, Jane Goodall, and Birute Galdikas in two significant ways: first, by representing distinctly individual animals as they lead rich and meaningful lives embedded within complex social and ecological communities; and second, by grounding those representations in scrupulous field observation and study (her own and that of others).

In the proposal and chapter outline she prepared for her first book, *One Day on Beetle Rock,* Sally Carrighar is precise about the stance she will take regarding her subjects and the beliefs that motivate her writing: "the animals should be presented realistically—yet with warmth. The scientists' efforts to avoid humanizing animals have succeeded so well that it has begun to seem as if people and animals have nothing in common."[14] Her own recent experiences with animals had taught her otherwise. And it was these experiences that had disclosed to her the possibility of a writing career.

In her autobiography, *Home to the Wilderness,* Carrighar recounts an experience in the mid-1930s with a flock of linnets (house finches) whom she had invited into her San Francisco apartment by spreading seed on a window sill and then across a table.[15] Soon, the birds were flying from room to room, perching on picture frames and the edges of vases; on one day she counted forty-three birds inside.[16] Recuperating from illness and depression at the time, Carrighar spent most of her day observing the birds from her bed. She soon noticed, however, that not only was the utter stillness that she adopted tiring for herself, it was also threatening to the birds. Predators are still and watchful, Carrighar reasoned. She realized that the birds would feel safer and go about their own business if she quietly went about her work, which included studying them.[17] She adds: "when there is no question of one hurting the other, as between the linnets and me, why shouldn't we be at ease together and even have some communal feeling in being here in the same environment?"[18] Carrighar reveals here, even at this point *prior* to the full inception of her plan to study and write about animals, the form her inquiry would take: She would approach her project in effect as an anthropologist engaged in participant observation.

Anthropology is of course premised on the belief that within the species *Homo sapiens* some level of commonality exists in spite of our diversity. This allows the anthropologist to establish meaningful contact with others across cultural borders even though these contacts may be partial and provisional. Our social and cultural discontinuities, though actual, are not total impediments to empathy and understanding.[19] The open question is whether such meaningful contact can cross species boundaries.[20]

In her autobiography, Sally Carrighar recounts a remarkable encounter with a lion on a Hollywood film set that might serve as a response to this question. She describes how one day on the set she felt a surge of sympathy for the lion whose "magnificent grace and strength" she felt was diminished "in this farcical setting."[21] She recalls thinking how fate had entered his life as unpredictably as it does the lives of human beings, pulling him out of his own world and trapping him in this impossibly trivial one. She remembers holding an image in her mind, drawn from a photograph perhaps, of the lion's "own high plateau, where the savannah was wide beyond the scattered trees." Immediately, the lion was off the set and moving toward her; he then sat down, gazing up at her. It happened again and again over the several days of shooting. Carrighar was told to stay away from work until the lion's part in the film was complete. She quit sooner than that and was even offered a job by the Selig Zoo as a lion tamer! She didn't accept it, though she felt disloyal to the lion. She also felt grateful. "What I had thought of as human truth," she notes, "had always made it disturbing to live in the false atmosphere of Hollywood, but the glimpses of nature's much deeper truth [offered by the lion] had made it impossible to stay."[22] Something had passed between them. Contact had been made.

In the early stages of articulating her project, Carrighar realized that what she proposed to do had not so far been done—"describ[e] minutely the activities, hour by hour, day and night, . . . [of] a particular limited group of animals."[23] She prepared for two years, making contacts in the late 1930s with established scientists in the San Francisco area, gaining access through them to museum collections, university libraries, and scientific societies in California, immersing herself in the work of naturalists and ethologists but without taking an advanced professional degree.[24] Finally, she was ready to begin her own fieldwork at Beetle Rock, a two-acre granite outcropping in Sequoia National Park in northern California. She stayed two months the first year, visiting again for extended periods over the following three years, and staying "from the snows of spring to the snows of the following winter" in one of those years.[25]

It was during that long stay, when her intention was to both watch and write, that she was able to assemble a group of birds and animals, the same species who frequented Beetle Rock, at her nearby cabin. She did this by providing a salt lick, sinking a dishpan for water in the ground, and offering them their natural foods, but in small enough quantities not to discourage their own independent foraging.[26] She then learned how to encourage the animals to stay all day: She adopted reassuring postures, low to the ground and relaxed; she crocheted, believing that the birds and animals enjoy "any continuous small motion that [isn't] threatening," and otherwise moved in smooth, nonaggressive ways; she talked "sense" to the animals, "in a normal voice not the high-pitched baby talk that is one's impulse."[27] In this way, Carrighar acknowledged that she was in fact the visitor seeking admittance into the animals' community.[28] She describes her own reasoning: "in making friends with the wild ones, I found, it is necessary to be absolutely sincere. I had to act in a friendly way for no other reason but that I cared about these companions; to seem like a friend because I wanted them to come near so I could watch them and write a book about them would never have been enough. It seemed that they had to feel a true sense of warmth, not sentimentality but concern. On the days when my thoughts were absorbed in some writing problem they went away."[29]

Carrighar reveals here that she believed her project's success crucially depended upon its unfolding in respectful terms with the voluntary cooperation of these other beings. She reveals as well the holism of her vision. For, not only did she presuppose the self-awareness of the animals about whom she hoped to write, she also recognized the contributions offered by her own interested engagement to the goals of her project. As she herself noted in another context, "a certain amount of subjective interest can call intuition into play, and intuition can furnish leads to more understanding."[30]

Fundamental to all of Sally Carrighar's writing is her thorough grounding in the new (at the time) biological discipline of ethology. Recall her two years of study and preparation before going to Beetle Rock and then her months in the field over four years there. Consider as well her ongoing contacts with contemporary naturalists and ethologists; her publication in 1965 of *Wild Heritage,* a synthesis of recent animal behavior studies; her own lemming study conducted while she was living in Nome, Alaska, in the 1950s.[31] It is primarily this ethological grounding, I believe, that supports her effort to accurately represent the "self-world" of each animal, that is, the surrounding world as constructed by the animal him/herself based upon information from the animal's own sense organs.[32] It is Carrighar's presumption of animal subjecthood, however, that encourages her commitment to represent each individual animal's experience of "inwardness" and to ground that inner self-awareness and integrity in the self-world constructed of sensory experience.[33]

In each of her books, Carrighar provides a vivid introduction to the physical setting and ecological character of the natural communities to which the animals she portrays belong. In *One Day on Beetle Rock,* we follow nine individuals—among them a weasel, a lizard, a coyote, a Sierra grouse, and a mule deer buck—on a June day. Each is the subject of a chapter, and in each chapter the same day unfolds in the same place from the unique perspective of a distinctly individual animal. *One Day at Teton Marsh* is similarly structured. Here Carrighar tells of the lives of fourteen animals on a single September day in and around a marsh near Jackson Hole, Wyoming. *Icebound Summer* is concerned with the animals, migratory and resident, whose paths cross in the brief Alaskan summer in and around Norton Sound. In *The Twilight Seas,* Carrighar tells the story of the life of a single blue whale from birth to death. In *Wild Voice of the North,* she tells the story of the husky dog, Bobo, and his life with her in Nome, Alaska. Always, her strategy is to focus on each animal's life as it is embedded inextricably in a network of relationships and interdependencies, its own individuality, however, never erased.

Carrighar gives her readers no alternative but to take these animals seriously as subjects of their own lives. Except in *Wild Voice of the North,* she offers no human narrator whose activities and experiences structure the narratives as plot. Nor does she provide a human mediating eye/I, creating an interpretive framework through which we view these animals as objects of study. It is their activities, experiences, responses, decisions, and occasional interactions with others as well as the natural events unfolding around them that create the narrative lines, the drama and tension running through each book and linking the parts of the books together. Carrighar's central concern in her books is to enter so faithfully into each animal's own world that, as she states it, her work would be "truthful as nothing I ever had written had been. At the start I made this test for myself: Would what I am saying about this animal seem true to 'him'?"[34]

In *One Day on Beetle Rock,* a thunderstorm blows up and weaves its way through each chapter, engaging the attention of each of the animals and prompting a range of sensory responses and awarenesses. For the Lizard, it is a lightening of his scales in excited reaction to the lightning strike outside his rock crevice.[35] For the Coyote, it is a mental response, a "link[ing] up [of] memories of other storms, images of the rain's end, of animals coming out of their holes at the return of the sun."[36] For the Mule Deer Buck, "[t]he storm was stimulating." Watching from beneath a cedar as it blew up and then passed, he "felt ready again for any encounter. The storm had renewed his energy, as rest had not."[37]

Some animals are prey or potential prey; others are predators. Others still are witnesses. In representing their awareness of one another and their encounters, their species-specific strategies for hunting, avoidance, or defense as well as their individual deployments of those strategies, Carrighar provides insight into the depth and complexity of their distinctive styles of inwardness. In *Teton Marsh,* the Cutthroat Trout attends to the tactile sensation of the unfamiliar strokes of an otter entering his pond, a rhythm unlike those he can identify as harmless or threatening. As Carrighar describes it, "it had more pulse than the Beaver's paddling or the striding of a moose. It was rougher than the swimming of a fish and heavier than the muskrat's sculling." For the Trout, "the lightest sounds were wave-beats in the pond."[38] Aware of the Osprey's three-times daily plunge into the pond for fish, "the Trout's good time-sense held him under cover when its strike was due."[39] As Carrighar portrays him, his perceptual capacities and instincts enable him to respond in creative and life-sustaining ways.

In *Beetle Rock,* Carrighar represents the more complex inwardness of the Mule Deer Buck, the now aging leader of his herd. She asks us to consider his perceptual capacities. On a night of unusual cloud-reflected light, he moves "the soft end of his nose up and down to sharpen the wind-borne scents." His flexible ears, capable of turning independently, respond "to each minute sound, quivering to enfold it more completely. They responded to sounds that human ears could not catch, perhaps to a cave-in behind an insect's burrowing, to a mouse's panting, or milk in the throats of little flying squirrels."[40]

But Carrighar does not stop here with perception alone. She invites her readers to enter into another animal's acts of reflection, contemplation, and decision. For instance, she describes how the mule deer herd, under the leadership of the Buck, places itself on this unusual night in the middle of a meadow. The Buck is uneasy: "There was no deer tradition to cover a night like this. The moon had not yet risen, but vapory small clouds had caught the approaching gleam and reflected it down on the grass. The leader could clearly see the herd's black tail-tips tight against the white rumps. If he could see them, so could a cougar. Should the deer go back under the trees?"[41]

The Buck continues to observe, to reflect upon his observations, and finally decides to lead the herd to an open grove where they are accustomed to spend moonlit nights.[42] In this passage and elsewhere, Carrighar represents other animals as creatures of intention and volition. She moves her readers toward a realization that animal knowledge may well be a complex accretion of instinct, observation, experience, and accumulated shared traditions different certainly from our own, but perhaps not completely inaccessible to the human who is willing to engage in imaginative inquiry across species boundaries.

The communities Carrighar describes are not in fact devoid of human presence, though they are communities she wishes to present as not defined by humans or their interests and concerns. Beetle Rock, for instance, is a place that humans visit; cabins stand nearby and campers and hikers pass through the animals'domain. But they seem odd somehow, not so much because they are out of place as because they are curiously distanced and out of scale. Readers experience them as if from within the"self-worlds," the perceptual spheres of the animals themselves; in other words, readers view other humans as if from over the animals' shoulders. The Weasel with her kits, coming across two sleeping campers—"gray mounds, strong with human scent"—seeks to evade a potential danger.[43] The Lizard, encountering the now-waking campers, "was so delighted to have something new occurring that he felt sharp and quick all over." Curiosity seems to prompt him to approach them, to allow a closer proximity. "But they did not even know of the exciting chase," Carrighar continues. "If they saw the Lizard, they gave no sign."[44] For the Deer Mouse in a cabin, the humans' voices are "tremendous," loud but not frightening, the higher-pitched woman's voice more audible than the man's.[45] The Bear with her two cubs, meeting two men and a boy on a trail, recognizes that the encounter could turn deadly—for herself and her cubs as well as for the humans. Carrighar describes the sow's response to the reaction of the man: "The man's face paled. The sharpening of his scent proved his fear. It was exciting to her! A sense of her great strength swept her, the knowledge that she could knock him from the trail with one stroke of her paw . . . The Bear raised her paw to attack. Only the presence of the second man restrained her. Might he reach the cubs? Even the Bear's pride meant less to her than their safety."[46] We experience humans not only as if we share the animal's scale and perceptual framework but also as if we share the animal's concerns and interpretive framework as well.

In *One Day on Beetle Rock,* because humans are minimally present and consistently perceived from an unfamiliar, even disorienting perspective, the notion of understanding and empathy crossing species boundaries is a possibility the reader contemplates in the abstract. While humans are only slightly more in evidence in *Icebound Summer* and *The Twilight Seas,* Carrighar is far more pointed in her presentation of them, their behaviors and motivations.[47] In these books, just as in the earlier ones, she moves her readers imaginatively into the "self-worlds," the perceptual spheres of other animals in order to consider the nature of animal self-awareness from a nonhuman vantagepoint. We look out upon and experience our *own* species from within these animal perceptual and interpretive frameworks. In the later books, however, the experience can be deeply distressing. For in these books, Carrighar compels her readers to contemplate in concrete terms our human capacity to establish or reject empathetic contact with other animals. In both *Icebound Summer* and *The Twilight Seas,* Carrighar presents the actual effects of the whaling industry on individual animal lives.

In *Icebound Summer,* she tells the story of an aging rogue walrus. At the age of six months, he lost his mother. She was killed by white men, whalers, interested in the ivory of her tusks. First hauled into the whalers' ship for amusement, the bewildered young walrus is at last thrown overboard. He finds his father, but is unable to feed alongside him in the clam beds on the ocean floor, as his tusks have not yet grown in. Facing starvation, the young walrus begins to kill and eat seals. Now feared by the others in his herd, the scent of a flesh eater about him, he becomes an outcast, and so is deprived of the companionship of his own kind for the whole of his life.[48] Carrighar describes his understanding of the experience and his lifelong response to humans: "If the event of his infancy did not remain in his memory as a clear picture, yet the sailors' treatment of him and his mother, their association with a dreadful and shocking experience, had touched off a hatred of men that later incidents had confirmed."[49]

In *The Twilight Seas,* the mother of the soon-to-be-born blue whale who is the subject of the book becomes aware that she has not heard her mate for some time. Carrighar enables her readers to extend themselves imaginatively into the whale's mind as she considers the sound she has heard: "had it really been a sound? Or was it only the memory that haunted any whale who had been at the feeding grounds? It was the most sickening sound on earth perhaps—the sound of exploding flesh. For four months she had heard it all day, every day . . . Suddenly in a panic she began swimming back and forth, constantly calling, calling. There was never an answer . . . Made bold by anxiety she sped directly toward the sound, surfaced—and saw her mate being towed away, belly uppermost. One of his severed flukes floated near."[50]

In these two instances, Carrighar is unsparing in her portrayal of human-initiated contact between ourselves and other animals. Whereas "natural" occurrences such as the lightning storm of *Beetle Rock* or the break-up of the beaver dam in *Teton Marsh* account for the drama and change in those animals' lives, here the forces of drama and change are purely destructive and solely human in origin.

Nonetheless, in these later books Carrighar also provides positive representations of human/animal contact. In *Icebound Summer,* she portrays a female loon who lingers in late winter on a city pond in Victoria because she has come to trust and take pleasure in the attentions of a human who comes each day to toss her shrimp. The loon is wary of humans, but feels confident in this one, "a slight, aging figure," "a human being as trim as a bird."[51]

Carrighar continues, "It was the courteous way that he walked . . . that made the loon feel at ease with him. For there was no aggressiveness in his gait; no secret wish to strike betrayed itself in his motions."[52] But one day her friend does not appear. We infer that he realizes that the loon's migration must begin; already she is late.[53] When she does at last arrive on her far northern river, she is captured by an Eskimo boy who makes a jess for her leg and attaches her to a stake on the beach. Panic-stricken, she has become his possession.[54] One person, however, a young Eskimo girl, is able to respond to the loon's predicament with understanding and empathy. Hearing the plaintive calls of other loons at night, in effect, hearing their stories, the girl imagines that they speak both for the captive loon and for herself. She imagines that the captive's life, her needs and desires are, if not the same as her own, then at least potentially congruent with them. Understanding the stories she hears, she releases the bird.[55]

Near the end of *The Twilight Seas,* Carrighar describes how the young blue whale and his female companion venture into waters unfamiliar to them. In a whale refuge established by the Argentine government, they encounter for the first time southern right whales, moving about with confidence, unafraid, in the company of human beings. Slowly approaching the blue whales, the humans "tactfully . . . touched various parts of the Whale's skin and he was startled but not frightened." As one of the men "drifted up close to an eye of the Whale and looked into it, . . . the Whale had an intimation that they were meeting in some mysterious way."[56] But "the Whale's past experience with killer men" thwarts the efforts of these humans to persuade the whales to submit to further encounters.[57] Carrighar emphasizes here how the long and violent history of whaling burdens human efforts at contact.

In her representations of animal self-awareness in these later books, Carrighar offers her readers opportunities to consider the real costs in the lives of animals of our insistence upon their objectification. She provides opportunities as well to consider what will be required if humans are to establish genuinely empathetic and understanding relationships with other animals. In *Wild Voice of the North,* where she recounts the story of the husky Bobo with whom she herself lived in Nome, Alaska, she explains what we will gain: "Any wild voice can stir something mysterious and exciting in us, and silent messages from a wild mind are even more moving. The human mind that receives them knows directly then what wildness is. The human being knows it because he himself is partaking a little of wildness. I realize that ideas like these, as well as examples of Bobo's insight, will seem unconvincing to many. To them, their doubts; to me, who had the experience, my knowledge of what can happen."[58]

Sally Carrighar's long-term project is a recuperative act: to return animals to the center of their own life stories in human representations of the natural world. Grounded in the ethological research on observable animal behavior in their natural settings gaining prominence at the time she was writing, her narratives take as a point of departure the "self-world" each animal constructs based upon his or her own sensory perceptions. In each of her books, the animal him/herself constitutes and experiences the world.

As an "animal anthropologist," Carrighar approached the animal communities about which she hoped to write from the perspective of the participant observer. She was thus able to put into practice her prior intuition that contact between humans and other animals is, however partial and provisional, fundamentally interpersonal in nature. In her early books, *One Day on Beetle Rock* and *One Day at Teton Marsh,* Carrighar concentrates on the inherent integrity and wholeness of animal lives, representing diverse forms of self-generated animal consciousness independent of human objectification. In her later books, *Icebound Summer* and *The Twilight Seas,* Carrighar goes further by considering as well the various kinds of contact that humans and other animals can establish. She shows on the one hand the world-shattering effects entailed by our human denial of animal self-awareness. On the other, she suggests the potential for genuinely interpersonal contact if humans acknowledge and respect animal forms of consciousness. Her story of her own relationship with the husky dog Bobo in *Wild Voice of the North* is a case in point.

Finally, how are we to judge the value and veracity of Carrighar's portraits of animal subjectivity in her fictional natural histories? One method of judging is suggested by critic Patrick Murphy. He argues that the authenticity of human literary representations of the world that animals themselves construct cannot be assessed in an abstract, context-free manner. As he puts it, "the test of whether such depictions seem accurate as renderings of non-human speaking subjects should be the actions that they call on humans to perform in the world. The voicing is directed at us as agents in the world."[59]

In effect, we must ask whether Carrighar's representations of animal subjectivity can contribute usefully to the real-world, real-life effort to establish understanding, respectful, and empathetic contact across species boundaries with the other animals with whom we share the world. I believe that they can.

In a related way, we can judge the authenticity of Carrighar's representations in a manner suggested by philosopher Mary Midgley: We can use our demonstrated practical ability to know what other human beings are feeling, extend those abilities to the animals we encounter, "and see where they take us."[60] In other words, we can, as Carrighar did, engage in animal anthropology. We can try to find out through our own participant observation if what Carrighar says about the Steller's Jay or the Cutthroat Trout or the Lemming or the husky dog bears out in our own experience. We can make contact with our own wild heritage.

Tamara Fritze

A View of Her Own

The Garden as Text

All landscapes tell us about the people who shape them and who are shaped by them. Every change that is made—or not made—is a result of a culture's values and way of living or an individual's attempt to reach her ideal habitat.[1] In the United States, the freeways we build are indicative of our concern with time and distance. The large number of single-family dwellings demonstrates our desire for privacy and our focus on the nuclear family. The concern with zoning within a town's city limits reflects the attempt by those in power to maintain their ideal habitat, especially within their own neighborhoods. By studying the changes societies and individuals have made to the land, scholars from a variety of fields are discovering that the landscape around us is a text that can be read in much the same way literature may be read. Reading landscapes provides us with primary evidence that can deepen our understanding of others and ourselves.[2]

However, up to this point, most scholars have examined only the changes men have made to the land.[3] They discuss the destruction of the environment through plowing, logging, or industrialization. Despite this focus, women also make changes to the land, and these changes tell us just as much about them and the society in which they live as the changes that men make. In rural areas of the West, women's transformations generally take place within the dooryard or barnyard, and their dooryard gardens are usually the most conscious attempt to alter the land, to culturize the land around them. For this reason, a study of rural western women's dooryard gardens reveals individual dialectics with nature and with the dominant culture and may serve as a way to overcome the perceived nature-culture binary that many scholars are debating.[4] After interviewing twenty gardeners and reading extensively from western women's literature, I have concluded that a single taxonomy may be created for rural western women's dooryard gardens. Such a taxonomy includes five taxa: gardens of domesticity, gardens of dominion, gardens of subversion, gardens of abundance, and gardens of abandonment. The same plants may be used in each, and in some cases, the layout may be similar. How the gardener responds to and works with the nature of the place distinguishes one taxon from the next.

A dooryard garden includes all that surrounds the home. It may include a kitchen garden, several flower beds, shrubs, trees, and garden architecture that includes walkways, patios, walls, birdbaths, fountains, and buildings. Gardens have been defined numerous ways, and undoubtedly mean something different to each gardener. However, I consider a garden a middleground—a place where culture blends with nature, where the gardener often, but not always, attempts to establish some order in nature and make it more bountiful. In turn, a garden provides the gardener with a greater sense of her place in nature *and* in her culture. The garden and the gardener work together, each nurturing the other.

The great variety of gardens reflects the variety of ways of living and responding to one's environment, and all of them serve several important cultural functions. The gardens ensure survival by providing necessary vitamins and minerals that may not be easily obtained elsewhere. Sale of the produce improves economic conditions, and the dooryard gardens themselves improve aesthetic conditions and bring not only physical but psychological sustenance to the gardener and the rest of her family. Often the dooryards serve an important spiritual role for the gardener, and provide a reason and place for fellow gardeners to gather together. Most important, these gardens help women come to know and understand the landscape around them and their place within it. These dooryard gardens should help us better understand what the gardeners perceive as their place in the West's large and potentially overwhelming environment. Because gardens do such a variety of cultural work, they must be evaluated not just aesthetically but as a place in which the gardener lives and works, as geographer J. B. Jackson suggests. The total meaning and purpose of the place must be analyzed.[5]

Gardens have historically been associated with domesticity, but the first taxon, the garden of domesticity, involves more than turning the space around the home into the place of the home. Gardeners of this taxon purposely extend the home boundaries so that the distance between themselves and their gardens is lessened. They know intimately what their garden contains, touching, smelling, and tasting the individual plants within it. They hold their gardens in their entirety close to themselves, embracing them with affection and recognizing them as home.

Anna Holbrook's garden in *Yonnondio* by Tillie Olsen defines a garden of domesticity. Forced to leave her country home, Anna feels disconnected from the land she loves and lacks a sense of homeplace. Although she is ill and too weak to work, she manages to plant a garden. As she nurtures it, it reaches out to her, providing not only biological sustenance but spiritual and aesthetic sustenance as well. The garden helps develop a sense of homeplace within her, and in such a secure environment, Anna is once again able to bury her own roots in soil.[6]

Danielle Klaveano of Cashup, Washington, grows a beautiful garden of domesticity, shown in figure 1. She has created an "old-fashioned" garden of perennials planted in drifts that flow together. She spends hours each day working in her

Fig. 1. Photograph by the author

garden, and the weeds are minimal—although she says they are there. She has learned through the years that her garden is not ever going to appear perfect; it is a living entity that makes many decisions on its own, so she uses the same philosophy in her garden that she used to raise her boys: "a lot of love and a lot of freedom." If a plant blows into a bed, she accepts it, telling it, "If you want to be there, okay." She tries to remain flexible, respecting the plants' growth habits and permitting them to make some decisions regarding their own location.[7]

Because the garden is an extension of her home, family members' response to her garden is deeply important to Danielle. She wants her grown children to enjoy what she has created since they left home, and through cuttings, she is helping her daughters-in-law create their own gardens. Her sharing of this extension of her home is not limited to the present. It goes both backwards and forwards in time. The home she lives in formerly belonged to her in-laws, and though Danielle has transformed the garden, and her mother-in-law is no longer living, Danielle still feels she shares the place with her. Danielle is certain "if she knows what's going on in her garden . . . she's just loving it." Similarly, Danielle is anxiously awaiting grandchildren old enough to enjoy the garden, and she recognizes that one day her eldest son will live in the home. Under his ownership the garden will again transform.

Danielle's dooryard serves as a source of spirituality. She explains, "Mother Nature is who I believe in, whether that's God or what." The garden brings Danielle close to this nature, permits her an intimacy with a power that she deeply appreciates. Because of this, her garden is a very private place, an intimate part of her home. According to Danielle, planting puts down roots, creates a homeplace, a sense of belonging. To be able to claim a place as one's own, a gardener must plant and nurture her plants, growing a private, domestic place in the process. This does not mean nature does not enter here. Danielle describes arriving home to discover her cherry tree lost to a whirlwind. Through her tears, she exclaimed, "I want to move!" The tree was as precious as any family heirloom, and its demise transformed a home she loved. However, she came to accept the tree's loss and to recognize that she must accept what nature offers, and this offering becomes part of her homeplace. Nature exists here in Danielle's culturized places, and it must be appreciated and respected. She, however, cannot remove herself from the garden and see it from such a distance that analysis is possible. Danielle realizes with Yi-Fu Tuan that seeing and thinking deeply about the garden distances oneself from it. Analyzing her dooryard would remove the enchantment, destroy the ideal place.[8]

Some women gardeners, however, wish to hold dominion over the soil and have a place where they are in control. For these women, gardens of dominion satisfy their need for ownership and a tamed nature. They do not see themselves at war with nature, nor do they believe their assertion of power is a desire for conquest. Instead, they feel great affection for their place and use their aesthetic of "neatness" to increase the perceived beauty there. They recognize that a garden must be vigilantly maintained or nature will creep in and destroy it. Plant choice is based on what the gardener enjoys, not what the environment requires, and no concern with the destruction of what was in the place prior to their own garden is felt, for they perceive "nothing was there." Whatever control is necessary to achieve the desired result is exerted.

Alexandra Bergson in Willa Cather's *O Pioneers!* demonstrates what a garden of dominion can be. Her dooryard is perfectly organized, fruitful, and beautiful. The road is lined with hedges, confining the grain within the fields and signifying the order of the place to which they lead. The home is surrounded by Alexandra's flower garden, which manifests the sense of control Alexandra asserts over the land. Through Alexandra's garden of dominion, Cather suggests that working with the spirit or genius of the land to achieve order is the only way to survive in the harsh western landscape.[10]

Donna Donnelly lives in Columbia County, Washington, and grows a garden of dominion. Her dooryard is carefully delimited by a steep bank and fences. She works hard to achieve a look of neatness and order, emphasizing that she does not like her garden to look "messy."[10] When she first began gardening in this place, she planted some herbs, but soon discovered that "they just take over everything, all over the yard," so she removed them, and she no longer plants invasive perennials, admitting, "I do control [the garden]. Any flower can become a weed if you

Fig. 2. Photograph by the author

don't take care of it. It just takes over, and I don't like that; I like everything in its place, where I put it." To achieve this orderliness, Donna plants in rows. As shown in figure 2, even the low-growing ground covers on the banks, stonecrop *(Sedum album)* and snow-in-summer *(Cerastium tomentosum),* are planted in rows with dirt paths in between so they can be easily weeded. In one bed, shown in figure 3, Donna has attempted to plant in drifts, without much success. She has left a large border of dirt around each drift and has even put a short, straight row of both the ground cover white alyssum *(Lobularia maritima)* and marigold flowers *(Tagetes patula)* amongst her drifts. Her concern for messiness has prevented the full and billowy bed that drift planting generally creates, and resulted in a bed that appears only half full. However, it is "neat" and free from weeds, conditions essential to Donna's aesthetic.

When Donna and her husband built their home, they located it to provide the best possible view of the backyard. The fact that they also have a very nice view of the hills in back of the house is incidental. This area beyond her dooryard she calls "the jungle," and she seldom goes there. She explains that she has "never liked bare hills," and although these have considerable brush on them, they lack the color and order that she appreciates. Her perception is not of a hill covered with plant and animal life but rather a place that is dreary and lacks color and life. The hill is naked; to be appreciated it must be clothed by the culture of the garden. As

Fig. 3. Photograph by the author

she studies the view from her back deck, her eye stops at the fence between her dooryard and "the jungle." Beyond the fence is chaos and disorder with limited color; Donna has no control in this area and does not concern herself with it.

Despite Donna's concern for control and order, she does not feel she is struggling against nature or at war with it, as was typical of the Euro-American male settlers. She feels no sense that she is conquering a powerful force. Instead, she feels great affection for the garden, and it is a source of peace for her. Working within her controlled space is relaxing, and she feels a love for her plants, singing and talking to them. While nurturing her plants, she finds herself thinking of her students and singing hymns of praise. Her garden provides a source of spirituality, and working in this place fulfills her. She explains, "You need to feel like there's something out there that you have control over or part of." Her garden empowers her, giving her a greater sense of herself and her accomplishments. She takes pride in the beauty of it all, but it is much more than beauty alone. The garden provides her with a sense of control over her environment, and the more she works this land, the greater her feeling of ownership and dominion.

The third taxon, the garden of subversion, is not just a garden that uses indigenous plants but is planted by a gardener who purposely desires to subvert the culturally accepted dooryard. She is willing to struggle against the cultural hegemony in order to celebrate "wild" nature and the natural landscape. The gardener

Fig. 4. Photograph by the author

of subversion is nearly always in conflict with city or county ordinances that attempt to dictate the plant life of her dooryard and the shape of each plant found there. As she attempts to give nature its rein, her culture interferes and insists that she make corrections to better control it.

The tension between the cultural hegemony and the gardener of subversion is demonstrated in the character of Marie Shabata in *O Pioneers!* by Willa Cather. Marie does not collaborate with nature to create her garden. Instead, she simply accepts what nature has created for her garden. She treasures the wild flowers, grasses, and herbs that have volunteered in her dooryard, but she is the only one who feels comfortable in the unkempt atmosphere. Eventually, Marie's garden is the scene of her and her lover's deaths. Marie's attempts to accept wild nature can, in Cather's view and that of the cultural hegemony, only be perceived as ominous and destructive.[11]

Despite this view, many gardeners of subversion work to celebrate the essential spirit of the land. Sandy Simmons of Walla Walla, Washington, is using native plants or cultivars of native plants to create a hedge between herself and her neighbors to provide greater privacy and because she desires a sense of naturalness around her home. She has been quite successful in her attempt, as shown in figure 4.

Originally, Sandy did not mow this lawn; she had a brush pile under the cherry tree for bird habitat, hoping to attract the pheasant and quail in the area. She thoroughly enjoyed the wild atmosphere, but it could not last. Her struggle to permit nature to work in this place resulted in a struggle with her society. The city sent her notification that the grass must be mowed and the brush removed. They were a fire hazard and were perceived not to lend themselves to the city beautification project. So she now mows and waters this part of her dooryard, but the lawn does not have a clean, tidy appearance nor is the grass evenly green.

Sandy is always conscious of the fact that she is tapping into ground water to care for her garden, but does her best to conserve the water she uses. Water conservation is a major consideration for her and one of the primary reasons she has a conflict with what the city requires. She suffers from what Yi-Fu Tuan calls the "modern dis-ease": a perceived conflict between the "beautiful" and the "good." Pesticides and artificial fertilizers destroy the soil and irrigating is decreasing the water tables at a frightening rate. However, she also recognizes that the "good" dooryard she has created is not perceived by the city as "beautiful."[12]

When asked why she gardens, Sandy explains, "Growing your own food . . . means more than going . . . [to] an office and pushing paper around. It gets back to . . . what's really important in life, which is . . . food, shelter, water, all those basic things you need to survive." She enjoys "bringing in the harvest," and although she does not believe she is influenced by the fact that this is a traditional role for women, she compares it to men who were supposed to "go out there and do that hunter business, kill the fatted calf." Her role is to go out to the garden and "bring in food . . . the fatted zucchini."[13]

The fact that this work is essential for life is not the primary reason Sandy gardens, however. As a gardener of subversion, she has a political purpose as well. By growing her own food, Sandy subverts the "hurry up and buy it" attitude of the cultural hegemony. Her work slows down the pace of living, making her appreciate the essentials of life. At the same time, she contributes less to corporate agriculture, which her culture supports, but which she believes is destructive to the earth. Gardening gives her the opportunity to discover what is "good" and to celebrate the "beautiful" of the earth within her own dooryard.

The fourth taxon, the garden of abundance, is one that is grown to celebrate growth and life, the process of gardening. There is no single end product, for the process is never ending. Gardeners of abundance will spend hours each day in their gardens, discovering what art they can create as they work with the nature of their homeplace and the nature they have chosen to introduce to that place. These gardeners are artists, and their gardens are their art form. This art is invoked and performs an essential function in the life of the gardener, helping her to celebrate the process of living and providing guidance as she progresses through her life.

In western women's literature, a garden of abundance is found in Jeanne Wakutsuki Houston's "Rock Garden." At the beginning of the story, Reiko, the young protagonist, holds herself apart from others and nature. A Japanese-American in an internment camp during World War II, she imagines herself a queen on her throne, watching other internees stand in the morning latrine line. Reiko's throne is a tree stump, cut down to build the barracks in which she lives. It exists as a result of destruction and is symbolic of the destruction internment causes in the lives of the Japanese-Americans.[14]

One morning, Morita-san, an old man who lives across the way, invites her into his garden. Reiko immediately recognizes that the stone garden is a sanctuary within this place of imprisonment. Here she feels a connection to her family, her heritage, and the homeplace of her grandparents. She accepts Morita-san's offer to teach her to garden and to meditate, and before long, the lessons she learns affect her daily life. She discovers that there are many ways of perceiving: the moss-covered stone she waters is a turtle who one day may get up and walk away; the carefully raked white pebbles undulate, making her seasick as she stands on dry ground. The garden begins to illuminate for Reiko a different way of living.

Morita-san's lessons in meditation include not only Buddha but Native American spirits as well. He demonstrates for the girl that she can connect to this desert land and its history, even while imprisoned. She is part of nature, Morita-san says, no different from the rocks, the wood, or the water. Reiko is a long time in understanding this lesson, but when Morita-san disappears while on a fishing trip, she has a vision of him and Native American warriors rising into the sky. She takes comfort from this and the lessons Morita-san's garden of abundance have taught. The garden provides Reiko with a way of perceiving that can help weave together disparate pieces of daily life. Binaries merge: nature and culture, life and death, the sacred and the secular. She recognizes a new way of perceiving time and the world around her.

When Morita-san's bones are found a few years later, Reiko has accepted the loss, knowing her own bones will gradually become the mountain, which will eventually become the desert sand. Reiko can no longer hold herself separate from others or from nature as she did at the story's beginning. She is the rock, the plant, and the people around her. All is one; all are the same. The reader can easily imagine that when Reiko grows up she will discover some form of garden of abundance in her own dooryard. The manner of perceiving this art form has provided her is so much part of her daily life, the process of the garden will be present in her dooryard whether she consciously grows it or not.

Nez Perce gardener Mary Maier was also a gardener of abundance, although she is no longer able to participate in the process because of the pain of arthritis. However, she and her husband used to raise all their own vegetables. Theirs was a conservative process that they learned from their parents and grandparents, and the division of labor within the garden reflects this. Mary assisted with the weeding and the food preservation and preparation. Her husband was responsible for the tilling, planting, weeding, and watering. Although they usually had subscriptions to one or two gardening magazines, Mary admits, "We never really paid attention to them."[15] The traditional family process was their guide. Their gardening bound them to their history, the way their parents and grandparents had shown them to live. The garden's process is closely connected to daily life. It connects the gardeners with family and with their family's lifeway.

Mary also defines the garden as "spiritual, just as everything that grows is spiritual." She emphasizes that "spirituality is part of the essence of a garden." This spirituality is based on her recognition that by caring for the earth and showing a regard for it, the gardener can create new life, and from this life, human life is sustained. "Nature doesn't do it alone," according to Mary. She believes an important part of the gardening process is the nurturing the gardener provides the garden. A gardener must put herself into the garden, and the garden is an expression of the gardener. Mary acknowledges that God has created everything, including the garden, but she is just as certain that the gardener and nature together have created the garden. This may at first seem a contradiction, but for many Christian gardeners, these two seemingly distinct belief systems work together. God is present and contributing to the life of all that is found on earth. He is a power who receives ultimate credit. However, Mary as a gardener is empowered by Him to work with nature, nurturing the seeds and plants that are nourished by the soil and water of her landscape, and together the gardener and nature create a bountiful garden. Neither could complete the task alone; the garden is a middleground for both.

In 1990, for the first time, Mary's husband planted red and yellow tulips in a ring around the garden. Before the year was out, he died, and with his death, Mary had to give up gardening. She simply was not capable of all the physical labor. However, the garden persisted for several more years because of the tulip bulbs, which she perceived as "a memorial" to her husband, and in this way, the garden gained another definition and remained a significant part of her daily life for several years after she was no longer an active participant in the art.

Now that the bulbs are almost gone and just a few tulips come up in the spring, the gardening process is nearly at an end. Mary does not really miss the vegetables that they grew; she just enjoyed the process of "watching the garden grow." It was an art form of "magic," rooted in her belief system and her history, shaping her daily life. Through gardening, she worked beside nature, and hand-in-hand they created an experience of gardening, not just a vegetable garden.

Mary's garden has been abandoned and is presently part of the final taxon, the garden of abandonment. Such gardens may occur through old age and death or when a new resident makes the decision not to garden, not to make a connection with the land of this space. Gradually, nature invades in the form of weeds and wildflowers; disease, then death overtakes many of the cultivars. Death and wild nature present in this place do not necessarily indicate that the gardener no longer cares about the dooryard, that it no longer holds meaning for her, however. Often, quite the opposite is true; the abandoned garden can have a complex meaning for the women who enter it. It can help women find the strength within themselves to make decisions and carry on with their lives or temporarily transport them to a world beyond the reality in which they live. The abandoned dooryard is still a work of invoked art, although the cultural hegemony does not perceive it as such.

Gardens of abandonment are also a result of the transience of Westerners. The first gardener leaves the garden when her family moves on, and another woman may move onto the land and continue to nurture and modify it. New gardeners result in new plants that must be included, old ones that are taken out or moved to a new location. The purpose and meaning of the garden changes. What was a garden of dominion may become one of domesticity or abundance. Janie Tippett has experienced this type of garden of abandonment. For the first twenty years of her marriage, Janie's husband moved from place to place, and she reluctantly followed. As soon as she had a nice garden established, they would move. "I left a trail of gorgeous gardens behind me."[16] She does not know what those gardens came to mean to the women who followed her, but undoubtedly some of them, at least, appreciated her efforts. The clearing of ground and careful weeding that Janie did would have made their own vegetable gardens easier to establish. The flowers that Janie planted in the dooryard must have been a welcoming sight to many of the new gardeners, and more than a few of them must have added to the flowers growing there. With the change in gardeners, the garden had to change in meaning as well. What was Janie's garden of abundance could have developed into a garden of domesticity or dominion. More than likely, some gardens were abandoned altogether.

The garden of my childhood is now a garden of abandonment. In that place, I learned how to garden and how a family works together to supply nutrition for the rest of the year. My older siblings taught me the pleasure of pulling a kohlrabi, peeling it with my teeth, spitting the thick skin onto the ground, and biting into the tender white globe smeared with dirt from my hands. I believe I still remember how it felt to lay my tiny child's body in the cool, moist soil beneath the tomato plants with the pungent smell surrounding me, although this could be a memory created from my parents' stories of their frantic searches for me when I could barely walk, and their discovery of me among the ground cherries or cherry tomatoes, gorging myself. My siblings and I began to participate in the gardening process well before we started school. I remember my parents demonstrating for us how far apart to plant the seeds, how deep to cover them with the rich black earth. Dad taught me to clean the garden by pulling the weeds out by their roots, not just picking off their tops, and we were regularly assigned a row or two of vegetables to weed every summer day. With the garden located beside the barn, I would weed awhile, spend some time with my horses, then return to weeding. In my memory, these days were sunny, long, and slow.

By the time I was in high school, I chose to spend time weeding the garden. Mom always looked surprised when I came up the hill and told her the beans were weeded, even though the task had not been assigned. By that time, she was so tired of raising kids, keeping house, cooking dinner, and working full-time, the garden was just another chore for her, food that had to be harvested, canned, and frozen. She no longer appreciated the smell of the soil and the feel of it between her fingers. She did not have the extra minutes to experience the slowing pace of time within the garden. She was tired of the labor that could connect her to the land and create a sense of place. She and Dad were connected; they knew the place. They built their home and their family upon this land. The labor was just labor; the enchantment was gone.

However, as often the case, the same garden held a very different meaning for me. As a teenager, I was just becoming conscious of the enchantment the land brought to us, just learning to appreciate the privileged life we had running up and down the hill between the house and the garden and the barn. Playing and working on this land, I knew it intimately but did not appreciate its meaning nor understand the sense of belonging it held for me until I was reaching out beyond it to the world that teens explore. I could not perceive the garden as my Mom did: more chores to do. For me, it provided a place to express the love of the land that I was discovering, a place to relax and enjoy myself and nature.

The land on which the garden of my childhood grew now belongs to my older brother, and he and his wife do not have time for a garden. This past fall, I went to visit and wandered out to the old garden plot. Now covered with a variety of grasses, deer beds can be found all over it. Bear and deer have destroyed most of the trees in the tiny orchard. Only the golden delicious is still alive, and while I was there, a young buck stood on his back legs and boxed down its largest remaining branch to get at the apples. Despite the changes and the ache of memory they cause, the garden still feels familiar when I enter it. Here, I see the teenage girl I was and hear Chib nicker over the fence to me. This place I know. It was part of me, within me, and I left part of myself within it. That day, though, the abandoned garden it has become offered me no refuge. I could only see that nature has reclaimed this place. The skeletons of plum and apple trees reminded me of the home up the hill where strangers now live.

However, my sister-in-law enjoys what the garden has become. She and my brother built their home in Chib's pasture, and they watch deer, raccoon, and turkeys come and go within the old garden plot. They are thrilled by the occasional bear or cougar. For several summers, a doe has slept nightly with her yearly "litter" of three fawns in this place, and Davonna loves to watch "the Bambis" grow. She has a compost pile that she admits is really for the deer who sneak in after dark to devour the lettuce leaves or leftover corn. Although each morning all summer long, Davonna goes out to her dooryard and discovers buds that were about to bloom lost and replaced by only deer prints, she does not mind too much.[17] She celebrates the nature of her homeplace, and the garden of abandonment permits that nature to thrive.

Perhaps these transformations are more a matter of perception than reality. Plant life still thrives in this place, probably due to the manure of those chickens eaten long ago. If I were to look with eyes open a little wider, I might discover the red and green grasses are no less beautiful than the neat rows of berries and vegetables. Both nature and culture are responsible for what the garden was and for what it has become. Although some gardeners may perceive them at odds, nature and culture function together, creating gardens, then destroying and recreating them. The process is long term, and the gardener working for the moment often leaves an unintended legacy. For gardens do not just work in the present. They evolve, changing daily, seasonally, yearly. Every new generation on the land creates a different garden from the bones of the old, and the culture that is introduced marks the nature of the place, just as nature shapes the cultural artifact.

This process helps us recognize the nature-culture binary as a cultural construct. In a garden of abandonment, it becomes increasingly difficult to recognize where the culture ends and nature begins. In my childhood garden, if it were not for the dead and dying fruit trees and the remaining rhubarb plant, one would not know a garden once lived here. The falsehood of the binary is reinforced by the gardeners of abundance. These artists explain nature and culture "are one and the same."[18] The gardener is of nature and creates culture with nature, so the binary cannot exist.[19] Similarly, gardeners of domesticity do not recognize this binary. Their gardens are so much a part of themselves that they cannot analyze them in terms of this cultural construct. The enchantment of their dooryards is a result of what nature has offered, and they know this nature so intimately, they can only perceive it as home.

Because gardeners of subversion struggle against the cultural hegemony in their attempt to bring nature into their garden, they are aware of the nature-culture binary as the hegemony has constructed it. However, although they may be in the midst of a struggle based on this binary, they deny the existence of it within their own dooryards. They recognize the falsehood of the binary and desire to create a middleground where nature and culture exist together to celebrate the essential spirit of the land.

The gardener of dominion is the only gardener who perceives the nature-culture binary as reality and attempts to maintain it. Interestingly, these are also the only dooryards that suggest themselves to the viewer; they have lines that are straighter, rows within the beds, and shrubs planted over a weed-preventing mulch. In order for these gardeners to create a beautiful garden, they cannot leave nature to itself. It must have a guide who is clearly separated from nature. Donna Donnelly desires to "look at it, not sit in the middle of it," not be part of it. By standing back, then reaching into nature and creating order, a gardener can make a garden emerge from chaos. This could not happen if she became part of nature or recognized herself as collaborating with nature. She would not feel in control, and the garden would be too "messy." By separating herself from the nature in her garden, she can better perceive the beauty that she has created.

This beauty that the gardener creates is appreciated by all gardeners in all taxa. A dooryard reflects the gardener's own aesthetic. Because it is the personal expression of an ideal place, it may be achieved through the planting of trees, roses, or sagebrush. All gardeners believe this place must be beautiful, must bring pleasure to its visitor. The dooryard is a work of art, an attempt to celebrate the aesthetics of the rural woman who created it.

As with other folk art, though, dooryard gardens do not only serve an aesthetic purpose. They also fill biological, economic, social, and spiritual functions. They can overcome differences created by class or ethnicity and create close friendships that last a lifetime. The garden nurtures, as well as needs nurturing, and for this reason, work within it is undemanding and heals rather than exhausts the gardener. Many of us turn to this art form because it represents what Pat Mora defines as "a quieter life." It gives us a means for "combining utility and beauty" and brings us delight when "we see pansies in hard, cold ground, the blooms delicate in the snow."[20] For this beauty alone it is worth growing a dooryard garden, but the garden does something much more as well. It helps to develop a sense of place, connecting us to the land upon which we live. As we work the land, we come to know it intimately, it becomes part of us, expanding within us. Our sense of self is clarified through the process of gardening. It provides us with a deeper understanding of ourselves and our role in both nature and culture.

Karen Cole

Tending the Southern Vernacular Garden
Elizabeth Lawrence and the Market Bulletin

Garden writer Elizabeth Lawrence's last and most passionate project was a history of southern market bulletins. First appearing around 1915 and usually subsidized by state agencies, these periodicals offered notices placed by farm men, who had land, livestock, produce, and equipment to sell; and by farm women, who had plants, seeds, and bulbs to offer. For the last decades of her life, Lawrence collected material and drafted sections for a book that she hoped would reflect, as she put it, "the rich sense of social history of the rural Deep South" and would "speak of times and customs that are long gone, or that are rapidly passing."[1] The bulletins, and the stories they told of farm women and backyard gardeners throughout the South, chronicle what we commonly think of as vernacular: Their gardens are local in design and indigenous in their uses of material. As Vera Norwood has pointed out in *Made from This Earth* (1993), these stories are important as documents of "the meaning of plants to rural American women and the power [these women] held in making the landscapes of their homes and of a region."[2] But Lawrence's study of these bulletins is valuable too for its insistence on a conversation about southern gardens and gardeners, a conversation in which the academically trained, the long experienced, and the amateur have equal say. The inclusiveness of her method ensured that the writing of these women would be saved, that their voices, their language would be documented along with their experience, wisdom, and love of the natural world.

As both collector and chronicler, Lawrence recognized the social importance of preservation. While the act of collecting itself (with bulletins from eight states and letters—and plants—from over one hundred gardening correspondents) was an important first step, her preservation of the talk about plants, about gardening and gardening ways, distinguishes her study. Her goal was not merely to preserve a female tradition but to understand the way that tradition was being preserved. At a moment when she perceived that the ways were being lost, she stepped forward to protect southern gardening by recording the voices of those who knew its ways best. Speaking from within this community and about it, she worked to ensure that its dynamic work would continue, and at the same time she sought to understand how the dynamics of a tradition maintain it. Lawrence's study of the market bulletins and the gardening women they served can help broaden our current understanding of the term "vernacular" as it is used to describe the garden, particularly the southern ornamental garden.

The outline of Elizabeth Lawrence's life and career suggests something about the place of the southern garden in the twentieth century. Born in 1904, the first daughter of a successful engineer and an equally successful gardener, she and her sister Ann spent their childhood in Georgia and North Carolina. In her only published autobiography, a brief four paragraphs published in *Herbertia* in 1944 (an issue dedicated to Lawrence by the American Amaryllis Society as its Herbert Medalist), she reveals how her love of gardening and her sense of place were passed down: "When I was a little girl my mother took great pains to interest me in learning to know the birds and wild flowers and in planting a garden . . . I remember the first time I planted seeds. My mother asked me if I knew the Parable of the Sower. I said I did not, and she took me into the house and read it to me. Once the relation between poetry and the soil is established in the mind, all growing things are endowed with more than material beauty."[3]

Her formal education—at a preparatory school, St. Mary's in Raleigh, and then as a student of literature and the classics at Barnard College—helped with the poetry, but it was her training in her mother's garden that made her an astute observer of growing things and what is behind them:

When I was twelve we came to live in Raleigh, in a house with an already established garden. It was fall when we came, and there was not much in bloom—only some old fashioned roses and chrysanthemums that the frost had not caught. But the first spring was like living my favorite book, The Secret Garden. Every day the leaves and flower buds of some plant that we did not know was there, would break through the cold earth . . . No other spring has ever been so beautiful, except the spring of the year I came home from college. That first spring in the South after four years in New York led me to choose gardening as a profession.[4]

Soon after, she enrolled in a course in landscape architecture that had just begun at then–North Carolina State College in Raleigh. Lawrence points out, through typically modest indirection, that she was the only young woman in this course, the first of its kind in the South. Even this professional training led her back to the soil: "I soon learned, however, that a knowledge of plant material for the South could not be got in the library, most of the literature of horticulture being for a different climate, and that I would have to grow the plants in my garden, and learn about them for myself."[5] Learning and writing about garden culture became her life's work.

By the time Lawrence started her study of the market bulletins, she was probably the best-known garden writer of the South. Her first book, *A Southern Garden* (1942), was in wide circulation among enthusiastic, middle-class gardeners throughout the region (as it is still today), and had gone through three editions. Two other works, *The Little Bulbs* (1957) and *Gardens in Winter* (1961), had gained her a national readership. As Sunday garden columnist for the Charlotte *Observer,* Lawrence formed long-standing correspondences with backyard gardeners, well-published writers, and experts in horticulture and garden design. The index to her correspondence includes the names of over two hundred frequent letter writers, ranging from Inez Conger of Arcadia, Louisiana, to Katherine White of *The New Yorker,* to John K. Small. Fostering conversation, tending it in print, had become as important to this hands-on gardener as the garden itself. All who had information and experience to add, from backyard gardeners, to garden clubbers, to columnists and professional horticulturists, were welcomed and encouraged.

From the first, Lawrence identified herself with the region in which she gardened. Like many of her generation, that of the Southern Renaissance, she was touched by the spirit of preservation. In a 1945 booklet on "Gardens of the South," published for the University of North Carolina Library Extension Service, she wrote succinctly of her method: "One of the best ways to study and preserve native plants is to grow them in the garden. Whenever I see the mutilation of a hillside where I once found *hepatica* or *trillium,* it is a consolation to me to know that a patch of the *hepatica* or *trillium* is now safely established in my own garden." For Lawrence, the southern garden was especially ephemeral: "When we begin a study of the gardens of the South, we too often find that while the houses connected with them are still standing, little is left of the original garden, for gardens are more perishable than houses."[6]

Yet her notion of preservation was something apart from that of those southerners for whom the garden was a static symbol, a monument to old ways not forgotten. Lawrence, dynamic and egalitarian, rejected that dangerous nostalgia. That concept of the garden as site and as justification of the Old South can be traced back to the literature of the early nineteenth century. Many literary and cultural critics have argued that the region's most formative myth is Arcadian, a return or preservation of the old times and simpler ways, marked by a working respect for nature and a practice of neighborliness. At the center of this Arcadian ideal is, of course, the garden. In *The Dispossessed Garden: Pastoral and History in Southern Literature,* Lewis Simpson demonstrates how, in the early nineteenth century, the tasteful and orderly formal garden frequently came to represent the potential of Southern patriarchy to cultivate gallantry, heroism, and European values, particularly among the "humble slaves of the Old Dominion." Lucinda Hardwick MacKethan characterizes the Southern pastoral as a literary device through which writers can "expose or rebuke, escape or confront, the complexities of the actual time in which they have lived."[7] Timeless and therefore static, the garden of the southern literary imagination frequently represents a desire to preserve the region itself.

Elizabeth Lawrence's professional training in landscape gardening in the late 1920s would certainly have reinforced this view. The textbook she used at North Carolina State, the 1917 *Landscape Design,* was authored by two Harvard professors, Henry Hubbard and Theodora Kimball.[8] It was a Harvard-trained landscape designer, Joseph Pillsbury, new to the faculty at State, who was apparently instrumental in developing its new curriculum in landscape architecture.[9] Hubbard and Kimball's text reflects the tastes and design principles of the formal garden, the estate garden, which dominated the American imagination, not only the southern imagination, between the 1880s and the 1920s.

As landscape historian Suzanne Turner has recently pointed out in her study of *The Gardens of Louisiana,* the trend in both houses and gardens alike harked back to the classical and traditional during this period, sometimes called the Country Place Era or the Golden Age. In the landscape writing of and about this period, attention is usually focused on nationally known grand estates, including North Carolina's Biltmore and the District of Columbia's Dumbarton Oaks. Although the South did not produce the industrial magnates primarily responsible for this fascination with the formalities of the grand estate, the region's pride in its own golden age welcomed this design heritage. As Turner so aptly puts it, "architecture, garden design, furniture design and decorative arts were the uncontroversial, nonpolitical survivors of the antebellum period."[10]

A vestige of this design heritage appears in one of Lawrence's few pronouncements on garden design, the first chapter of "Gardens of the South." There she offers two "indispensable rules": "There must be a central axis, and the garden must be enclosed."[11] Her own garden in Charlotte follows these principles. Yet formal garden design seems never to have been her priority. In one self-description, she makes several important distinctions: "For the record: I design gardens but cannot bear to be called a Landscape Architect; lecture and write about gardening but cannot bear to be called an expert. Cannot bear to be called an amateur, but like to be taken seriously as a gardener and a writer . . . Cannot bear for people to say (as they often do) that I am better at plant material than design: I cannot help it if I have to use my own well-designed garden as a laboratory, thereby ruining it as a garden."[12]

Certainly, it was the history of ornamental garden plants that drew Lawrence to the market bulletins—a living part of the rural South's material culture. Plenty of horticultural information can be found in the market bulletins, recorded and then re-shaped by Lawrence's wonderful hand. This discussion of the geraniums offered through the market bulletin of Mississippi is typical:

"Spice-smelling geraniums" of all kinds are offered in the market bulletin, as is the strawberry-geranium (or begonia), which is not a geranium (nor a begonia). It is a *Saxifraga stolonifera,* mother-of-thousands or roving sailor, which is hardy with me and very useful in dark, neglected parts of the garden. Another begonia that is not a begonia is *Justicia carnea,* the plume plant, which Viva Mae Pipkins offers as the pine-burr begonia. "It blooms from the last of May through early winter," she says, "and has rose-colored flowers shaped like pine burrs."[13]

All of the Lawrence characteristics are here: the delight in plant names (both Latin and common) and their descriptions; the practical advice straightforwardly given; the modest claims of her own experience; and, above all, the reverence for a fellow gardener. Mrs. Pipkins' voice is as strong and clear in this passage as horticulturist B. Y. Morrison's a few pages earlier and garden writer Ernesta Drinker Ballard's on the following.

But it is the conversation about gardens, the language of the vernacular garden, the letters of the gardeners themselves that became her central interest. The particular contribution of the market bulletin study to our understanding of the vernacular garden is its tribute to these farm women and the network of correspondents she met through the bulletins. The participants are almost always mentioned by name and quoted generously. Their efforts are celebrated on each page—not merely their efforts as gardeners, but as writers as well: "The farm women are great letter writers, and usually answer (delightfully and often at length) if a stamped and addressed envelope is enclosed . . . Questions are answered freely, willingly and with love."[14] The care they took in handling and shipping is duly noted: Mrs. Ottice Breland sent explicit directions for planting Chinese pink morning glory seeds (which involve moistening a cloth covering for them every day until they come up). Miss Nancy Holder paid the extra postage to pack her Ladies eardrops in deep moss. Their work in the garden, after a day of work on the farm, is memorialized in Lawrence's study: "Most of all," wrote Lawrence, "I like to think about the hard-working farm women who are never too tired, when their farm work is done, to cultivate their flower gardens. They always find time to gather seeds, to dig and pack plants, and to send them off with friendly letters."[15]

Lawrence credited Eudora Welty with introducing her to the southern market bulletin, an exchange that seems to have taken place just after the 1942 publication of *A Southern Garden*. In the preface of her manuscript, she claims that "ever since the Second World War, I have been in correspondence with the country gardeners who advertise their flowers in the southern market bulletins, and who garden for love."[16] Then in 1944, as she explains in the preface, Caroline and Ruth Dormon of Louisiana told her of the Louisiana, Alabama, and Florida Bulletins— which put her on the trail of other publications and the recognition that these bulletins are "largely a southern institution." Because of the market bulletins, she remarks: "I have written many a letter, and gotten many a letter in return, from people like the Dormons, from professional botanists such as Dr. B. Y. Morrison . . . and Dr. Frederick Meyer . . . and from plant-lovers such as Kim Kimery and Weezie Smith and Ethel Harmon. I treasure these letters, have kept them, and have much drawn on them in this book, as is, I hope, obvious."[17]

In a passage on the blue wonder lily in her chapter on the Mississippi bulletin, many of these players—Welty, the Dormons, as well as others—come into play. For that reason, tracing its history in her unpublished correspondence as well as in her manuscript helps illustrate her method of working—of getting information, of networking gardeners, and of making them present in her text. The broad outline goes like this. In 1942, Ruth Dormon, the sister-in-law and business partner of Louisiana garden writer Caroline Dormon, wrote to Lawrence in praise of *A Southern Garden.* As the owner and nurserywoman of Felicity Wild Gardens, Ruth also offered additional information about the Louisiana iris to which Lawrence had devoted a section of her southern manual. The information in *A Southern Garden,* she wrote, had been "good and accurate" but "does not go far enough." Ruth Dormon and her sister had been collecting species from swamps and hybridizing for years.[18]

This letter initiated a correspondence that lasted, off and on, until Ruth Dormon's death. In it is recorded a typical relationship, one of hundreds Lawrence maintained. She sent silver bell bulbs to Ruth Dormon, who in turn sent a list of the native plants she offered through her nursery. The two of them exchanged hymenocallis, an exchange that was followed by correspondence on both sides in the next year when blooms appeared, filled with careful descriptions and measurements.[19] Then, in the fall of 1943, Lawrence wrote to ask for help: "Eudora Welty sent me a "Blue Wonder Lily" that she got from an add [*sic*] in the Miss. Farm Bulletin. Did you ever hear of it? It is *not* agapanthus. It has foliage like a crinum, but narrower, & shorter, & it lies flat on the ground Makes fall foliage that dies in summer—It hasn't flowered yet. Eudora's hasn't either and she says she is still wondering . . . It is all very confusing."[20]

Ruth Dormon responded quickly: "Back to your wonder lily. Perhaps it is camassia found in old gardens here called 'seven year hyacinth' because it is said to bloom every seven years. I have not heard of your lily called 'wonder.'"[21] A paraphrase of this statement appears in the manuscript, along with a description of a slide of a scrape of the *Scilla hyacinthoides* sent by "another friend" and a brief history of its cultivation, from the Mediterranean, to England, to the United States.[22]

Ruth Dormon's name appears four times in *Gardening for Love;* references to her help appear indirectly throughout. She emerges as one of the "characters" in Lawrence's garden history. To be sure, the bulletins, like Welty's fiction, offer a social history. But in capturing that social history, Lawrence also shaped it. Her gardeners are polled, networked, and acknowledged in her text. The question Eudora Welty poses to Elizabeth Lawrence becomes a link to Ruth Dormon, then to a network of Louisiana gardeners. Finally, the information is filtered back to Lawrence, who records at least a portion of the transaction in her manuscript. The complications of Lawrence's method, fostering a wide-ranging network of voices, more than aging or ill health, suggest to me why this labor of love was never, could never be completed.

The appeal of Lawrence's garden writing is in this voice, modest yet authoritative, personal yet inclusive. Emily Wilson astutely observes that it is in her language, rather than in her biographies, that one learns best of Elizabeth Lawrence. As Lacy points out in his introduction, she is the kind of garden writer more akin to the poet and novelist. Her own voice is clear and consistent; she has a strong sense of character and story. In the preface to *A Southern Garden,* she makes both her method and her rhetorical strategy clear: "Although I have gathered as much information as I can from all sources reliable and unreliable . . . I must necessarily depend upon my own records for what I write about plants. Any one person's experience can be taken only as an indication of what plants will do in a given locality." She offers an account of what has grown in her garden against the "recorded experience of many gardeners."[23] Throughout her writing career, from *A Southern Garden,* through her columns in the Charlotte *Observer,* to the meditative pages of *Gardens in Winter,* her awareness of gardening as a collective experience, as communal and cultural, is quite clear. But it is in the market bulletin project that she most completely devotes herself to the "recorded experience of many gardeners."

Behind every plant is a story of a gardener and her efforts. This social history is the larger goal of Lawrence's study. "Preservation of plant material," she claims early in her manuscript, "depends upon individual effort, and it is only in private gardens, in lonely farm yards, or around deserted houses that certain plants, long out of commerce, are still to be found."[24] It is the work of the gardener, the dynamic interaction with the green world, which Lawrence hopes to engage. Through all of her works, Lawrence contributed to this preservation effort; in fact, southern garden books of the last decade (Allen Lacy's *Home Ground,* Steve Bender and Felder Rushing's *Passalong Plants,* Scott Ogden's *Garden Bulbs for the South,* Welch and Grant's *The Southern Heirloom Garden*—to name a few) necessarily pay tribute to Elizabeth Lawrence.[25] Her work to preserve southern plant material has been fully recognized, and in that sense her contribution to southern vernacula is widely appreciated.

Recent studies of the vernacular American garden have brought keener awareness of the important role women in particular have played in our national garden history. Vera Norwood's *Made from This Earth* was an important first step in documenting the long history of American women and the natural world.[26] In *Grandmother's Garden,* May Brawley Hill documents what she calls an "indigenous national gardening tradition," despite some expert commentary that has claimed that there has been no such thing. An art historian, Hill finds her evidence of the images of the garden presented in the poetry, fiction, gardening books, photographs, and especially the painting of American women at the turn of the century.[27] In particular, her work points to southern women who helped preserve their region's gardens. As Hill argues, by the turn of the nineteenth century, women had turned the acceptable leisure pursuits of painting, photographing, writing, and gardening into successful careers. Watercolorist Alice Ravenel Huger

Smith produced two preservationist works of her native Charleston: *Twenty Drawings of the Pringle House on King Street* (1914) and *The Dwelling Houses of Charleston* (1917). Similarly, Blondelle Malone returned to Charleston after years of study abroad to paint the gardens of her family and neighbors. These visions of the garden as art helped preserve the region's designs just as the work of other women recorded its decline. The photographs of Frances Benjamin Johnson made picturesque the old and untended sites throughout the impoverished South of the early twentieth century, much as did Eudora Welty's photographs of the thirties and forties. In *A Woman Rice Planter* (1913), Elizabeth Pringle's account of her struggles to maintain her own plantation gave voice to the hardships faced by rural women of all classes. As Lawrence recognizes in *Gardening for Love,* Eudora Welty's fiction, too, had played a significant part in documenting women and their gardens.[28]

Hill's study of a particular vernacular style, what she calls "grandmother's garden," offers visual, and overwhelmingly persuasive evidence of the role that women played, not only as gardeners but as its chroniclers and designers as well. This design—a garden closely related to the house, usually enclosed by a fence or hedge, with flowers arranged in rectangular border beds—was passed on in a variety of imaginative ways. At the close of the twentieth century, as we come to terms with the water and other chemical necessities of a well-manicured lawn, the "new" cottage garden has re-emerged as a sustainable design.

Yet, our understanding of the term "vernacular" as it applies to the garden has changed in the last two decades. No longer does the label merely apply to regional styles, even one as wide-ranging as "grandmother's garden." Nor can it rest upon the use of indigenous plants; swapping of plants across regions and continents is as old as gardening itself. Most of the plants Lawrence identifies in her market bulletin study, traditionally "southern" as they seem, are non-natives. Lawrence makes this point subtly yet frequently throughout her text.

In his study of African-American garden traditions in the rural South, Richard Westmacott has broadened the concept in some ways helpful to an examination of Lawrence's study. He argues that the vernacular garden is not only an aesthetic choice but a utilitarian one; that it is a place of leisure but also of work; and that the vernacular garden is dynamic, a place where tradition and adaptation are not contradictory.[29] For the correspondents Lawrence gained through the market bulletins, the garden was all of these things: a place to earn money; a source of aesthetic pleasure as well as pleasurable work; an experimental plot where old plants grow next to new hybrids. The work and the workers are Lawrence's focus; how they represent their work and their gardens is her passion. Whether through the poetic language of a notice in a market bulletin or through a description or tip generously offered in a letter, the voices of her market bulletin correspondents are preserved right along with their experience, wisdom, observation, and love. It is through their language, through open and shared conversation, that Lawrence hopes to preserve the dynamics of southern gardening.

Westmacott cautions against equating "vernacular" with "traditional"; landscape historian J. B. Jackson expresses a similar reservation about the term because it derives mostly "from the mid-nineteenth-century exploration of rural life by antiquarians and those dissatisfied with urban, industrialized life."[30] His examination of the vernacular landscape, especially in the twentieth century, seems particularly applicable to Lawrence's project, one which is preservationist without being traditionalist.

In his re-thinking of the political dimensions of our use of the concept vernacular, Jackson maintains that in the twentieth century we have inherited a relationship with the natural environment that began to take shape at the end of the fifteenth century. It is a landscape aesthetic and ethic of beauty and order, culminating in the United States in the monuments of our capitol and the grid system on which most of our cities have been planned. If that relationship has taught us to lavish care upon the environment and to recognize that "the contemplation of nature can be a kind of revelation of the invisible world and of ourselves," it has also impressed upon us a "notion" of one kind of landscape, one "identified with a very static, very conservative social order." He calls for a new relationship, for nowadays "we no longer live in the country, we no longer farm, we no longer derive our identity from possession of land." These are, of course, the changes in the rural south to which Lawrence responded. These changes in our relationship to the land cause changes in our relationship to people: "We derive our identity from our relationship with other people, and when we talk about the importance of *place,* the necessity of belonging to a *place* . . . [we] mean the people in it, not simply the natural environment." Out of this new relationship, a dynamic and egalitarian community can develop, one where a single symbol of permanence can indicate "a history ahead."[31] What both Westmacott's study and Jackson's theory help to identify is the problem Lawrence had recognized between the lines in the market bulletins. The rural South was a widespread region undergoing the changes of a new economy as well as rapid population growth and urbanization of post-war years. The old order, based on working the land and owning the land, was rapidly changing. Westmacott's and Jackson's questions— how to maintain a sense of region, of community, of place amidst these changes? How to keep a tradition alive rather than pay tribute to one that has died?—are Elizabeth Lawrence's questions.

The answer for Lawrence, although she would not have used the language, is in keeping alive a discourse community; her term was "the society of inveterate letter writers." Much of her life's work aims at just that stewardship: from the idea of the garden center to the cultivation and collection of her letters to the inclusiveness of voices in all of her published work—from columns through garden books to this final, culminating project.

This is where Lawrence's work is most significant. Instead of depicting the southern garden as an emblem of tradition, a monument of the aristocratic Old South (as did many of the garden books and literary works of the period),

Lawrence's study of the market bulletins promotes an idea of the southern garden as a site of conversation—not imagined but lived, cultivated, preserved through long-standing correspondences. She returns the emphasis on language to the connotation of vernacular. In the gardens described in her study, correspondences are as carefully tended as plants. Place is identified by the names of its gardeners as well as by the names they give their plants.

Elizabeth Lawrence's study of the market bulletins opened up the conception of the southern garden, making it inclusive of both gardening as process and gardener as worker *and* speaker.

Her twenty-year project attempts to do for this "Friendly Society of Inveterate Letter Writers" what she claims that they have done for gardening: "Among them they keep in cultivation many valuable plants that would otherwise be lost."[32] Hers is, as she called theirs, a "saving love" and talk is its medium. It is a conversation upon which we can build a new landscape history, a vernacular tradition both dynamic and egalitarian.

Rena Sanderson

Linda Hasselstrom

The Woman Rancher as Nature Writer

During the last twenty years, Linda Hasselstrom has gained recognition as one of the most-published nature writers of the Great Plains.[1] Between 1978 and 1994, she produced eight books of ranch-centered nonfiction and poetry, including *Windbreak: A Woman Rancher on the Northern Plains* (1987), *Going Over East: Reflections of a Woman Rancher* (1987), *Land Circle: Writings Collected from the Land* (1991), and a massive regional historical study, *A Roadside History of South Dakota* (1994).[2] In addition, her essays and poems have appeared in numerous anthologies and periodicals.[3]

Her perspective, she stresses, is that of someone rooted in the land, someone with first-hand knowledge gained from *working* the land. She writes to correct the misconceptions of those who, out of touch with the primitive, insist on the sanitized nature depicted on Sierra Club calendars, and she aims to correct the Romantic myths of the Wild West. In a style of unflinching realism, at times especially startling in a woman writer, she records both the hardships and the rewards of the rancher's participation in a cyclical, regenerative nature.

Drawing selectively on Hasselstrom's nonfiction and poetry of the last twenty years, I will examine her rancher's perspective on the natural world. In the course of her career, she has written from both the insider's and the outsider's viewpoint. Hasselstrom was raised, since the age of nine, on her stepfather's cattle ranch located in southwestern South Dakota between the Black Hills to the west and the Badlands to the east. As an adult, she has divided her time between ranch work, the publishing and editing of regional literature, environmental activism, and writing. After working for years side by side with her second husband, George Randolph Snell, on her parents' ranch, Hasselstrom experienced major changes in her situation (George's death in 1988; her departure from the ranch in 1992 as the result of a disagreement with her stepfather; and her stepfather's death in 1992). Nevertheless, she has continued to feel inspired by the region and has continued to represent it as lecturer, teacher, and writer.

The subtext that emerges in Hasselstrom's writings confirms theories, such as those formulated by Hector St. John de Crèvecoeur and by Hyppolite Adolphe Taine, that stress a region's formative influence on character.[4] As this study aims to show, Hasselstrom has defined her values, her identity, and her purpose primarily through her relationship to the land and its legacy. In other words, like other women writers of the West (such as Terry Tempest Williams of Utah and Mary Clearman Blew of Idaho), Hasselstrom has found herself in her region's landscape, people, and history.[5] But in addition, unlike other regional women nature writers, Hasselstrom has derived her sense of identity to a great degree from *working* the land *as a woman* within a male-dominated field.

Hasselstrom has said: "On the ranch my life became a circle: . . . writing, riding, branding, and gardening—taking care of the land—coincided with writing. The arid soil nourished my body and my work."[6] Like other nature writers, she records the beauty she finds outdoors—"the scent of lavender and yellow mustard . . . white daisies, phlox and star lilies," the flight of hawks, the song of meadowlarks.[7] Also, in the tradition of the Romantic poets, she finds philosophical consolation in nature, but the lesson that nature imparts to her is filtered through ranch work.

Hasselstrom is not shy in insisting that such work experience "may confer authentic knowledge missing from the work of writers who spend more time inside than out."[8] She notes that she has attended conferences for "rural" writers where she was the only writer making a living off the land.[9] Distancing herself from the "visiting experts," who she fears have "frequently overrun" the West, she aligns herself with "writers who have rooted, been battered and tested by the terrain, the weather, and the people . . . folks who know the country and write from love of it," including Teresa Jordan, Mary Clearman Blew, Terry Tempest Williams, and Dayton Hyde.[10]

In one of her finest essays, "The Price of Bullets," Hasselstrom recalls her initiation as a fourteen-year-old girl into the ways of ranch life. After her stepfather gives her a .22 rifle for her birthday, her parents leave her in charge of the ranch and she has to shoot a sick bull. The incident teaches the young girl that the rancher must be self-reliant, skillful, and tough enough to accept harsh reality.[11] For forty years, she later recalled, she learned from her stepfather, "eager to prove" that she was "as good as a man."[12]

The rigors of the life she embraced are most powerfully depicted in *Windbreak: A Woman Rancher on the Northern Plains.* The book, written in journal form, chronicles one year of Hasselstrom's life with her husband George on her parents' ranch and details the hard work and challenges posed by a nature that is both destructive and regenerative, thus offering a close study of the human place in the natural order. While it documents the author's close familiarity with her region, the localized also assumes larger, symbolic significance.

Windbreak, like Thoreau's *Walden,* follows the order of the natural seasons (with sixteen poems carefully placed—one before each season and one after each month), its structure thus reflecting the coherence of the yearly cycle. Each season dictates specific chores, but in South Dakota, "Mother Nature has a talent for nasty surprises" that make routine tasks more difficult.[13] Extreme weather is common: In early November, winds blow at sixty-five miles an hour; by December, below-zero temperatures, high snow drifts, and winds test the physical limits of both humans and animals. Feeding the cattle becomes a challenge: "My arms are practically pulled out of their sockets when I stagger through the drifts with thirty pounds at the end of each arm."[14] Late March brings snow blizzards; by May, highs are in the nineties, but night-time lows still kill off much of the garden; in July, a hailstorm destroys the remainder of the garden as well as the pasture and alfalfa. Summer temperatures above one hundred degrees bring drought, thunderstorms, fire hazards, and grasshoppers that "bounce into the truck windows by the voracious hundreds, thump against our shoulders and our knees, dig their sharp claws into our bare skin until we stop flinching."[15] For Hasselstrom, the weather on the Great Plains is "almost an intelligence" symptomatic of a nature that needs to be outwitted:[16] "Everyday we get up in the gray light of dawn . . . and go out to battle nature again. Only it's not really a battle; it's a war of nerves, of tactics. If we considered it a fight we could only lose; instead we try to outmaneuver her, to survive, to keep the cattle alive."[17]

Accidents and injuries of humans are routine. One day the truck's tailgate drops on her finger. George uses "his favorite method" of treatment: "he heated the end of a paperclip in a flame and pressed it into the center of the fingernail. When the nail burned through, blood and fluid shot five feet in the air, but the instant of pain was worth the relief. In a moment the finger didn't hurt at all."[18] A couple of days later, the truck's motion causes Hasselstrom to fall off the tailgate while she is feeding the cows, and she lands on her "hands and knees in front of an astonished bull." Her "hands are full of cuts from the ice, with a little manure worked in, both knees are skinned" and one leg "is swollen more than usual, but at least I didn't land on my head."[19] In fact, these are minor scrapes. As her poem "Rancher Roulette" points out, ranching can be a deadly occupation.[20]

Injuries and losses of animals are routine as well. In the spring section of the book, the frequency of the fatalities may seem especially surprising. Hasselstrom finds injured wildlife, such as the meadowlarks that either froze or starved to death during a storm in late April. Spring is also, of course, the season of calving that exposes the rancher to recurring life or death situations. In one particular description, Hasselstrom employs parallel syntax to stress the close proximity of life and death: "Last night around midnight George and I pulled a calf and walked back to the house under the stars, happy to have saved one. Tonight, around the same time, we had to call the veterinarian, and still lost one."[21] As Geraldine Sanford points out, "the incidence of birth gone awry recurs over and over again in Hasselstrom's pages."[22] Within two pages, two dead calves have to be cut up inside the mother cow. In the first instance, the veterinarian, using "a flexible saw inside, cut off the calf's head" and divides the rest of the body so that the pieces can be removed; in the second case, the Hasselstroms themselves perform the procedure.[23]

During another birth, the calf's hooves get pulled off. When the calf dies two days later, Hasselstrom comments that she is "glad. He was so brave trying to walk."[24] She clearly admires courage, and her many detailed depictions of grim and tragic reality may reflect her wish to confront bravely what clearly is not an easy topic but rather a preoccupation for her.

While traditional literature tends to celebrate spring as the most hopeful season of romance, Hasselstrom never lets the reader forget that death and life are intimately interconnected in nature. A light-hearted treatment of this theme is given in one of her most popular poems, "Mulch," which opens the spring section of *Windbreak*. The gardener in this poem discovers that tokens of past losses (old love letters, dead kittens and baby chicks, "plans I made for children of my own, / photographs of places I've been / and a husband I had once; / as well as old bouquets / and an occasional unsatisfactory lover") make excellent compost.[25] In a piece entitled "Journal of a Woman Rancher" that she wrote for *Life Magazine* after George's death in 1988, she confirmed her acceptance of regenerative nature in a more serious way: "Spring always brings this tart mixture of life and death, and I am oddly heartened to see it again, to know I am part of the cycle."[26]

In her depiction of the destructive and restorative cycles of nature, Hasselstrom resembles Hemingway, whose instructions for shooting a horse she quotes and whose influence she has acknowledged.[27] Like him, she insists that she is telling the truth by presenting in an unflinching and, at times, aggressively shocking manner the unpleasant realities—violence, pain, loss, death. Finally, also like Hemingway, she believes in the larger continuum of an earth that abides, and she promotes the myth of a regenerative, cyclical nature. Her philosophy, however, puts an unusual spin on traditional definitions of the natural order since she reserves in it a special place for the work of the rancher.

In the essays collected in the two books *Going Over East: Reflections of a Woman Rancher* and *Land Circle: Writings Collected from the Land,* Hasselstrom fully explores environmental issues pertaining to ranching. In effect, she asserts that ranching is so well suited to the environment of her region that it illustrates human adaptation to the natural order: "[T]he steady rhythm of night turning to day, spring to summer, birth to death, the progress of the moon and sun, the sweep of wind and rain—those natural cycles determine how we arrange our lives."[28]

She suggests that ranch life at once fits smoothly into the ecological circle and symbolizes self-sufficiency: "[t]he circle of our world lies within a mile radius. Here lie our homes, the garden whose products supplement our own beef to feed our bodies, the wintering and birthing grounds for our cattle, the hayfields that feed them, the boneyard where they slowly return to earth, the junkyard where dead machinery becomes spare parts, and the garbage dump where we get rid of what we cannot use."[29]

Writing from the perspective of a "second-century South Dakotan who has worked most of her life at the business of ranching," she perceives herself as a "custodian of the land":[30] "I believe that many of our advantages and disadvantages, our triumphs and defeats, have sprung from the way successive populations have treated the land and natural resources, as well as one another. The prairie, for all its apparent toughness, is fragile and could be lost to us if we don't sensibly choose among varied choices of development. Those opinions affect everything I write and do."[31]

She identifies the West as a "third frontier" that "could become a battleground, with a money-rich, resource-poor urban society pitted against the center of the nation, traditionally the Breadbasket."[32] Herself an environmental activist, she has defended her region against those who "seem to regard the plains as an 'empty quarter,' fit only for the disposal of garbage," and she has protested various proposed environmental threats, from uranium mines and mills and a national hazardous waste disposal site to strip mining and heap leach processing operations.[33] Indeed, at times, locals have resented the fact that her activism helped to prevent the influx of new industries and jobs.[34] Nevertheless, Hasselstrom sides with the locals in defending the family ranch and its endangered way of life against such threats as interstate highways, snowmobiles, pollution, and corporate ranching, as well as against well-meaning newcomers and environmentalists who have no ancestral ties to the land and no commitment to the local economy. She speaks as one who has a stake in the local land-centered economy of ranching, but she wants that ranching done in an environmentally sensitive way.

Hasselstrom believes that the survival of the rancher justifies certain measures reflecting the kind of clear-eyed practicality she has been raised to admire, but some of her practical measures may offend some environmentalists. For example, she readily, and sometimes in graphic detail, describes other wildlife fatalities that are unintentional and regrettable but nevertheless part of ranching. Riding her tractor, she mutilates and kills blackbirds, deer, rabbits.

Her defense of killing prairie dogs illustrates her position. Since prairie dogs eat the vegetation, once a prairie dog colony reaches a certain size, "it destroys pasture and endangers cattle and horses by harboring rattlesnakes as well as creating leg-breaking traps"[35] Environmentalists, she says, simply don't understand that the rancher's survival depends on certain unpleasant but necessary steps. "Only survivors of the job should speak to the problem," she believes.[36] In her poem "Shooting Prairie Dogs," she identifies herself as a former peace activist ("who marched behind Martin Luther King, / who stood the taunts behind a sign that read 'Silent Vigil Against the War,'"), but she nevertheless shoots the prairie dog, and the closing, disturbing image of the poem suggests that her shot rips out the animal's heart, which "lies on the edge, / quivering."[37]

Killing coyotes, a common practice with other ranchers, is more difficult for the Hasselstroms. In *Windbreak,* Hasselstrom explains that she and George enjoy the spectacle of a pair of coyotes that regularly play in the field. One day, however, they find a calf "dead and partly eaten" surrounded by tracks that suggest an attack by a coyote. That evening, George shoots at a coyote that attacks another calf.[38] Yet one month later, when Hasselstrom identifies a three-legged coyote as the one George shot, she is "glad he's alive, for reasons I'd rather not have to explain."[39] Generally though, the Hasselstroms kill neither snakes nor coyotes. She explains the principle: Those animals "serve their purposes here, cleaning up mice and other vermin. We give them the same respect we give responsible humans, and don't bother them if they don't bother us."[40]

Overall, she insists, responsible ranching can be environmentally friendly, with cattle doing "no permanent damage; in fact, their manure and their feeding habits encourage plants that might not survive elsewhere . . . What's good for cows is usually good for wildlife, so they can coexist if the number of cows isn't excessive."[41] She suggests not only that western South Dakota's grasslands are ideally suited for livestock grazing but that cattle ranching can benefit the land and the people who live on it.

Hasselstrom's belief in this ideal fit between the Great Plains and ranching activity illustrates her commitment to bioregionalism, the idea that each region should draw on its own strengths, utilize its own resources, and import a minimum from other areas. As a bioregionalist, Hasselstrom aims to foster local authenticity and diversity as an alternative to the uniform culture promoted by the media and consumerism. Her defense of the family ranch as an economic unit is thus part of a cultural resistance as well.[42]

Her staunch support of ranching takes unusual and memorable form in "Beef Eater," a poem that functions as an exclamation point at the end of *Land Circle*. The speaker in the poem cooks and eats, with much pleasure, a beef heart and consequently becomes cow-like:

> My friends have begun to notice my placid air,
> which they mistake for serenity.
> Yesterday a man remarked on my large brown eyes,
> my long eyelashes,
> my easy walk.
>
> I switched my tail at him
> as if he were a fly,
> paced
> deliberately
> away.[43]

Hasselstrom ascribes to her favorite type of cow, the Hereford-Angus, the characteristics of autonomy and vigor. To survive, the rancher must be as hardy and independent as the Hereford-Angus. Hasselstrom endorses, once again like Hemingway, a certain code of "grace under pressure," but hers is the rancher's code. It prescribes physical and mental strength, courage, pride in one's work, and a certain philosophical stoicism. Her writings demonstrate that she herself has developed those very qualities and that she admires those qualities in others.

After her husband's death and after she was displaced from the ranch in 1992, Hasselstrom had to formulate for herself a new identity. After forty years of hard work on the ranch, she regarded herself "pledged soul, body and brain to a few thousand acres of buffalo grass on the arid high plains, to a ranch" and yet found herself suddenly living "in exile" in urban Cheyenne.[44] In her subsequent writings, she tells how much she misses life on the prairie, but she also draws comfort from her vivid memories. Informed by a familiarity with the prairie recollected in tranquility, this work of her maturity seamlessly combines the educational with the symbolic as information on local geography, vegetation, and fauna is interwoven with narratives rich in psychological, philosophical, spiritual, environmental, and political significance.

Although Hasselstrom celebrates self-reliance and autonomy, she also expresses her sense of connectedness with those, both living and dead, who have worked the same land and been shaped by it. She has celebrated historic and mythic figures, such as local settlers and legendary mountain men. Early in her career, she edited and introduced James Clyman's *Journal of a Mountain Man,* and in 1994, she published a massive 467-page volume *Roadside History of South Dakota* that combines social history with geography, topography, and ecology.[45]

As early as 1978, Hasselstrom had already expressed an interest in women's unique historic role in the West, in particular, when she edited and published *Next-Year Country: One Woman's View,* a collection of photographs taken at the turn of the century by a Nebraska farm woman named Alma.[46] Since then, and especially during the last decade, Hasselstrom's sense of connectedness with western rural women has grown and taken many different forms. After the warm reception of her second book *Going Over East,* she came to see her role as a writer with a new gender awareness: "Abruptly, I saw myself representing a breed many people did not know existed: the western woman. Hesitant about speaking for the cantankerous women I knew, I also believed we needed to be recognized as part of the West as well as the environmental movement."[47]

In the third and last section of *Land Circle,* significantly entitled "A Woman's Covenant," several pieces examine women's relationship to the land. The poem "The Only Place" and the essay "Prairie Relief" are light-hearted explorations of a woman's bathroom troubles out in nature.[48] In the essay "Confessions of a Born-Again Pagan," Hasselstrom more seriously examines the Christian tradition that teaches dominion over the earth, and she embraces ecofeminism as an alternative.[49] On a more personal note, in such poems as "Planting Peas" and "Digging Potatoes," she expresses her special kinship with her grandmother; and in various writings she has acknowledged the formative influence of her strong women friends, Jo and Margaret.[50] "Strong and independent, they all grew up with men whose habit it was to belittle women. Their work quietly proved such men wrong, but they carried no banners."[51] Hasselstrom admits that "imitating" her father, she grew up "sneering at women's work," but that she eventually came to respect the work of prairie women who "relish a freedom" from traditional gender restrictions "largely unknown elsewhere."[52] Her writings suggest that the demands of working the land transform Western women into a special breed, a third gender, that combines traditionally separate male and female strengths.

Like Willa Cather almost a century ago, and like her contemporaries Teresa Jordan and Mary Clearman Blew, Hasselstrom celebrates the strengths of western women and alerts the public to the unique situation of western women. In the tradition of Tillie Olsen, Meridel Le Sueur, and Adrienne Rich, she has made it her task to give voice to women left out of history as she has reclaimed for them and their experiences a place in the history of the American West.[53] "I felt some of those women seize my hand to write what they could not write."[54] In "Tapestry," a poem dedicated to Meridel Le Sueur, Hasselstrom celebrates women's part in the creation of the West:

> Ripe grain and rough brown silo walls
> carpet the earth, a tapestry ancient as Europe's finest.
> This mat was not woven by the tender white fingers of
> virgins, but by the scarred, broken
> hands of farm wives. It was created, warp and woof,
> of their blood and bones.[55]

And recently she collaborated with Gaydell Collier and Nancy Curtis in editing *Leaning into the Wind: Women Write from the Heart of the West* (1997), a collection of writings by contemporary western women "who do, or once did, physical labor connected to the land."[56] Their lives, she feels, serve as inspirations for other women whose circumstances perhaps don't provide them the same freedom to realize their potential.[57]

In his study *Beyond the Frontier: Writers, Western Regionalism, and a Sense of Place,* Harold P. Simonson (using sexist language typical of his generation) describes the psychological, philosophical, and artistic function of the connectedness experienced by the literary regionalist: "The artist feels [what it means to be part of a region] because he merges his own experience of place with the people and events that have come before, and with the fuller symbolism into which they have expanded. The artist recapitulates all these things within his psychic depths . . . What emerges is the artist's voice, as complex and universal as the human condition."[58]

Hasselstrom is what Simonson has called a "true regionalist," someone who equates home with "wholeness, centeredness, and connection."[59] Her beloved Great Plains have not only shaped her values but have made her life meaningful by providing her with an identity, with a sense of belonging, and with a calling. She, who has always celebrated the redemptive power of a person's intimacy with nature and place, continues to function as mediator between the range and the world at large. She has finetuned her message that the local is the universal, that the "center of the universe *is* South Dakota" and that its problems and lessons are everyone's.[60] She has, in other words, successfully mythologized her life.

Karen E. Waldron

The Land as Consciousness

Ecological Being and the Movement of Words in the Works of Leslie Marmon Silko

> The earth is your mother,
> she holds you.
> The sky is your father,
> he protects you.
> Sleep,
> sleep.
> Rainbow is your sister,
> She loves you.
> The winds are your brothers,
>
> They sing to you.
> Sleep,
> sleep.
> We are together always
> We are together always
> There never was a time
> when this
> was not so.
>
> —Leslie Marmon Silko, *Storyteller*[1]

Nature writing, writing nature—such terms reflect an emerging category of study in American literature, itself a broadening category that over the last few decades has been seen to encompass far more than the literature of the American Adam or Eve in a new world garden.[2] Between the dawn of the conservation movement, the travel literature and regionalism that accompanied westward expansion, modern environmentalism, and American studies, Henry David Thoreau's *Walden* has been variously but repeatedly considered the touchstone of a body of writing defining and expressing the American experience of and with land and landscape, wilderness and nature. Thoreau initiated a way of writing about what is seen in nature that mirrors the approach of the landscape school of painting, having its roots in European and Christian fantasies of a "new world" alternately perceived as howling wilderness and bountiful garden by English settlers.[3] Whether located and writing from the Massachusetts Bay Colony or Virginia, English writers of landscape and botanical literature who prefigured Thoreau lay out the primary ground of what nature writing has been considered to be with their European and Christian perspective on art as representation of nature. Yet as Carolyn Merchant details in *Ecological Revolutions,* landscape by definition remained outside their personal consciousness. Nature was a separate force and kind of being from the human and the divine. What Thoreau and modern nature writers such as Annie Dillard have brought to this framework includes ecstatic, focused, nature-based individual attention. Such nature writers' consciousness of landscape involves recognition of personally transcendent realities based on and in what Stephen Trimble, himself a literary naturalist, identifies as "journeys *into* the landscape."[4] What their style of writing fails to do—to actually accomplish the reconnection of the viewing self with the natural world, which their expressions of "the naturalist's trance" or "ecological way of seeing" express as desire—comes automatically from the writings of Leslie Marmon Silko whose work details *the land,* not the self moving into landscape, as consciousness.

In her oft-anthologized essay, "Landscape, History, and the Pueblo Imagination," Silko directly challenges "landscape" as one of the fundamental terms through which nature and the land have been perceived by European-based writers and artists: "So long as the human consciousness remains *within* the hills, canyons, cliffs, and the plants, clouds, and sky, the term *landscape,* as it has entered the English language, is misleading. 'A portion of territory the eye can comprehend in a single view' does not correctly describe the relationship between the human being and his or her surroundings."[5] In developing a theory of nature writing, Lee Schweninger pointedly notes that to call the essays and fiction arising from Silko's perspective "nature writing" is to impose a European perspective on the reality and necessity of Silko's written expression, the very *nature* of her work's textuality; there is no Keresan or other Native American writing that does not, in essence, encompass nature.[6] Helen Jaskowski adds that however central landscape and place are to Silko's work, her writing points to the limitations of the written word.[7] Nevertheless, Silko's writing contains inherent lessons for nature writing and nature writers. Silko's poems, essays, and novels manifest the relationship between the human being and his or her surroundings as one of being rather than viewing.

The natural world Silko's works express offers no wilderness concept. Nature has no existence apart from the people: "Viewers are as much a part of the landscape as the boulders they stand on."[8] To deny the impact of humans on the landscape is to misunderstand the nature of the relationship. Unlike nature writing that highlights a separated natural wilderness through the vision of an isolated observing consciousness, Silko's writings show consciousness and vision of the relationship between humans and their surroundings—a phenomenon more like insight than sight—arising from the containing sphere of the natural world. The core relationship of all being derives from the land itself, not the self-authorizing eye. Although clearly cultural and autobiographical, Silko's work is not about the "I" in the usual or Western sense.[9] In Leslie Marmon Silko's work, the land—and often a particular kind of land, the southwestern portion of the North American continent—witnesses to and carries spiritual insight. What collectors and critics of nature writing like Trimble have called "the ecological way of seeing" comes from the mountains and plains, ice and snow, the river and the desert—not the human eye.

Silko's essay, like all her nonfiction, vividly describes what such a perspective might mean to human consciousness: "Landscape thus has similarities with dreams"; "You are never the first to suffer a grave loss or profound humiliation"; "The arroyo demands from us the caution and attention that constitute respect"; "the Hopi elders are grateful to the landscape for aiding them in their quest as spiritual people."[10] The poems and fiction, however—*Ceremony, Storyteller, Almanac of the Dead,* and *Gardens of the Dunes—manifest* the egoless quality that marks Silko's major contribution to the understanding of nature through writing. Silko's different kind of vision, one in which "nothing is overlooked or taken for granted," illustrates and creates the actual sensation and knowledge of a land-based consciousness, its shape and method of working with words, and its illuminative potential and necessary reorientation of what is commonly considered nature writing. In particular, the fiction demonstrates how stories that relate to the land from this perspective create what I am calling an egoless kind of narrative and textual consciousness, one in which individual achievement, including of understanding—the typical vehicle of plot even in nature writing—gets absorbed by the larger, containing whole of the transcendent communal and natural mind. Silko's stories are of the land before they are of the individual or ego; the stories relate the diminishment of ego through awareness of the land. Thus, as Jaskowski notes, the storytelling vehicle remains primary while Silko's interviews and essays serve to emphasize "that self, land, and community are united by presence at a place whose meaning is expressed and internalized *through* story."[11] Or as Silko herself says, "The narratives linked with prominent features of the landscape . . . delineate the complexities of the relationship that human beings must maintain with the surrounding natural world"[12]; and "The whole feeling of the place, the whole identity of it was established for me by the stories I'd heard, all the stories."[13]

Thus although Silko's interviews and essays *say* how important the landscape is to the fiction, how basic the fact of stolen land is to tribal peoples, and how native writing must respond to self-centeredness, fiction provides the *means* of avoiding egocentric individualism and cultural appropriation while creating land as consciousness.[14] The nature of story itself, along with the egoless nature of Silko's story writing—the modes by which she develops characters, for example—articulate most fully the true vitality of Silko's ecological vision and importance for nature writing. Structure and form, in Silko's fictional narratives and poetic texts, lead her works to transcend their apparent agendas (the fact that she has agendas, a common trait of nature writers, being a matter often criticized) and the challenges of audience to create a narrative consciousness that knows and shows how and why the land can witness.[15] Silko's mode of writing nature articulates a consciousness born in focal points that could be labeled Keresan, Laguna, Native American, western regional, and female but extend beyond and yet through multiple, overlapping concepts and experiences to reach all peoples who seek to listen to the earth, not as a bearer of news or vision, but of spirit and consciousness. Leslie Marmon Silko's novels create a transformative experience of land and nature for non-native readers at the same time that they generate recognition of core values for Indian peoples. Her storytelling frame and its vivid expression of the consciousness of humans within the landscape manifests a nonlinear, metaphorical, egoless lesson for nature writing. Silko's stories compel the reader to see, feel, and know the land.

Silko's work accomplishes this compelling call to vision with fiction and poetry that describes the Southwest, Alaska, and all places they touch with loving, respectful, intensely realized and emotionally challenging detail. The land dominates her characterizations, always at heart becoming the central figure in her fiction. Descriptions of action, although taking shape differently in each work, all formulate themselves around place. Each of Silko's fictional works also provides a titular and structuring metaphor through which the story and its consciousness, meaning, and expressed relationship of humans to their surroundings, must be read: ceremony, story, almanac, garden. While critically distinct, the central metaphors all express the nature of the containing relationship of the land for the people who dwell within its consciousness. All describe a form of ritual or repeated activity connecting humans to their surroundings and reminding them not only of the function of story but of the nature of abundance as spiritual rather than physical. For example, in Silko's first novel, *Ceremony,* Tayo discovers a sustaining consciousness and identity when he realizes that just by being Tayo he is part of the ceremony that grows out of and responds to the Indian lands of the Southwest—dry, impoverished, home of pueblos, hogans, deer, elk, mountains, mesa, desert, alcohol problems—and uranium. Silko's second work, *Storyteller,* a compilation of fiction, photographs, and poetry framed autobiographically, highlights the nature of story as spiritual expression. Each element shows exactly how place manifests both story and character, communicating the spirit of place by centering movement, experience, and identity in the natural world, its only context.

Almanac of the Dead weaves itself around a constantly referenced but always extra-textual collection of records and revolts that ultimately point to land theft and desecration, an addiction to excess defined by capitalism and its sense of the land as source of raw material. The novel begins with a map, stating at its outset that it is not only about but framed by place and that consciousness of place will provide understanding of the story. With the map acknowledging time as well as place, Silko moves to extend the historical consciousness of her work, complicating the vision of *Ceremony* and *Storyteller* even further. The being and consciousness that *Almanac of the Dead* articulates through Sterling, its subtly central Native American character, stretch over a much vaster landscape than that of *Ceremony,* adding layers of difficulty and loss as well as immense historical detail and present complexity to the narrative of the Southwest. *Gardens in the Dunes* extends further over centuries and landscapes, reclaiming the contested metaphor of the garden for the Southwest from a Euro-Christian perspective that would deny the land consciousness and being. In all four works, the central metaphor both expresses the nature of being and provides the storytelling vehicle. Story structures delineate the relationship of humans to their surroundings, and the importance of honoring that relationship in sacredly conscious ways.

In the context of American nature writing, critical attention has been paid most directly to *Storyteller* and *Ceremony;* these novels present metaphors of the environment and healing visions of the land that are both accessible and hopeful to Euro-American readers. In *Ceremony* and *Storyteller,* lyrical descriptions of the natural world surround and reinforce themesof appreciation and preservation. *Ceremony,* a novel that is itself a ceremony, includes extensive narrative description of the Southwest and teaches a right relationship of humans to their surroundings. *Ceremony* and *Storyteller* also show how the vehicle of the short story and poem combined into a longer, more sustained work operate to centralize and lyricize the spirit of place. Yet despite these novels' accessibility to romantic interpretations, their important lesson for nature writing derives from the way they insist that the non-linear, egoless whole of the text and the world it narrates—dismissive of the centralizing eye and consciousness of both some fiction and "the naturalist's trance"—is always greater than the sum of the textual, descriptive parts.[16] Understanding this movement of words in Silko's early works leads the reader inevitably to appreciate the developments of land as consciousness in Silko's more complex and wide-ranging later plots.

Ceremony

The descriptions of place in *Ceremony* dramatize how its movement of written words away from ego prefigure *Storyteller* and the later fiction as nature writing. Early Silko scholars Paula Gunn Allen and A. LaVonne Ruoff both immediately highlighted how the place Silko comes from, interpreted geographically, socially, spiritually, and in terms of gender, articulates the relationship of the native peoples and beliefs of the Southwest to the landscape and words about landscape that these texts create. Allen, for example, notes that *Ceremony* manifests the nature of Indian time as arising from the interconnectedness of all life and a consciousness that "assumes the individual as a moving event shaped by and shaping human and nonhuman surroundings."[17] The work of critics such as Allen and Ruoff highlights how *Ceremony* articulates a kind of nature writing, forms a core of commentary and guidance for understanding the import of all Silko's fiction as writing derived from and responding to the natural world.

Within *Ceremony,* the protagonist Tayo, facing a life or death struggle with illness, addiction, and witchery, must ultimately realize—by listening to the land—both what has happened to his home and the egoless, active role that he must play in restoring it to balance. Restoring the land creates both ceremony and novel. Like any reader who is not already conscious of the land's being and able to center all understanding there, Tayo has to see both that the land has been violated and that alcoholism, bloodlust, and other forms of what western medicine calls addictive illness arise from over-consumption and a desacralized relation to the earth. To give in to the ego of desire—whether for property, power, wealth, fame, or forgetfulness—when desire has been shaped by excess into a force destructive of spirit and being, means personal and communal death. All Tayo's understandings come from the same, sacred place, his home land. To convey the point, *Ceremony* opens with a story of storytelling and then the beginning, "sunrise"; the framed story starts after the invocations with Tayo's return to the reservation.[18] Movement between dramatically different natural and cultural worlds, to fight the Japanese in the Philippine jungle in World War II, has left him spiritually and fraternally bereft. Most importantly, Tayo's dislocation has made him seriously ill with constant vomiting—but not in any way that the western psychiatric medicine that has been applied, with its focus on ego control, can address.

As Silko's prose insists that the reader understand Tayo's predicament culturally, it weaves in descriptions of the landscape and its meaning that gradually build the sense of the land as both consciousness and ground of insight. Tayo's Uncle Josiah becomes a vehicle for this understanding, as when Tayo worries about the drought and recalls the jungle rain in the Philippines: "It was there that Tayo began to understand what Josiah had said. Nothing was all good or all bad either; it all depended" (*Ceremony* 11).

Tayo's thought processes, weaving together past and present in each moment of the text, voice the intimate relation of land to the movement of words: "He damned the rain until the words were a chant, and he sang it while he crawled through the mud to find the corporal and get him up before the Japanese saw them. He wanted the words to make a cloudless blue sky, pale with a summer sun pressing across wide and empty horizons. The words gathered inside him and gave him strength" (*Ceremony* 12). Tayo also recognizes that his illness has to do with losses his home and the whole earth have sustained: "he cried at how the world had come undone, how thousands of miles, high ocean waves and green jungles could not hold people in their place" (*Ceremony* 18). When he hears another reservation veteran, Emo, talking about post-war racism, Tayo understands that his fellow Indians "blamed themselves just like they blamed themselves for losing the land the white people took" (*Ceremony* 43). And he remembers Josiah saying: "'This is where we come from, see. This sand, this stone, these trees, the vines, all the wildflowers. This earth keeps us going'" (*Ceremony* 45). From the beginning, the novel establishes the land as the center of health and being.

Silko's plot develops the import of the land's relevance to story. When the medicine man Ku'oosh comes to see Tayo, he reminds him along with the reader that "this world is fragile." The text elaborates again by explaining the relation of words to the earth's natural reality: "The word he chose to express 'fragile' was filled with the intricacies of a continuing process, and with a strength inherent in spider webs woven across paths through sand hills where early in the morning the sun becomes entangled in each filament of web. It took a long time to explain the fragility and intricacy because no word exists alone" (*Ceremony* 36). Tayo begins to reconnect, to understand, by listening to Ku'oosh's words, remembering his Uncle Josiah's stories, and allowing the natural world to resurface in his memories. Tayo's entire story and development as a protagonist exist to show how and why human beings have their consciousness *within* landscape, the ground of being, spirit, and love.

Ceremony reverberates with examples of Tayo's and the text's gradually elaborated understanding of both healthy and destructive relations between human beings and their surroundings. One series of characters, egoless in that they are more concerned with community and the land than with themselves, share the story and hope of healing with Tayo. Others, struggling with ego and addiction like Harley and Emo, share story and ceremony through despair and disease. As the ceremony proceeds, invoking the many elements that contribute to Tayo's illness—his cousin Rocky's dismissal of the old ways, Auntie's susceptibility to shame, Uncle Josiah's dream of raising Mexican cattle able to tolerate the heat and drought of the reservation, Harley's and Emo's beer drinking, the way European culture treats land, self, and other—the novel's words show Tayo's consciousness formed upon and by the world that *is* nature and its gift of being, word, and spirit. For example, Tayo realizes that the conflict with Christianity comes from ways of naming and highlighting individual action:

The fifth world had become entangled with European names: the names of the rivers, the hills, the names of the animals and plants—all of creation suddenly had two names: an Indian name and a white name. Christianity separated the people from themselves; it tried to crush the single clan name, encouraging each person to stand alone, because Jesus Christ would save only the individual soul; Jesus Christ was not like the Mother who loved and cared for them as her children, as her family. (*Ceremony* 68)

The novel supports its egoless thesis with lyrical descriptions of the immanence of nature and its elements: "The cottonwood trees had not lost their familiar feeling with him. They had always been there with the people but they were much more than summer shade" (*Ceremony* 103); "There was something about the way the old man said the word 'comfortable.' It had a different meaning—not the comfort of big houses or rich foods or even clean streets, but the comfort of belonging with the land, and the peace of being with these hills" (*Ceremony* 117). Silko's ceremony and inherent lessons for non-native nature writing proceed through Tayo's simultaneous recovery of the meaning of words and land.

Ceremony's allocation of blame for illness conveys the danger of seeing only objects, not land as ground of being. A prophetic tale forms a central part of the ceremony Tayo performs with another medicine man, Betonie:

> Caves across the ocean
> in caves of dark hills
> white skin people
> like the belly of a fish
> covered with hair.
> Then they grow away from the earth
> then they grow away from the sun
> then they grow away from the plants and animals.
> They see no life
> When they look
> they see only objects.
> The world is a dead thing for them
> the trees and rivers are not alive
> the mountains and stones are not alive.
> the deer and bear are objects
> They see no life.
>
> (*Ceremony* 135)

Betonie tells Tayo that he has been watching for the transition in the constellations, then shows him how knowledge rests in the natural world. Himself watching for the stars, Tayo meets Ts'eh and realizes for the first time that it is all right for him to be alive (*Ceremony* 181). Still following the cattle, he ascends Mount Taylor, the "most sacred of mountains to the Keresan peoples," now occupied and claimed by ranchers.[19] Each step Tayo travels through the ceremony and over the reservation reminds him of the power of what he is fighting. Caught between egolessness and ego, health and illness, he frequently gives in to despair. But the land, the words, the ceremony, keep returning—textually and thus by implication to Tayo. As he confronts death, Mount Taylor shows him his connection to the ground of being:

He was aware of the center beneath him; it soaked into his body from the ground through the torn skin on his hands, covered with powdery black dirt. The magnetism of the center spread over him smoothly like rainwater down his neck and shoulders; the vacant cool sensation glided over the pain like feather-down wings. It was pulling him back, close to the earth, where the core was cool and silent as mountain stone, and even with the noise and pain in his head he knew how it would be: a returning rather than a separation. He was relieved because he feared leaving people he loved. But lying above the center that pulled him down closer felt more familiar to him than any embrace he could remember; and he was sinking into the elemental arms of mountain silence. (*Ceremony* 201)

Though Mount Taylor allows Tayo to make the choice to live and struggle, supported by the land's power, hope, and guidance, Tayo still must show the natural world that he is not one of the destroyers (*Ceremony* 203). Yet the earth itself provides strength and insight to continue the ego-removing process of the ceremony. Tayo's surrender of the ego that interprets loss as personal and his release of the eye that delineates only one kind of vision comes increasingly from his natural surroundings:

The buzzing of grasshopper wings came from the weeds in the yard, and the sound made his backbone loose. He lay back in the red dust on the old mattress and closed his eyes. The dreams had been terror at loss, at something lost forever; but nothing was lost; all was retained between the sky and the earth, and within himself. He had lost nothing. The snow-covered mountain remained, without regard to titles of ownership or the white ranchers who thought they possessed it . . . The mountain outdistanced their destruction, just as love had outdistanced death. (*Ceremony* 219, emphasis added)

After being embraced by the mountain, Tayo stops relying on his eyes to know what the land is telling him. Tayo realizes that "this feeling was their life, vitality locked deep in blood memory, and the people were strong, and the fifth world endured, and nothing was ever lost as long as the love remained" (*Ceremony* 220). As he comes to know, "The strength came from here, from this feeling. It had always been here." The novel dramatizes that nothing that damages ego damages the essence of self and being (*Ceremony* 237).

The challenge that remains for the reader with Tayo is to *always* hear the land as a mind holding thoughts together, even when "the steel and glass closed out everything" (*Ceremony* 241). Tayo must recognize all that the land can tell him at the uranium mine: "There was not end to it; it knew no boundaries; and he had arrived at the point of convergence where the fate of all living things, and even the earth, had been laid" (*Ceremony* 246). But Tayo almost loses the awareness in his rage at Emo, who is still caught in the witchery. When Tayo turns away from violence, refusing to harm Emo, he completes the ceremony and shows the earth's consciousness redeeming all the persons in the text, and more importantly the larger world surrounding the writing, from disaster. The novel ends with "Sunrise / accept this offering / sunrise"—a sense of peace, recovery, and healing that situates consciousness firmly in the larger world. The ceremonial story's fulfillment lies in Tayo's release of ego needs and desires to the land which is his home.

Storyteller

Silko's *Storyteller* demonstrates formally, rather than through the ceremony of character, the release of ego needs and desires, showing exactly how place manifests character. While the poems describe place and the photographs document it visually, the tales set amongst and between these use them and their own obviously provisional textuality to center movement, experience, and identity in the natural world as fiction's only possible context. The natural world contains the human world, of course—including the history of contact of Native American and European peoples. Contact with European peoples creates both writing (from an oral tradition) and the still orally based style of writing that notes that there is always a surrounding consciousness and story.

In a tale such as "Yellow Woman," however, contact remains within the Keresan imagination. From the opening "My thighs clung to his with dampness," "Yellow Woman's" textual consciousness, though narrated in the first person, moves immediately into the landscape: "And I watched the sun rising up through the tamaracks and willows. The small brown water birds came to the river and hopped across the mud, leaving brown scratches in the alkali-white crust. They bathed in the river silently. I could hear the water, almost at our feet where the narrow fast channel bubbled and washed green ragged moss and fern leaves. I looked at him beside me, rolled in the red blanket on the white river sand" (*Storyteller* 54).

The relationship of the narrating Yellow Woman to Silva, the man beside her, belongs to the land and the river in particular. Thoughts and feelings come from and through the surrounding sky, mountains, wind: "I wondered who was over there to feel the mountain wind on those sheer blue edges—who walks on the pine needles in those blue mountains (*Storyteller* 57). With the unnamed narrator herself a character created by landscape and story, the consciousness of the text—certainly of the narration—arises so profoundly from the environment that the reader cannot separate characters and characterization from place. Yellow Woman doubts both her own and Silva's identity as separate minds or mere humans (*Storyteller* 56). Her egoless first-person narration leads the work to a mythical or transcendent level.

Scholars have explicated the mythical qualities of *Storyteller*'s tales as a natural element of Keresan storytelling and have done important work documenting its means of negotiating multiple audiences.[20] All such textual elements contribute to Silko's lesson for and form of nature writing. As writing, "Yellow Woman" lives in the space created by Silko's overall textual structure and its own narration as a story. The narrator who has been Yellow Woman but also, given the architecture of the text as a whole, acts as an implied projection of identity into landscape, can constantly remind the reader of the surrounding oral tradition, as in: "I decided to tell them that some Navajo had kidnaped [*sic*] me, but I was sorry that old Grandpa wasn't alive to hear my story because it was the Yellow Woman stories he liked to tell best" (*Storyteller* 62). What matters is not the writing but the experience of listening, of becoming part of a story, grounded in the land for that time. "Yellow Woman" illustrates how tale and character come from the land throughout *Storyteller*. As "Yellow Woman" tells of the marriage of person with surroundings and shows how the relation originates in the river and valleys, it also compels the reader to listen; the words made textual give the land consciousness. "Yellow Woman" is more about how place creates the imaginative than about the natural world as view or backdrop or projection of the imagination.

"Yellow Woman" and its companion tales, including the titular "Storyteller," which highlights the question of narration and identity, also explore the nature of the "I" in story by asking a fundamental question: Who or what are persons? In both *Storyteller* and *Ceremony,* the experience of one person in the environment gets enhanced by meetings with others, often sexual others, who are also spiritual beings with medicine or a profound connection to the earth—a relationship Patricia Chase Smith and Paula Gunn Allen analyze in showing just how completely humans interact with the landscape. In *Ceremony,* Tayo has a deeply spiritual encounter with Ts'eh, his Uncle Josiah with the Night Swan. Yellow Woman *is* Yellow Woman because of the man she met by the river, who is both Silva and a ka'tsina or mountain spirit (*Storyteller* 59). Other sections of *Storyteller* illuminate the question of who or what persons are, dramatizing the danger of reliance on a human ego. The tales and poems offer little distinction between animals, parts of nature, and humans: "But you see, he *was* the Sun, / he was only pretending to be / a human being"(*Storyteller* 63).

"Cottonwood" and other poem-stories also highlight the related Silko theme of the role stories play and the necessity of reading and writing and hearing stories from the entirety of the landscape. Transcribing, by telling the land's stories, diminishes the power of assumed "I," moving human consciousness away from egocentric and human-centered concerns:

> Before that time, there were no stories
> about drastic things which
> must be done
> > for the world
> > to continue
>
> Out of love for this earth
>
> > cottonwood
> > sandstone
> > and sky.
> > > (*Storyteller* 65)

Reality comes from the environment, but not as a matter of weather and crops, landscape, or what the eye can see. Dependence on surroundings spiritualizes the relationship of humans to place, showing place as the origin as well as the ground of being. Part one of "Cottonwood" ends with: "So much depends / upon one in the great canyon" (*Storyteller* 67).

The places that give birth and consciousness to *Storyteller* offer themselves clearly and honestly (without ego), whether as Southwestern mesas or Alaskan ice. Lack of abundance creates a profound sense of gratitude and wisdom for how mother earth works. The land's honesty generates humility and respect among its people. Part two of "Cottonwood" returns to Yellow Woman as poem and prose merge to tell the same extensive, inclusive, story of place. The text shows how the profound relationship with the earth that leaves "Yellow Woman" and its characterization egoless, archetypal, about patterns and realities not personalities, comes from a dry land:

> It was one of those times
> one of those times when
> there had been no rain for months
> and everything was drying up.
>
> It was at this time
> long ago,
> Kochinkinako, Yellow Woman went searching
> for water to carry back to her family.
> She went first to the spring
> near the village
> but the water had dried up
> the earth there
> wasn't even damp when she touched it.
>
> (*Storyteller* 68)

Poems expressing the spirit of place in *Storyteller* become an essential part of the framework and sensibility of the short fiction. Included in a work of stories by and about storytellers, the poems themselves are particular kinds of stories in which the land is alive, a character, the basis for and of being. "Poem for Myself and Mei: Concerning *Abortion*" illustrates the prevalence of the natural world not only *in* every story but as the ground and reason for every story. Without the title, many readers would fail to understand the human significance of the last stanza about butterflies, along with the ultimate insignificance, or larger context of, abortions:

> They die softly
> against the windshield
> and the iridescent wings
> > flutter and cling
> > > all the way home.
> > > (*Storyteller* 123)

Although marked as a poem for a self, "concerning abortion," the lyric moves from the morning sun's release of "yellow light / butterflies tumbling loose" to a vision of them that has the "clarity of ice" to "butterflies are yellow mustard flowers / spilling out of the mountain" (*Storyteller* 122). The entire experience of abortion, the expression of profundity "for myself and Mei," comes through the landscape and natural world. Ice and sun and butterflies determine the nature of sight and ultimate repository of abortion's meaning.

Another poem in *Storyteller*, "for the students of the Bethel Middle School, Bethel, Alaska," comes as close as Silko ever does to lyricizing the process of the observing eye. Yet it addresses a collective, not an "I," and transforms seeing into a process of shared recognition of the natural world's being and aliveness, rather than an act of the speaker or any singled out individual member of her audience:

> You see the sky now
> colder than the frozen river
> so dense and white
> little birds
> walk across it.
> (*Storyteller* 177)

Even a fragment of a letter, written much as any nature writer might:

The purple asters are growing in wide fields around the red rocks past Mesita clear to the Sedillo Grant. This year there has been more rain here than I have ever seen. Yesterday at Dripping Springs I saw a blue flower I had never seen before. (*Storyteller* 170)

rapidly turns into a comment about the way such events create stories changing entirely the role of the eye and I:

I remember the stories they used to tell us about places that were meadows full of flowers or about canyons that had wide clear streams. I remember our amazement at these stories of lush grass and running water because the places they spoke of had all changed; the places they spoke of were dry and covered with tumbleweeds and all that was left of the streams were deep arroyos. But I understand now. (*Storyteller* 170)

Each element of *Storyteller's* textual collage shows the understanding that has been reached is not only *about* the transformation of the landscape by rain but *of* the power and truth of stories that give the natural world a life and history beyond and constitutive of any interacting human history.

Almanac of the Dead

Celebration of *Ceremony*'s achievement as a unified and healing vision of the land and *Storyteller*'s illustration of the particularity of place has somewhat obscured both recognition of the works' movement away from ego and critical appreciation of Silko's more recent, far-ranging, and (to some) darker novels, *Almanac of the Dead* and *Gardens in the Dunes*. Criticized for their political agendas, these works challenge non-native readers to see the land without the romanticization of Native American environmentalism that Eurocentric nature-writing critiques can tend to embrace in the works of Silko, Lopez, Hogan, and others. More significantly, these novels further develop the egoless expression that Silko's approach produces, frustrating the reader who seeks a recognizable ego in character development or, as in the case of Tayo, character healing. According to Silko, *Almanac of the Dead* expresses hope to its native readers, though many non-natives have seen the novel as a vehicle for despair.[21] The novel certainly challenges non-native readers with its scope and extension beyond the reservation. Egocentric white characters rage under the forces of addiction and capitalist over-consumption while native and mestizo characters turn away from ego and individualism— sometimes toward revolution. While the native characters in *Almanac of the Dead* build on the example of Tayo, they are more like Yellow Woman; they not only do not need western medicine, they show no appreciation for the egoistic concepts of property and individualism that underlie European identity in law, medicine—and fiction. *Almanac of the Dead*'s story of the Southwest becomes a native history of the Southwest, constituted as a map of character, economic, and territorial interactions that prophesy the land's future.

Almanac of the Dead creates its text and textual consciousness of prophecy from a map of the Americas that stretches over time as well as place and has its locus in Tucson, Arizona. Significantly, the almanac and its prophecies, like the oral tradition, operate always from outside the written text. The map that prefaces the novel remains primarily outside as well, pointing beyond its borders to indicate immediately its extension beyond the page and into past time, the five hundred years of European occupation of the Americas and their results for not only the land of the Southwest but for the earth and her people. *Almanac of the Dead* also directly makes the case about the European artistic tradition's rendering of landscape through its disturbed photographer David, who recognizes that "so-called still lifes and landscapes" are only "analogues for the artist's perceptions and emotions."[22] Like David, who demonstrates his own analogue and literal "still life" through brutal photography of a lover's suicide, many of its characters suffer and/or inflict suffering by focusing on their own perceptions and emotions. Silko connects addiction and loss in *Almanac of the Dead,* though in a more extremely egocentric form than in *Ceremony,* to a damaged or absent relation between humans and their surroundings.

With a text that maps relations and connects stories all over the world, Silko extends the experience of disconnection far beyond the reservation. *Almanac of the Dead* presents a complex matrix of characters whose lives and relations to the land have been interwoven by economic and political forces. Chapter headings "transcend the normal geography and history and lead us into the fourth and fifth world," a completely different way of seeing the earth than that of global capitalism, national boundaries, and maps that control economic landscapes.[23] Most significantly, and as documented by the map with which the text begins, all of the characters move about physically, their mobility expressing dislocation from the various and vivid landscapes around them. Patterns of settlement, exploitation, and loss in the Southwest haunt every description of place, ultimately dramatizing each disconnection's fragmentary part of worldwide displacement.

Almanac of the Dead's story of the Americas shows Silko quite aware of the fact that it is in the Americas and in particular North America that nature and landscape writing have emerged. Each of the novel's descriptions of the natural world challenges the romantic vision of the nature writer reconfiguring self and landscape. For example, the Yaqui grandmother Yoeme tells her granddaughters Lecha and Zeta why she chopped down a cottonwood tree: "I must always return to what the white men kept hanging in all the lovely cotton wood trees along the rivers and streams throughout this land. Swaying in the light wind, rags of clothing flapping the shrunken limbs into motion" (*AD* 129). Yoeme and the novel insist the cottonwood be seen as a hanging tree used to kill and display brown people; as the text continues, no image of the land escapes similarly communicating the sensibility of the dispossessed. *Almanac of the Dead* dramatizes immediately that Tayo's ceremony has not been enough, for dislocation and destructive forces act on the reservations, too. When a Hollywood film crew comes to the Laguna Pueblo, the character Sterling's job is to keep them "away from sacred places and from stepping on sacred land" (*AD* 90). But Hollywood cannot comprehend sacred land: "as far as the movie people were concerned, the reservation was rented, too."

Perceived most deeply by the novel's native and dispossessed characters, destructive loss of and from the land extends to white characters as well. The ex–strip dancer Seese, who has spent a good part of her life high on cocaine, stops using the drug after her friend Eric's death. Nevertheless, she cannot remember "seeing the hills and trees or the ocean after Eric's suicide." Literally unable to perceive the land's being or sense its consciousness—"They had done a lot of traveling after that, but she had no memory of it"—Seese manifests a disturbing kind of egolessness as the inability to find the ground of being. Other characters throughout the novel dramatize the origins of the overwhelming loss Seese feels only when her infant son is kidnapped. Multiple, overlapping losses create the text's web of awareness, as a kind of prophecy, of what an earth-centered consciousness might have to communicate.

Almanac of the Dead claims that the history of European conquest has created present-day Tucson (built by whites who made their money off the Apache Wars). Moving outward from Tucson, the novel relentlessly portrays characters—from the Yaqui Calabazas' perverse capitalistic revenge of drug dealing, supported by Zeta; to her sister Lecha's occupation as a psychic who finds the bodies of murder victims; to the cocaine-addicted Seese who started out as an exotic dancer and has now lost her child; to her former boss and partner's lover, Beaufrey, who makes his money selling snuff and abortion films; to Sterling, the quiet Indian who has been kicked off his reservation for failing to prevent Hollywood filmmakers from sniffing cocaine on sacred sites—whose lives all play out their environmental disconnection in settings that emphasize a profound imbalance of humans and their surroundings. Calabazas, a drug and arms dealer, voices the central issue as "the war for the land" (*AD* 178). Beaufrey's household shows how disconnected white people might experience this war for the land—when Seese lived there and had nightmares affecting her vision of humans and the natural world: "The blue sky and puffy clouds seen through the deadly jade water of the nightmare pool was a message about the whole of creation. The loss of the child was another, more final message" (*AD* 43). At the novel's center, vicious attack dogs articulate the war's effect on Lecha and Zeta, Sterling and Seese, by surrounding them on Zeta's isolated Tucson ranch.

Sterling, a native character whose loss, like Tayo's, disrupts his relation to his surroundings, provides the novel's central respondent to the war. When Sterling—clearly part of a ceremony—arrives at Lecha and Zeta's ranch, he sees the landscape through television cartoon lenses, noting "giant cactus you always saw in cartoons with Mexicans in big hats sleeping under them" (*AD* 25). Yet as he recognizes and muses on the value of traditional knowledge, Sterling shows the novel's potential for reaching the land's consciousness: "The old-timers had been dead set against ripping open Mother Earth so near to the holy place of the emergence. But those old ones had been dying off and already were in the minority. So the Tribal Council had gone along with the mine because the government gave them no choice, and the mine gave them jobs. They became the first of the Pueblos to realize wealth from something terrible done to the earth" (*AD* 34).

Silko's other native characters reinforce the focus on the land's consciousness. Calabazas, for example, makes the reader attend carefully to the significance of the novel's unconventional map: "'We don't believe in boundaries. Borders. Nothing like that. We are here thousands of years before the first whites. We are here before maps or quit claims'" (*AD* 216). And the novel's central map and almanac metaphors, with the native characters, challenge the reader to view ecology with something other than the eyes and to remember the history of mapping as a history of visual appropriation. The novel's mapping tells of loss as well as gain—and that the capitalist history of loss and gain disrupts the conscious in-grown relation of humans to their surroundings. As the novel claims mapping as an activity that carries the European quality of observation, its native characters critique that kind of eye as insufficient, marked by "inability to perceive unique details in the landscape," to distinguish one hill or rock or canyon from another (*AD* 225).

From such beginnings, *Almanac of the Dead* goes on to develop a set of intricate relationships among money-making, drugs and arms dealing, and consuming forces that cross the United States, Latin America (particularly Mexico), and Africa. Connections between native peoples and places comes abundantly clear—the places themselves are unwell, perceived as countries and markets rather than locations of relationship between humans and the earth on which they depend. Drug use, addiction, and the crime that feeds them keep the people from the spirit and consciousness of the earth, which might otherwise instigate rebellion. But while loss haunts the novel, so does a particular kind of hope that has to do with people connecting in and about the land. Consciousness erupts periodically through different narrative focalizations, all forming part of a story that shows that no one character can see all that needs to be seen. But native peoples at least realize the need for change in land use and ownership. Sterling and Seese, Lecha and Zeta, remain at the center—narratively because the novel starts with them, spatially because they are in Tucson, spiritually because their losses and humility, their egoless longings to do the right thing somehow, surround the traffic in ego desires and ever-present but never textual almanac of the dead. While Lecha realizes that the almanac contains a living power "that would bring all the tribal people of the Americas together to retake the land," characters at both the center and margins, like lines on the map, repeatedly articulate what the earth's consciousness might mean (*AD* 569).

With disabled persons, Vietnam veterans, Yaqui gun-runners, half-breeds, Mexican Indians, black Indians, and the white woman Seese seeking personal or collective responses to dispossession and oppression around Zeta's ranch, the novel's forces seem to be gathering for a cataclysm—"All across earth there were those listening and waiting, isolated and lonely, despised outcasts of the earth. First the lights would go out—dynamite or earthquake, it didn't matter" (*AD* 513). Stating and restating that there have been crimes against tribal histories, ways of telling the story of the land and its people, ways of writing nature the novel dramatizes a sense of collapse and revolt coming from the earth's knowing. Elements such as Tucson and its ranch of oddly but critically connected dispossessed characters, the almanac of the dead, a lost child symbolizing all losses as losses of essential being, and constant interspersed desires to exploit or retake the land, articulate the central consciousness of what is now a desert while the surrounding characters keep moving, stirring, and thinking. Sterling and Seese passively maintain the emotion of loss, but other characters, including Lecha and Zeta, have plans to act. When a natural-healing convention comes to Tucson, the novel suggests that a climax will come.

Egoless writing, however, removes climax. Silko's insistence that the land's being and consciousness—generating the vitally alive and working magic of the almanac—remain at the true center of the narrative manifests most profoundly through the anti-climactic actions and reactions of Sterling and Seese. Sterling's and Seese's inability to meet the needs of the tribes and families they have belonged to brings consciousness of the dead into living relationships, with a sense of what a global community of the earth might encompass. There is nothing left to do but let go of all ego concerns, share, and survive. Sterling and Seese can't, of course, help each other; yet the novel shows that they somehow do, Seese by continuing to live for her lost child and Sterling by finally recognizing that being and consciousness come from the earth. Caring for more than their ego desires, they act out an essential part of the possibility, articulated by the more activist characters, of there being "plenty of space, plenty of pasture and farmland and water for everyone who promised to respect all beings and do no harm"(*AD* 518).

Through Sterling and Seese's small, significant recognitions of the fact that "no one being stood above another" in their own lives, Silko's rendering of a rebellion against dispossession becomes a non-event for the novel, a reminder that the land's consciousness outlives that of humans and forms the text's containing narrative subject (*AD* 519). The novel communicates consciousness and prophecy of change, the land's knowledge, rather than changes themselves. *Almanac of the Dead* offers none of the dramatic and individualized knowledge of ego-based desire that forms the plot development of much western fiction. As the text finally rejects the naming of any one action as primary, it demonstrates how that would attempt to claim and realize consciousness for individuals, rather than as the land's possibility. To highlight events would be to bring an end to the story and the consciousness it both expresses and makes.

The other reason for the non-event comes from the limitations of the I and eye, even in the most visionary of human thinking. Many of the native characters' visions involve destruction, even on an ecological mission to restore the earth. Although almost optimistic, the action-oriented, revolutionary visions of the black ex-con/homeless veteran Clinton and the native Mexican Indian post-Marxist Angelita La Escapía invoke revolution and violence. Clinton, just wanting "to set his feet on the soil where the spirits of three continents had been manifested," realizes: "Nothing could be black only or brown only or white only anymore. The ancient prophecies had foretold a time when the destruction by man had left the earth desolate, and the human race was itself endangered. This was the last chance the people had against the Destroyers, and they would never prevail if they did not work together as a common force" (*AD* 747). Destruction shadows Angelita La Escapía's revolutionary vision as well. Although Angelita's optimism for tribal recovery is enormous and ecological, it involves revolutionary violence: "Now it was up to the poorest tribal people and survivors of European genocide to show the remaining humans how all could share and live together on earth, ravished as she was . . . All hell was going to break loose" (*AD* 749).

The vision of those who become warriors cannot help but reside in the "I" if they are to understand the Destroyers emblematized by Leah Blue, her Mafia family, and Beaufrey's household. Leah, turned real estate tycoon, dramatizes the egocentric human-centered vision of the land as material for human desire, a vision that comes from divorcing landscape from being. Leah has plans to build a Venice in Arizona, because "what possible good was this desert anyway? Full of poisonous snakes, sharp rocks, and cactus!" (*AD* 750). Leah also aims to make money by intervening to provide what the land cannot and transforming the desert mechanically into a garden. Clearly, her work with the land comes from the same habit of exploitation, denial, and misuse that her family and Beaufrey get rich on.

Between Clinton and Angelita as revolutionaries and Leah and Beaufrey as Destroyers, Sterling, at the center, has a completely different kind of realization, one that involves consciousness and being rather than action and vision, the products of the ego and the eye. As Lecha, Seese, and Sterling zoom out of Tucson, during the last real event of the novel, it becomes clear that Seese will never find her child. Sterling, in a parallel acceptance, goes to that most profound of all places, home. "When Sterling caught a glimpse of the distant blue peaks of Mount Taylor, his throat tightened and tears ran down his cheeks. Woman Veiled in Rain Clouds was what the old people had called the mountain. Sterling was home" (*AD* 757). The novel's concept of egolessness comes to light in Sterling's need for and recognition of the true meaning of his questions, but "What was 'home'?" Sterling spends time "alone with the earth" and wishes that he had listened more closely to his aunt so he might have better understood "the connection between human beings and ants" (*AD* 757, 758). Then Sterling walks to the uranium mines, where a giant sandstone snake has appeared.

Sterling's quiet epiphany, as he remembers and pieces together all the stories he has heard and returns consciousness to the land, is also the novel's. The consciousness begins as Sterling's surroundings remind him of the ancient stories, that a long time ago, "before the Europeans, the ancestors had lived far to the south in a land of more rain, where crops grew easily. But then something terrible had happened, and the people had to leave the abundance and flee far to the north, to harsh desert land" (*AD* 759). The desert then teaches Sterling its story: that blood sacrifices had to stop when the people reached the desert, for "every drop of moisture, every drop of blood, each tear, had been made precious by this arid land" (*AD* 760). When he sits beside the stone snake at the end of the novel, Sterling has to trust the being-ness, the spiritual quality, of the earth instead of his own eyes and mind: "Sterling sat for a long time near the stone snake. The breeze off the junipers cooled his face and neck. *He closed his eyes.* The snake didn't care if people were believers or not; the work of the spirits and prophecies went on regardless. Spirit beings might appear anywhere, even near open-pit mines. The snake didn't care about the uranium tailings; humans had desecrated only themselves with the mine, not the earth. Burned and radioactive, with all humans dead, the earth would still be sacred. Man was too insignificant to desecrate her" (*AD* 762; emphasis added). As in *Ceremony*, the central moment of *Almanac of the Dead* comes with a character feeling the land in solitude with his eyes closed. *Almanac of the Dead* uses its wandering, dislocated characters, to make this point historical and prophetic; the earth's being has vision and consciousness far beyond that of any human. The novel ends with Sterling *seeing* the sandstone snake, looking prophetically south, toward the direction from which the people will come (*AD* 763).

By emphasizing the earth's spirit and being as the location of vision, *Almanac of the Dead,* like all Silko's novels, asks a vital question of all writing, not "nature writing": What do you see/feel/know of the land? Also, *how* do you know? And:

What does it mean that you know? Where and how do you carry this conscious-ness / knowledge? Where does the knowledge go? How does writing embody or express what is known, seen, recognized? *Almanac of the Dead* moves the ques-tions forward from those of *Ceremony* and *Storyteller* to show that Silko's works are concerned with both recovery from past consequences and understanding of future directions; they intend as prophecy to act on the world, but to reform vision. Rather than centralize the consciousness of a mind meeting nature, *Almanac of the Dead* shows that everything depends on how humans are in the landscape and develops understanding of this relationship, not visual perception.

And in terms of visual perception, every character sees part of the whole from which *Almanac of the Dead* builds the land's vision, so that the profound descrip-tive beauty of its passages creates a haunting awareness not of the human capacity to comprehend place but of human smallness in relation to the earth's ecological being. And any character can see, for a while, showing Silko's ability to stun the reader with lyrical descriptions of the natural world:

To the south in Mexico, Angelo could see the pale blue ranges of mountains, like layers of paint growing progressively paler. The distance and the space did not seem to end. Not ever. The colors changed rapidly after the sun set. The sky ran in streams of ruby and bur-gundy, and the puffy clouds clotted the colors darker, into the red of dry roses, into the red of dried blood. The dunes of the horizon were soaked in the colors of the sky too; then the light faded and the breeze slashed at the ricegrass and yuccas. The cooling brought with it deep blues and deep purple bruising the flanks of the low, sandy hills. (*AD* 370–371)

But Silko's novels and poetry, show that poetry—not the individual eye—conveys the consciousness of the land (*AD* 713). *Almanac of the Dead* simply broadens the story and poetic frame outward, as an encounter with the narrative frame of west-ern storytelling at the meta-level of the map. With its map and a story that begins and ends in prophecy, not ego-fulfilling or destroying climax, *Almanac of the Dead*—like *Storyteller* and *Ceremony,* demonstrates narratively and structurally that the lyric vision comes from the land and earth, the mother being that commu-nicates its spirit to the poet, the seer, and the sufferer who seeks to understand.

Gardens in the Dunes

Gardens in the Dunes, Silko's most recent novel, develops a strong theme of con-nection between the earth and the feminine, another persistent and vital thread of her lesson for nature writing. Silko, clearly a Keresan, Native American, *and* woman writer, proclaims a particularized and feminine experience of the land's stories and their import.The arguments of her essays and novels gender the land female and articulate complex forms of its rape and destruction by persons vari-ously described as "destroyers," "witches," and "hybridizers." Gender as well as culture—gender in the context of culture—underlies the nature of the earth's being and its poetic communications. Yet males can be healers, grandfathers are as vital as grandmothers, and females can be destroyers in Silko's fiction. Both genders understand the meaning of feminine symbolization as just one profound element in the spiritual apprehension of the land's fullness. Silko's complex gendering of fiction thus comments on the tradition of American nature writing, one that derives historically from a separated gaze often characterized as male. Her project documents recovery from the westernized and often masculinized world view in which gender, nature, and land are all disassociated states of being.

Gardens in the Dunes develops the theme of gender as a subtext that shows how it remains a dominant thread of European-based egocentric cultures. Woman-centered views enter the novel both through the Mother Earth concept of the Sand Lizard and other Indian peoples and through the meeting of Indigo, a Sand Lizard daughter, and Hattie, a privileged but discontented white woman from Oyster Bay. As in *Almanac of the Dead,* multiple and multiply gendered characters and plots intertwine to weave together the novel's central set of messages about the fate of the earth, particularly in the Americas. Spanning historical time, *Gardens in the Dunes* similarly tells the story of places and the relations human beings need to have to them. But in this novel, egolessness takes a still different form. Silko directly and sometimes even from within the ego shows the fate of characters relying *on* ego, individualism, desires for more, the pursuit of monetary value, scientific classification and reduction—what European-based cultures have for years called civilization and characterized by masculine power and identity. Spiritual themes abound in the text, where there is a meeting and joining of indigenous religious beliefs about ego and those of "civilized" Christians. But as always in Silko's nature writing, the ground of the joining remains the egoless perspective of the natural world.

The novel first genders the land's story by opening with Indigo, Sister Salt, their Mama, and their Grandma Fleet leaving Needles (Arizona territory) in order to avoid being hounded onto a reservation. Grandma Fleet, profoundly connected to the people's history with the land, snatches the girls away from the white men who would send them to school and instead takes them to the "old gardens" in the dunes. The old gardens, with their grandfather snake guarding the spring, and with multiple other threads taking up the space often given by nature writers to the Garden of Eden creation myth, immediately establish the alternate and egoless ground of the novel in tribal experience. Readers used to the European style might find the opening unsettling, not only because of its challenge to the Christian myth, but because there seems at first to be no plot and little characterization. Sister Salt and Indigo do not display the selfconsciousness of individual distinction or desire for achievement typical of characters in the Anglo-American novel. Nor does the novel's narrative consciousness operate to replace their individualism with its own, as in many texts narrated in the first person and in much nature writing. Instead, the novel opens with a joyous female earth energy as the two "lay side by side with their mouths open and swallowed raindrops until the storm passed."[24]

Silko builds on the connection of female energy with an immediate flashback as Sister Salt remembers how they got to the old gardens with Grandma Fleet and their Mama. But the prolepsis does not take typical Anglo-American form; it does not signal Sister's Salt's consciousness as particularly important or relevant to her development as a character/ego. In the interruptions that signal the presence of white people and European contact with Sand Lizard stories, the novel moves back and forth to reclaim interpretation of the garden—as that of a nurturing mother earth with her guardian grandfather snake. The girls learn to plant seeds and tend the plants, caring for them as if they are babies (*GD* 16). Grandma Fleet teaches them and the reader that "the old gardens had always been there." And the gardens show gender as fluid, interconnected, without power struggle: "The old-time people found the gardens already growing, planted by the Sand Lizard, a relative of Grandfather Snake, who invited his niece to settle there and cultivate her seeds" (*GD* 16–17).

Contact with European people and stories comes inevitably, haunting even from the novel's early pages in a thematic reprise of Silko's continuing, containing story of the land. The metaphor of the garden controls the movement of words, establishing the relation of humans and their surroundings through an egoless understanding of gender as natural. The land remains the primary character even with the novel's range of white and Indian characters: Indigo, Sister Salt, and their other Indian friends; Hattie, her husband Edward, parents, Aunt Bronwyn, and friend Laura. *Gardens in the Dunes* directly counteracts the nature of the originating metaphor of most white American nature writing, the Garden of Eden as a bountiful wilderness, by gendering the land female without making it powerless or subject to violation as in Leo Marx's *The Machine in the Garden* or Annette Kolodny's *The Lay of the Land*. As in Kolodny's later *The Land Before Her,* Silko also emphasizes the role of women in making gardens. But for Kolodny and the women writers she discusses, the land exists as a frontier, a wilderness to be tamed and reclaimed from the fantasies of men. *Gardens in the Dunes* makes a more fundamental point that wilderness does not exist as such; the gardens have always been there, nurtured by a spring and a snake who guards them wisely and well. What Europeans perceived as a desert wilderness was a garden—an ecological being—land husbanded and populated by the people and their ancestors.

The omnipresent garden metaphor, underlying and surrounding every character and event in Silko's novel, makes the profound point that land and the food that comes from the land are basic to human life. Land as consciousness manifests as the creative, word-working force that novels as well as gardens bear witness to. Novels, like gardens, exist as fertile meeting grounds of multiple beings, spirits, and minds—of the creative force and the cultivator. Although gardens abound in Silko's text, the old gardens in the dunes remain the home place, the center repository of the novel's communication about the land's consciousness. As titular metaphor, the gardens in the dunes also provide background for the way that the novel explores politics of gardening. And the gardens are the memory and motivation for Indigo as she travels to Europe and visits gardens with Hattie and her botanist husband.

Hattie's meeting with Indigo in a garden provides just one demonstration of how *Gardens in the Dunes* develops concepts of egolessness and ego surrender in characters. The novel consistently locates moments of consciousness and change in gardens that comment on ego. Beautiful, landscaped gardens and the drive to collect rare orchids demonstrate dominating, ego-driven perspectives toward the natural world. Hattie, whose thesis proposal on the Gnostic gospels and feminine principle has been rejected by her graduate committee at Harvard Divinity School, and who never quite fits into the "garden-as-display" culture of some of her Long Island neighbors, finally marries Edward—much to her mother's relief (*GD* 80). Terrified of childbirth and as a result not inclined to sexual congress, Hattie believes that she has found a good companion in Edward, a collecting botanist who has no difficulty with either her physical distance or rejection of standard theology. But both Edward and Hattie, although cultural outsiders, nevertheless remain bound by the cultural framework and habits of individual ego. Edward takes Hattie to his family's neglected gardens and home in Riverside, California, which the reader discovers when Indigo escapes from Indian School and hides in a grove of tall trees that signal the presence of water (*GD* 72). Indigo's hiding place and reason for choosing it expose the very different nature of how Hattie and Edward see it. Hattie sees the garden as an expression of ego—of planning and human control: "She paced off the width of the grassy area and noted the measurements on one of the note cards she carried in her pocket, a habit left over from her days of scholarly research into early church history. Of course, to Edward, the garden *was* a research laboratory . . . She saw evidence of some sort of breakdown in the neglect of the orchid house" (*GD* 75).

Indigo has no difficulty appreciating the garden, but possesses no framework to see it as an extension of either herself or anyone else. She wonders where the functional garden lies, the one with beans and corn (*GD* 86). Other questions from her mind provoke the reader's consciousness: "Where did they get all the water?" Indigo's approach and understanding differ especially from those of Edward, who sees local and regional agriculture as sources of plants with commercial potential (*GD* 88). Edward's vision of plants has no husbandry in it, no sense of the limits of natural resources or presence of ancestors. Driven by the ego-bound concerns of capitalist and individual success, Edward cannot see or hear the land's consciousness. The novel shows that plant knowledge comes from natives, not from science or from the land that Edward objectifies even in his appreciative exploration.

Hattie's interest in the Gospel of Mary and her gradual realization that Indigo's knowledge has a very different source than her own makes her slowly changing perceptions of the nature of gardens and gender central to the novel's "plot." *Gardens in the Dunes* focuses ironically on Hattie's ego, her development as a character, but in order to move her from the place of individual and self-centered concerns (whether to pursue her thesis, what others think of her, how she "should" relate to

Edward, what she "should" do for Indigo) to a realization that the earth and its gardens, rather than scholarship, offer the consciousness she seeks and needs. Hattie insists that Indigo come with her and Edward on their planned trip to European gardens and Corsica, both to prevent Indigo from getting sent to the reservation and in recognition of Indigo's knowledge of and proximity to nature. When Edward calls Indigo an animal, Hattie understands the prejudice. Indigo, who doesn't listen for such comments, is no victim of patronization; she goes along on the trip because Edward's monkey—a person—needs her.

Through Hattie's, Edward's, and especially Indigo's journey, *Gardens in the Dunes*, like *Almanac of the Dead*, develops an extensive politics of the land. The garden metaphor, referring both to the garden that has always been there and the garden constructed from imports, controls and delivers the novel's consciousness of the earth's centrality to meaning. Each time Hattie and Indigo visit a garden, Indigo's presence and observations automatically signal the need to move away from the ego-based and dominating artistic and capitalist perspectives. For example, in Hattie's father's garden, Indigo wonders "Hello, Old Man Yucca, how did you end up here?" (*GD* 180). Edward as scientist, on the other hand, serves to point out the constant themes of use and fashion that dominate the garden culture of Long Island and the market he serves. Periodic returns of the text to the Arizona territory show both the unfair allocation of scarce fertile reservation land (to regular churchgoers) and the difficulty of agriculture in a populated desert (*GD* 206). The novel also demonstrates a continual conflict between native peoples who use the desert wisely and whites who plan and act destructively to dam the Colorado River. In their desire to supply water to the growing Los Angeles, whites leave "the earth blasted open, the soil moist and red as flesh" (*GD* 213).

In Europe, Hattie and Indigo meet others who understand the power of the land and garden in pre-Christian terms. These visits reframe the ego-driven vision the novel attempts to counteract. Aunt Bronwyn knows the worship of the great mother in Europe and that stones are beings and can move. Her gardens recognize the difference of landscapes, separating indigenous English plants from those from the Americas, Africa, and Asia—including the corn that again brings Indigo's recognition to the plot (*GD* 245). Aunt Bronwyn tells the story of Europe as a story of the land's violated spirit and prophesies, as does *Almanac of the Dead*, that the day of reckoning is not far off (*GD* 255). Similarly, Aunt Bronwyn's professor friend Laura, in Italy, insists that monuments of earth-worship stay in the garden rather than be removed to a museum. While Hattie struggles with the egoless and uncommodifying nature of such concerns, Indigo experiences constant recognition, manifested both through her conversations with her parrot and monkey and in her growing collection and understanding of seeds. Meanwhile, Edward disconnects further from the land as natural garden; not only are his plans to illegally import citron slips foiled, but he leaves botany behind for total resource extraction when he meets a doctor who wants him to invest in an Arizona mine. Hattie keeps seeing visions and determines that she must return Indigo to Arizona, but unlike Edward, she returns to Arizona trying to respect the earth's consciousness.

When Edward and Hattie, agreeing to separate, move to Arizona with their different purposes, they arrive in the midst of what *Almanac of the Dead* called a war for the land. Sister Salt's story has proceeded in the context of the building of the dam, the need for soldiers, and the exploitation of labor. Multiple brown-skinned characters move into and out of the text, all part of the struggle to return the land to native peoples. But as with *Almanac of the Dead, Gardens of the Dunes* delivers prophecy rather than climax, mood and intention rather than result. Destruction and rape of the land threaten Sister Salt, Indigo, and all the women—including Hattie, who insists on moving about without escort and gets left for dead by an exploitative attacker. Found and cared for by Indians and Indigo, Hattie recovers—but not to have a romantic reunion with Indigo. Instead, Hattie goes back east with her father and ends up in England with Aunt Bronwyn, where she may learn her own mythology and landscape. Indigo and Sister Salt, who has borne an old black grandfather baby, move back home to find the grandfather snake slaughtered and Grandma Fleet's apricot trees destroyed. But the land prevails; shoots spring up from the stumps and "Old Snake's beautiful daughter" moves back home to guard the spring (*GD* 479). *Gardens in the Dunes* shows the encounter with destruction and misunderstanding of the land generating the possibility of rebirth and ego-surrender for those who remember or learn the old stories. *Gardens in the Dunes* not only reclaims the metaphor of the snake and garden from a native perspective but shows the original garden of nature writing, too often viewed as an Adamic Eden, with an alternative, feminine, prophetic, and pre-existent consciousness.

Lessons for Nature Writing

The passionate, deliberate, self-authorized spiritual reconnection that gave birth to nature writing as a genre clearly responds to a need for greater wisdom about the land. But the framework of vision remains that of the separated consciousness, even in the critical assessment of "nature writers." As Trimble details in his introduction to the collection *Words from the Land:*

These writers make journeys into the landscape; they enter "the naturalist's trance." They weigh their journals against their research, spend long hours in libraries, and talk to the experts. Then they place themselves in a second trance and translate all this into lovely prose. Within and between the lines lies a wealth of natural history, basics and esoterica, that the reader absorbs painlessly—instructed, invisibly—carried on the buoyant leading edge of the writer's curiosity, skill, and enthusiasm.[25]

Not all of the writers Trimble anthologizes consider themselves naturalists. Nevertheless, the naturalist model and idea of "the writer's curiosity, skill, and enthusiasm"—the writer's consciousness—focuses most assessment of their work. Even when they include Native Americans and women, such compilations as *This Incomperable Lande, Words from the Land,* and *Made from This Earth,* which all highlight the centrality of the land to nature writing, fail to include the scope and import of Silko's fictional communication. Norwood's *Made from This Earth,* which highlights Silko's articulation of the relation of woman to nature, does challenge the "continued romanticization of American Indian environmentalism."[26] But, in addition to the gendered and native perspective, Silko's work offers a fictional vehicle that comments on the nature of storytelling as critical to its means of writing nature and communicating ecological being. Silko's fiction offers a slightly but critically different perspective from that of other nature writers, that of the land as surrounding consciousness. The distinction of fiction in Silko's case gets magnified by her Pueblo heritage so that the work of a novel like *Ceremony* or *Gardens of the Dunes* cannot be simply read as the transmutation of place into language. What Silko does, instead, is work to produce a containing consciousness and natural history, a surrounding vision rather than observing eye. Rather than the naturalist model that Trimble sets authorial comments and practices against, Silko is an author of already grounded fiction.

Tiffany Ana López with Phillip Serrato

A New *Mestiza* Primer
Borderlands Philosophy in the Children's Books of Gloria Anzaldúa

Latina/o voices are strikingly absent from most anthologies on nature writing, a conspicuous void given the abundance of narratives that historically focus on the centrality of geography—from fences to borders to nation states—and of a physical connection to the land, to deserts, fields, and islands. A litany of texts immediately comes to mind: Rudolfo Anaya's *Bless Me, Ultima;* Tomás Rivera's *And the Earth Did Not Devour Him;* Christina Garcia's *Dreaming in Cuban;* Julia Alvarez's *In The Time of Butterflies;* Pat Mora's *The Desert Is My Mother.* In each of these works, images of nature and issues of environmental justice are inseparable from thinking about ethnic identity and national culture. Yet, only recently has Chicana/o literary criticism explicitly discussed nature writing. In his groundbreaking volume, *Chicano Culture, Ecology, Politics: Subversive Kin,* editor Devon G. Peña explains the significance of a culturally specific critique inclusive of environmental issues: "By rooting identity in a 'sense of place' as much as in gender, class, ethnicity, and other constructions of difference, identity politics and the environmental movement are deepened and made more inclusive. They therefore become more subversive."[1] Comparative readings that locate Latina/o texts in conjunction with works more clearly labeled as nature writing keenly underscore the critical relevance of the genre to conversations about American literary and cultural studies.[2]

In *Borderlands/La Frontera,* Gloria Anzaldúa asserts that speaking about a geographic place demands thinking about emotional, political, and psychic spaces as well: "The actual physical borderland that I'm dealing with in this book is the Texas–U.S., Southwest/Mexican border. The psychological borderlands, the sexual borderlands, and spiritual borderlands are not particular to the Southwest. In fact, the Borderlands are physically present wherever two or more cultures edge each other, where people of different races occupy the same territory, where

204

under, lower, middle and upper classes touch, where the space between two individuals shrinks with intimacy."[3]

While Anzaldúa speaks directly about the United States–Mexico border, she also demarcates a common ground for her readers based on shared feelings of alienation, marginality, and otherness. Elsewhere, Anzaldúa has described the borderlands in even more utopic terms as "a place or state of awareness where we could all listen and talk to each other, where divisions may be breached, perhaps even healed."[4] Speaking of its possibilities to inspire a fully multicultural political movement, Anzaldúa's co-editor of *This Bridge Called My Back,* Cherríe Moraga, writes, "Nature will be our teacher, for she alone knows no prejudice."[5] Significantly, the space where writers most represent land as a common ground for cultural healing is the province of children's literature. It is precisely the intersection between children's writing and environmental criticism that informs this essay's reading of Gloria Anzaldúa as a nature writer.

Everything I Ever Needed to Know I Learned in the Borderlands

Both of Anzaldúa's borderlands children's books, *Friends from the Other Side* and *Prietita and the Ghost Woman,* focus on the coterminous concerns of social justice and personal, political, and environmental healing embodied by the exemplary figure of "the new *mestiza*" introduced in her landmark collection, *Borderlands/ La Frontera:*

> The new mestiza copes by developing a tolerance for contradictions, a tolerance for ambiguity. She learns to be an Indian in Mexican culture, to be Mexican from an Anglo point of view. She learns to juggle cultures. She has a plural personality, she operates in a pluralistic mode—nothing is thrust out, the good, the bad and the ugly, nothing rejected, nothing abandoned. Not only does she sustain contradictions, she turns the ambivalence into something else.[6]

Following this prescription for Chicana empowerment, Anzaldúa's young protagonist learns to negotiate a variety of borders and navigate towards an ethos of justice hinging specifically on the healing of people and their lands. In *Friends from the Other Side,* Prietita turns to the *curandera* to help her young friend Joaquín escape the border patrol and heal the wounds on his arms. The eponymous heroine of the sequel, *Prietita and the Ghost Woman,* continues her apprenticeship with *la curandera,* who sends her into the fenced-off borderlands of the King's Ranch to find the only healing herbs that will cure her mother's illness. As in so many fairy tales and mythological stories, Prietita's transformation must take place within the realm of nature. Prietita's maturation into a fledgling version of the new *mestiza* requires that she cross a variety of borders in order to work through her feelings of ambivalence and make difficult but decided choices. Will she confront the bullies that torment her friend Joaquín and successfully find him and his mother sanctuary from the Border Patrol? Will she risk her own physical safety to find the herbs necessary for healing her mother's wounds?

Beginning with the preface and opening illustrations, Anzaldua's books invite readers to consider more complicated, sensitive conversations about the border, immigration, and immigrants. As she does in *Borderlands/La Frontera,* Anzaldúa draws powerfully on the autobiographical to explain the personal, cultural, and political significance of this landscape. *Friends from the Other Side* begins: "I grew up in South Texas, close to the Rio Grande River, which is the Mexican-U.S. border. When I was a young girl I saw many women and children who had crossed to this side to get work because there was none in Mexico. Many of them got wet while crossing the river, so some people on this side who didn't like them called them 'wetbacks' or 'mojados.'" Accompanying illustrations by Consuelo Méndez present images of the land, flora, and fauna particular to the Southwest, the desert landscape populated by a mesquite tree, a lizard, chaparral, and cacti. Notably, this introductory image does not include people, a curious omission given Anzaldua's title emphasizing friendship. Rather, the artist represents human inhabitants of the landscape by three short posts of what used to be a barbed-wire fence, a key detail that clashes with the otherwise unadulterated picture of flora and fauna. This purposefully charged picture of barbed-wire fenceposts calls attention to their role as markers of "two very dissimilar countries" separated from each other by a river, an element of nature made to function as a political border.[7] In this way, Méndez visually introduces the icons of the border as clearly disruptive of the natural landscape. The role of people in this borderlands region is left undefined at this critical moment in the book, leaving open for question whether fences will be rebuilt or removed.

Unlike in her essays for adult readers, Anzaldúa's employment of the autobiographical in her children's books never comes across as self-righteous or politically strident. Her reflections about growing up serve a calculated pedagogical function, addressing those concerns articulated by educator Oralia Garza de Cortes in "Behind the Golden Door: The Latino Immigrant Child in Literature and Film for Children": "[W]hat are young people to think about the issues that confront them in daily newspaper headlines and on prime-time news? How can young people begin the educational process that will enable them to become more empathetic, rational, and reasonable about such a complex issue as immigration? How can they begin to develop an understanding of the struggles and human suffering of immigrants and of undocumented persons who have no legal stature in this country but who wish to participate more fully?"[8] At the outset, by explaining and problematizing the rise of the term "wetback" Anzaldúa exposes it as both an ignorant and malicious construction. In the process, she debunks the notion that undocumented Mexicans come to the United States in order to take undeserved advantage of "all the benefits of the welfare state."[9] She succinctly counters anti-immigration hostilities and exaggerations by drawing sympathetic attention to the complex motivations Mexicans have for migrating.

In the process of making friends with Joaquín, Prietita learns how he and his mother have found themselves displaced by an unstable economy in Mexico and then placeless in the United States, where they must watch out for the Border Patrol and deal with anti-immigrant hostility. Their "tumbledown shack with one wall missing" is merely a provisional residence on the margins of society, its location and condition indicative of their social, political, and economic disenfranchisement, a point made clear when the Border Patrol van enters the area and forces Joaquín and his mother to cower in hiding behind a bed. This is the only time in the entire book that readers see Prietita and Joaquín indoors. Previous illustrations show them outdoors playing *lotería* (a kind of bingo game whose cards feature images of plants and animals), climbing trees, visiting neighborhood yards, and gathering herbs for *la curandera*. The scene of their flight from *la migra* is particularly powerful because Joaquín and his mother are so clearly morally upstanding people. They embody none of the "negative values" of criminality and moral depravation that political analysts such as Leo Chavez find overwhelmingly associated in the popular American imagination with undocumented Mexican immigrants.[10] Rather, Anzaldúa directly comments on the bitter irony of Joaquín and his mother being labeled as "illegal aliens" on "land [that] was Mexican once."[11] In response to the *migra's* question, "Does anyone know of any illegals living in this area?" a neighbor woman, "pointing to the *gringo* side of town," replies, "yes, I saw some over there."

To further counter definitions of Mexican immigrants as "aliens," Anzaldúa encourages empathy for Joaquín and his mother. When she first meets Joaquín's mother, Prietita is told, "We had to cross the river because the situation on the other side was very bad. I couldn't find work and Joaquín was in rags." This explanation emphasizes misfortune and avoids maliciousness, in the process humanizing those who have been predominantly labeled as somehow outside the range of common human experience. The representation of their poverty deliberately contains an element of pathos, communicated in Anzaldúa's description of the shack's interior: "In place of the wall was a water-streaked tarp." She then adds details about how Prietita sits "on a straw mat that covered a dirt floor." Méndez's accompanying illustrations portrays the shack as a dilapidated structure with walls made of old wood and absent of any comforts of the typical storybook home. The interior is horribly bare except for a few religious articles meant to support the idea that Joaquín and his mother are *buena gente* or *gente decente* (good people). These eloquent pictures show the poverty of the immigrant experience, yet also the dignity and humanity of the poor working to better their conditions, as beautifully encapsulated by Prietita's thoughts when offered something to eat: "She saw pride in their faces and knew that they would offer a guest the last of their food and go hungry rather than appear bad-mannered." In her interview with Donna Perry, Anzaldúa states that "If we can make the empathetic leap to the other side and listen, we could come to a deeper understanding of each other's positions."[12] Anzaldúa's children's books thus valorize empathy for its capacity to enable cultural healing and facilitate community building.

While illustrations play an important role in any children's story, they play an even more pivotal role in Anzaldúa's children's books because they provide a counter-narrative to the dominant culture's most prevalent images of Mexican immigrants. Throughout *Friends from the Other Side,* although she draws attention to the inherent beauty of the border region, Méndez also uses visual details to emphasize Anzaldúa's larger political project. Subtle background images of a shack, clotheslines, and a broken chair poignantly suggest the poverty that Mexicans living in that borderlands space experience.[13] The sense of hardship and dispossession that charges Méndez's vision of the landscape actively disrupts the idealization of people of the Southwest as somehow closer to nature or God, the kind of musings that have, more often than not, characterized nature writing about indigenous peoples. However, on a certain level, Anzaldúa does participate in such idealized portraits, albeit in a more complicated way, in her consistent representation of Prietita living outside in communion with the natural environment. In one pivotal scene, Méndez's illustrations show Chicano boys who use the term "wetback" to taunt Joaquín. While Chicanos and Mexicanos historically have much in common in regard to their relationship to the land, Chicanos are nevertheless part of the people "on this side" who demonstrate their ignorance of the many intersecting factors informing Joaquín's life. The Chicano boys have somehow escaped this part of their cultural history—a narrative turn that makes Anzaldúa's translation of her borderlands philosophy into children's literature all the more poignant, most especially in light of the genre's performative component: that of an adult reading to a child, literally passing on history and culture to the next generation. In Spanish, *historia* is the same word used for what English separately calls history and story.

The concerted ways in which Anzaldúa and her illustrator work together show how reaching a common ground depends largely on shared political goals more so than any presumably shared ethnicity. As Anzaldúa points out in *Borderlands,* within the Chicano/Mexicano body women are oppressed and abused; homosexuals are subject to "being abandoned by the mother, the culture, *la Raza,* for being unacceptable, faulty, damaged"[14]; darker bodies are disparaged; Mexicanos call English-speaking Chicanos *"pochos"* and cultural traitors; and Chicanos call Mexicanos "wetbacks" who come from "North America's rubbish dump."[15] In *Friends from the Other Side,* the sores on Joaquín's body continue Anzaldúa's figuration of the U.S.-Mexico border as an unhealed *herida abierta,* a "1,950 mile-long open wound/dividing a *pueblo,* a culture."[16] Physical wounds signify the larger ruptures that exist within the Chicano/Mexicano social body, the intracultural divisions between Chicano/Mexicano, legal citizen/illegal alien, boy/girl. It is the gracefully simple yet philosophically complicated gesture of empathetic outreach in Anzaldúa's work that heals the most painfully cleaving differences.

Fig. 1. Illustration reprinted with permission of the publisher, Children's Book Press, San Francisco, CA. Illustration for *Prietita and the Ghost Woman* copyright © 1996 by Christina Gonzalez.

When Joaquín and Prietita meet for the first time, the Chicano/Mexicano split is captured by the gate that separates them (see figure 1). Like the construction of the Rio Grande as separator between the United States and Mexico, the gate exemplifies natural resources transformed into unnatural borders; in this case, trees have been made into gates and fences to keep people divided from each other, as well as in and out of certain spaces. When Joaquín visits Prietita's yard to sell firewood, the gate represents class divisions as well. Importantly, however, built into the gate (as with the Rio Grande as a border) is the wonderful possibility of its own transgression. This occurs a few scenes later when Prietita's cousin Teté and his chums taunt Joaquín. Teté yells, "Look at the *mojadito,* look at the wetback!" Another boy adds, "Hey, man, why don't you go back where you belong? We don't want any more *mojados* here." Prietita responds in anger, to the point of physical discomfort: "Prietita felt her body go stiff. She had known Teté and his friends all her life. Sometimes she even liked Teté, but now she was angry at him. She felt pulled between her new friend and her old friends." When one of the boys bends down to pick up a rock—another instance of nature being put into the service of social divisions—Prietita steps in front of Joaquín to protect him from injury. Before she can do so, Prietita must first pass through the gate and step beyond the comforts of her yard. When she does this, she makes an empathetic leap, one that has drawn applause from one teacher of Latina/o children's literature for the story's message about cultural wounding: "And by Prietita's example, the reader learns how to take a stand against prejudice and cruelty."[17]

When Prietita stands up to her cousin and his friends, she engages in a courageous intracultural confrontation of *familia* and *familiares.* In *Borderlands,* Anzaldúa specifies that the new *mestiza* must be able to respond critically to the world around her, and this demands a willingness to challenge that which is most familiar, in all senses of the word: "She is willing to share, to make herself vulnerable to foreign ways of seeing and thinking. She surrenders all notions of safety, of the familiar."[18] Armed with the ability to critically understand the situation, Prietita realizes that justice demands that she stand up for Joaquín. Hence, she breaks with *lo familiar,* including *familia* (her cousin Teté) and the safety contained therein, represented by her passing through the gate that separates her family's yard from the larger public space where she and her friend Joaquín are most vulnerable. Interestingly, in the moments before her intervention in the confrontation between Joaquín and the boys, Prietita contemplates seeking the *curandera's* help in healing Joaquín's arms. Rather than celebrate *la curandera's* healing abilities or valorize a mythic return to nature via traditional medicinal ways, Anzaldúa turns the focus to Prietita's development as an agent of social change.

Before she can become a full apprentice to *la curandera,* Prietita must demonstrate her ability to confront patriarchy and engage in social conflict, a test metaphorically represented by the gate through which she must pass. In response to a commotion that breaks out, Prietita makes her way through the gate to the scene of conflict and dismantles the boys' motivations for antagonizing Joaquín. In the same ways that Anzaldúa upbraids the othering of Mexican immigrants as "wet-backs," Prietita censures machismo in a manner that deconstructs it. She tells the boys, "What's the matter with you guys? How brave you are, a bunch of *machos* against one small boy. You should be ashamed of yourselves!" Prietita exposes their macho impulse to be *número uno* as a sign of cowardice and a behavior that needs to be unlearned. Admittedly, Anzaldúa's portrait of Joaquín's affliction risks coming across as excessive, with his being made to carry the burden of too much otherness. In addition to his status as a poor, undocumented Mexican immigrant, he also has a grotesque condition that the other boys find both unattractive and unmanly. Anzaldúa represents Joaquín as morally superior, abject, and wounded—all of which put him in a feminized position and paradoxically work against a larger project of blurring boundaries and embracing contradictions. However, one can also read Joaquín as representing the new Chicano male in the same way that Prietita represents the new *mestiza*. Here, one may recall that Rodolfo "Corky" Gonzales in his nationalist epic poem, *Yo Soy Joaquín (I Am Joaquín)* portrayed the young male as the most ardent political voice of the Chicano social body.[19] Anzaldúa directly draws upon this historical icon, yet redefines the social body as positively engendered female; Anzaldúa's feminized Joaquín is as much an icon of political agency as Prietita even though she certainly has greater access to power. Cherríe Moraga also gives the name Corky to her lead character in *Giving Up the Ghost,* a theatrical feminist engagement with Gonzalez's poetic anthem for the Chicano movement that portrays Chicanas as its most critically savvy leaders.[20]

Anzaldúa closes *Friends from the Other Side* with *la curandera* formally confirming her apprentice: "Prietita, I am going to show you how to prepare these herbs in a paste you can use to heal Joaquín's arms. It is time for you to learn. You are ready now." For this scene, Méndez presents an image of Prietita, Joaquín, and the *curandera* together in nature gathering herbs for the *remedio*. This image suggests a coalition-building based on shared respect for nature. A few pieces of fencing in this final image conspicuously stand amidst the otherwise open landscape. The borderlessness that characterizes the scene reflects the new vision of community building. Yet the presence of fence pieces also reminds readers of the borders that continue to exist or can be made, prompting us to keep firmly in mind the impulse to foreclose private spaces—geographical as well as emotional, political, and psychic.

La Historia and Its Influences on Growing Up New Mestiza

In the sequel to *Friends from the Other Side,* Anzaldúa continues the story of Prietita's apprenticeship under the direction of *la curandera,* here also identified as Doña Lola. As with the first book, Prietita must endure a test and prove herself as a skilled border crosser and healer with the critical savvy necessary to navigate even the most dangerous terrain. Additionally, like *Friends from the Other Side,* the illustrations in *Prietita and the Ghost Woman* show Prietita outdoors in connection with nature, save for a single scene when she enters *curandera* Doña Lola's house to obtain a sketch of the rue plant, a type of evergreen shrub with bitter leaves needed to complete a *remedio* for healing Prietita's mother, who has fallen victim to "the old sickness." After giving Prietita the drawing, *la curandera* explains the risks involved: The only rue plants in the area grow behind the barbed-wire fence of the King Ranch, where they "they shoot trespassers."

Anzaldúa's references to the King Ranch provide an important avenue by which to further explore her translations of borderlands philosophy from *Borderlands/La Frontera* into her children's books. In the earlier work intended for adult readers, Anzaldúa speaks of the lands taken from her own family in the wake of the Treaty of Guadalupe Hidalgo. Because her ancestors did not speak English, they were unable to enter the contract negotiations required by the treaty and, as a result, their lands were taken away from them under an act fully protected by new land laws published only in the colonizer's language. The sole piece of property left to her family was a small parcel of land, the family cemetery located adjacent to the King Ranch, eventually viewed as a monument to injustice: "But there was a fence around the cemetery, chained and padlocked by the ranch owners of the surrounding land. We couldn't even get in to visit the graves, much less bury her *[abuela]* there. Today, it is still padlocked. The sign reads: 'Keep out. Trespassers will be shot.'"[21]

Here we might consider Malia Davis's argument in her essay "Philosophy Meets Practice: A Critique of Ecofeminism through the Voices of Three Chicana Activists." She writes, "the domination of women, economic exploitation, racism against indigenous peoples, and the domination of the earth are all rooted in Western patriarchal ideology, theory, and practice."[22] Anzaldúa weaves together these various concerns of environmental injustice using the trope of *la historia,* history and myth, specifically the history of the King Ranch and the mythic story of La Llorona. When Prietita enters the King Ranch to find the rue, she loses sight of the fence and becomes lost. Not only must she locate the plant, she must also safely navigate her way home. None of the animals she encounters help her, and rocks and stars refuse to function as legible maps. As it grows dark, Prietita comes across a woman dressed in white floating alongside the lagoon who casts a glowing light that illuminates a path to the rue plants and back to the main road where Doña Lola and the search party greet her.

Like other fairytale plots featuring characters who get lost in the woods, Prietita cries when she finds herself on completely foreign ground without any sight of familiar roads and landmarks. Yet, her desire to find the ingredient necessary to heal her mother eventually calms her fears and refocuses her attention to her task: "She began to cry, but then she said to herself, 'Don't give up, Prietita. You have to find the rue for Mami.'" It is shortly after this affirmation that La Llorona comes to her rescue. Notably, Anzaldúa's portraits of Prietita's mentoring relationship with Doña Lola emphasize her encouragement of Prietita's critical awareness and self-reliance, both demonstrated by her declaration of responsibility to her family (and, in the first book, community). It is at such moments in Anzaldúa's story that illustrations play perhaps the greatest role. Christina Gonzalez's drawings for *Prietita and the Ghost Woman* employ a distinct set of shared features to emphasize the personal and political ties between the trio of female healers—La Llorona, *la curandera,* and Prietita—that Anzaldúa portrays as new *mestiza* role models (see figure 2). In this illustration, Prietita's eyes are cast down, reflecting her introspection and isolation as she enters the King Ranch. Each is brown skinned with long black hair, a broad nose, and full lips. All have the same large, almond-shaped eyes of the darkest brown intensity, the artist accentuating the eyes as each character's most predominant feature, perhaps reflecting their desire to view the world with eyes wide open. Gonzalez's dynamic illustrative style includes bright yellow squiggles at the feet of each character and near the borders of several pictures, capturing the movement and progression of Prietita's journey. Gonzalez's exciting use of such bold lines and bright colors also accentuates physical and spiritual strength, the curlicues and wavy designs surrounding each character like an aura.

Anzaldúa's version of the La Llorona myth revises the story's most antifeminist elements, notably the representation of women working against one another to the benefit of patriarchal power. A typical version tells the story of a woman who discovers her husband having an affair with another woman (often of the ruling class) and responds by drowning her children, henceforth wailing along the shore mourning her irrecoverable act.[23] Anzaldúa's Llorona wanders the woods to protect Prietita within the dangerous confines of the King Ranch, literally shedding light on the problem by guiding Prietita to the rue and back through the woods. After she emerges from the woods and joins the search party, Prietita shares her story of an encounter with the mythic figure. Her cousin Teté exclaims "But everyone knows she takes children away. She doesn't bring them back." Much like his role in *Friends,* Teté functions here as a kind of intracultural heckler. In Anzaldúa's first children's book, he calls Joaquín a "wetback." In the sequel, he reads La Llorona as a threatening figure. In both works, Teté uncritically accepts cultural myths that portray border-crossing figures as disruptive in only negative ways. Doña Lola counters Teté's remark: "Perhaps she is not what others think she is." Her comment has great resonance, for like La Llorona, the *curandera* is typically feared as a kind of witch.[24] Then, in Latino culture, girls are not allowed the same status or freedoms as boys. Prietita bravely enters the woods despite threats of physical harm and legal punishment, acts that defy historical representations of proper behavior for young Chicanas. Indeed, Prietita demonstrates a courage and wisdom beyond her years. As in *Friends,* her actions are motivated always by a desire to help others. While this helping impulse is certainly gendered behavior, Anzaldúa presents it not as the norm, but as an exemplary act of community building, a vivid strengthening of socially, economically, and

Fig. 2. Illustration reprinted with permission of the publisher, Children's Book Press, San Francisco, CA. Illustration for *Friends from the Other Side* copyright © 1993 by Consuelo Méndez.

physically weaker members. Hence, Doña Lola's final words to Prietita: "I am proud of you. You have grown up this night."

Several Chicano cultural practices reinforce the privileging of patriarchal power, but none are as powerful and lasting as the explicit hierarchy between daughter and son roles in relationship to the mother. On her own awareness of this practice and the pain it elicits, Anzaldúa writes: "Though she loved me she would only show it covertly—in the tone of her voice, in a look. Not so with my brothers—there it was visible for all the world to see . . . Her allegiance was and is to her male children, not to the female."[25] Other Latinas have written similar testimony. For example, more closely analyzing her relationship with her mother as a means by which to better understand the cultural reinforcement of patriarchal privilege, Cherríe Moraga writes: "Ask, for example, any Chicana mother about her children and she is quick to tell you she loves them all the same, but she doesn't. *The boys are different.* Sometimes I sense that she feels this way because she wants to believe that through her mothering, she can develop the kind of man she would have liked to have married, or even have been. That through her son she can get a small taste of male privilege, since without race or class privilege that's all there is to be had."[26]

In both of her children's books, Anzaldúa emphasizes the importance of breaking with the familiar and developing an ability to stand up to acts of injustice, no matter how normalized. In *Friends from the Other Side,* Prietita must stand up to her cousin Teté and confront his racist bullying ways. In *Prietita and the Ghost Woman,* she must transcend the ways the world has been ordered and demarcated for her. To begin, she has to physically leave the familiarity of her neighborhood and more local plants to find the rue needed to complete the healing recipe. Then, she has to emotionally reject her culture's gender norm practices, which traditionally do not sanction a young girl's independent excursion into the woods. (As a kind of obligatory aside, Doña Lola reminds Prietita, "It is not safe for a little girl.") Finally, Prietita has to reject the mythology of La Llorona in order to follow her guidance without reservations or fear. This rejection of the familiar allows her to find the rue plant and, in the end, share her revisionist account of the La Llorona story—such transformative powers of passing on *la historia* are represented by the book's final illustration of Doña Lola and Prietita embracing Prietita's little sister Miranda.

Opening Pages at the End of a Book

Curiously, there are no direct equivalents in children's literature to the type of nature writing marketed to adults. Publications for young readers that do treat issues of the environment tend to feature the subjects of garbage, pollution, and recycling, with the Amazon rain forest as a paradigmatic case study.[27] While a book like Susan E. Goodman's *Ultimate Field Trip: Adventures in the Amazon Rain Forest* spotlights many of the same issues found in Anzaldúa's children's books—culture clash, economic exploitation, environmental injustice, human-animal relations, herbal remedies—placing those concerns within the context of a field trip to a foreign country allows readers to imagine a border between themselves and the issues presented in the text, which become conveniently compartmentalized as part of a school project and a field trip from which one can always return.

In both *Friends from the Other Side* and *Prietita and the Ghost Woman,* Anzaldúa artfully unfurls meaningful stories, raises relevant political issues, portrays children as agents of social change, and—most importantly—constructs open-ended conclusions that push readers toward thinking about the future deconstruction of borders on multiple fronts: cultural, linguistic, geographic, economic. In "Ecopolitics and the Literature of the Borderlands," one of the few critical essays on environmental literature to discuss the pivotal role of Latina writers, M. Jimmie Killingsworth and Jacqueline S. Porter conclude, "What the borderland writers dramatize for us is that the story of justice can no longer be told without reference to the earth's brown body. The land and the people have suffered together, and together they must be healed."[28] In memorable fashion, Anzaldúa ends her children's books with the theme of unfinished work, which can be read as a result of her activist commitment. For, as several critics of children's literature have noted, happy-ending closure does not encourage a sense of the need for change or a sense of the child's possible agency in effecting change. Anzaldúa's trope of healing in both *Friends* and *Prietita* emphasizes each individual's responsibility for the healing of social divisions and preserving the environment, all a part of what it means to growing up—and not only as a new *mestiza*.

Although children's literature has increasingly gained attention over the last decade as an area of serious scholarly inquiry, work by Latina/o children's authors has been largely ignored within both American and Latina/o literary studies. Currently, none of the existing critical works by such leading voices as Ramón Saldívar, José David Saldívar, Alvina Quintana, or Tey Diana Rebolledo includes a discussion of children's literature. The handful of texts available are mostly limited to book recommendations and project ideas for teachers of grades K–12. All of this seems a glaring omission, given that the major authors of Latina/o adult fiction—Gloria Anzaldúa, Rudolfo Anaya, Sandra Cisneros, Luis Rodriguez, Francisco Alarcon—have recently begun publishing children's stories and to much critical acclaim. In the context of today's political climate, most notably characterized by legislative action (particularly in the Southwest) against affirmative action, bilingual education, and youth of color generally, we can read Latina/o children's writers as strategically cultivating a future readership and increased literacy in Latina/o communities—groups that studies show are the most at risk for failure in the school system at all levels. In these politically and economically trying times, children's literature takes on a spiritual and philosophical dimension because of the optimism that necessarily characterizes the genre.

Jenny Emery Davidson

Stalking a Prayer

Crossings of the Hunter and the Shaman in
Pilgrim at Tinker Creek

At the beginning of Annie Dillard's *Pilgrim at Tinker Creek,* a tomcat jumps through the sleeping narrator's window and lands on her chest. The cat is bloody and dirty, and as he paws her—"as if sharpening his claws, or pummeling a mother for milk"—he leaves red footprints all over her body. When the narrator looked at her painted self in the mirror the next morning, she asked, "What blood was this, and what roses? It could have been the rose of union, the blood of murder, or the rose of beauty bare and the blood of some unspeakable sacrifice or birth."[1] The question haunts the rest of the narrative, as well as many of Dillard's later works, because she, like the tomcat, paws the chest of the world. She, too, is a hunter. As her narrative persona stalks the banks of Tinker Creek and the pages of scientific volumes, striving to get closer to the world around her, she collects blood roses: paradoxical images of violence and beauty. She watches a frog disintegrate into a bag of skin as a giant water bug sucks out its guts.[2] She witnesses a mosquito drawing blood through the scaly skin of a copperhead snake.[3] She touches the surface of the world, and when she pulls back her hand, she discovers the mark of blood. "I am the arrow shaft," she writes, "carved along my length by unexpected lights and gashes from the very sky, and this book is the straying trail of blood."[4] The bloody traces that we then read repeatedly beg the question, what blood is this? Is it the blood of violent confrontation and consumption? Or of the more reciprocal act of consummation—even of communion? Are the boundaries between entities transcended or transgressed as separate bodies—the cat and the narrator, the writer and the world, the reader and the text—touch, and leave a track of blood in the wake of their encounter?

The ambiguity of the blood roses takes us to the heart of Dillard's Pulitzer-prize winning memoir, and leads us to central issues in the field of environmental literature in general. Nature writing is engaged in a paradoxical project: It strives to encourage an environmental ethic, while transforming the environment into a product—a text—to be consumed. Nature writing repeatedly risks a kind of scientific objectification, rather than intimacy, as it prods the reader to look more closely, more attentively, at nature. And, in the other extreme, nature writing also risks romanticizing and simplifying nature's complexity by juxtaposing it too starkly with the modern world of human technology. Too often nature writers pick and choose from nature to create a human fantasy that hinders an honestly reciprocal relationship with the environment. So we must ask ourselves, as we read and write about nature, are we achieving some sort of communion with nature, or are we engaging in a violent confrontation, bending nature to our will and to our aesthetic categories? How do we reconcile our human desires with the reality of the natural world? And how do we reconcile—through language—our human bodies with the living flesh of the world around us? The space between our individual selves and the selves of Others—both human and non-human—holds the potential for intimate contact or violent collision, and as we struggle to mediate that fathomless space with language, our words shape the kind of contact that will occur.

A common trope that has been used in literature to mediate this space between humans and nature is the hunting myth. From frontier writers such as James Fenimore Cooper, who preceded Dillard by more that one hundred years, through Jack London, Ernest Hemingway, and William Faulkner, to more contemporary nature writers such as Richard Nelson, a long tradition of primarily male writers have used the story of the hunt—a symbolic ritual of violence—to represent the relationship between humans and nature. Women nature writers, on the other hand, have more often written from the perspective of the garden or the house, often engaging more domestic tropes for describing an intimate relationship with the land. But Dillard, with the publication of *Pilgrim at Tinker Creek* in 1974, ventures into the terrain of the hunt, employing its rhetoric while also challenging its conventions. According to the conventions of the myth, the hunt is a means for merging with the non-human world. By learning the ways of the wilderness, the hunter gains an intimacy with it. The successful hunt signifies a marriage contract, a joining with the wilderness, as the flesh of the wilderness is consumed by the hunter and transformed into his own flesh and spirit. But the dynamics of this mythic marriage to the wilderness are problematic, as *consumption* becomes the ultimate means of *consummation* between humans and the wilderness: The wilderness is devoured by the dominant hunter. The relationship between humans and the wilderness thus parallels other common binary oppositions, such as: society/nature, active/passive, masculine/feminine, predator/prey, violator/victim. These oppositions clearly do not represent an equal relationship; rather, they suggest a loaded power dynamic with dangerous consequences. So when Dillard enters the realm of the hunting myth by engaging in the rhetoric of hunting, she enters an already bloody ring—a trope that refigures a violent act as consummation, and a genre that consumes the very thing it wishes to save.

However, throughout the course of *Pilgrim at Tinker Creek,* Dillard both exposes the consequences of the hunting trope and revises it. As she does so, she slyly slips out of the role of the hunter and assumes the role of another important mythic figure: the shaman or the magician. Unlike the hunter, the shaman assumes a more submissive role in relationship to the environment as he or she straddles the boundary between the human and non-human worlds. The philosopher and ecologist David Abram offers a provocative characterization of the shaman in *The Spell of the Sensuous:*

The traditional or tribal shaman . . . acts as an intermediary between the human community and the larger ecological field, ensuring that there is an appropriate flow of nourishment, not just from the landscape to the human inhabitants, but from the human community back to the local earth . . . [H]e ensures that the relation between human society and the larger society of beings is balanced and reciprocal, and the village never takes more from the living land than it returns to it—not just materially but with prayers, propitiations, and praise . . . [T]he shaman . . . is the exemplary voyager in the intermediate realm between the human and the more-than-human worlds, the primary strategist and negotiator in any dealings with the Others.[5]

The shaman, then, exists on the edge of the human society, or, say, at Tinker Creek rather than in Pittsburgh, and there monitors the health of both the human and the non-human communities by "ensuring . . . an appropriate flow of nourishment" between them. Through the narrative personae in *Pilgrim at Tinker Creek,* Dillard assumes the role of such a spiritual figure as she negotiates between the worlds of various creatures and monitors the exchanges between both human and non-human entities.

Although the shaman figure offers an alternative to the masculine hunter trope, *Pilgrim at Tinker Creek* is *not* the story of the narrator's development from an aggressive hunter figure to a passive shaman-like figure. Dillard never abandons the rhetoric of the hunt, even as she also engages the rhetoric of the shaman. Perhaps we must then consider the possibility that these two roles—the hunter and the shaman—are not mutually exclusive, but in fact are intimately intertwined and dependent upon each other. In fact, this interconnectedness is evident in many indigenous cultures. As Richard Slotkin notes in *Regeneration through Violence,* "In the Indian way of life, and in the mythology, these figures often appeared paired and balanced with each other, since the hunter's courage is as essential to the shaman as the latter's spirit of kinship with nature is to the good hunter."[6] The hunter and the shaman each fill important roles for the life cycle of the tribe: the hunter provides physical sustenance, while the shaman provides spiritual nourishment; the hunter takes a gift from the environment with each kill he makes, and the shaman gives something back to the environment through offerings and sacrifices. Both figures negotiate the boundaries between the human and non-human worlds, the hunter through aggressive engagement, the shaman through acceptance and endurance. By incorporating both roles and both ethics into the narrative of *Pilgrim at Tinker Creek,* Dillard sustains a natural cycle. This is not to say that Dillard strives to advocate or emulate indigenous practices, but that the shifting roles of her narrative persona make sense in many non-Western cultures. "The power we seek," she writes, "seems to be a continuous loop . . . [T]he spirit seems to roll along like the mythical hoop snake with its tail in its mouth."[7] The cycle encompasses both beauty and violence, life and death, and so, too, does the narrative. Dillard never averts her eyes.

And she forces us to open our eyes, too. Seeing is essential to the narrative's project, because Dillard's hunt is a vision quest—a shaman's spiritual journey. Dillard relates two ways of seeing and confronting the world, two ways of encountering the Other. One is the mode of the hunter, willful and aggressive, and the other is the way of the shaman, meditative and receptive. In her more predatory mode, she explains, the act of seeing involves purpose and determination: "I analyze and pry. I hurl over logs and roll away stones; I study the bank a square foot at a time, probing and tilting my head. Some days when a mist covers the mountains, when the muskrats won't show and the microscope's mirror shatters, I want to climb up the blank blue dome as a man would storm the inside of a circus tent, wildly, dangling, and with a steel knife claw a rent in the top, peep, and if I must, fall."[8] Here Dillard sees the world through the eyes of the hunter, engaging in a kind of scientific predation. She strives for an intimate look at the world by prying it open and analyzing it, aggressively hurling away obstacles and forging her own trail for knowledge. She aims at the world as though she is aiming a camera, or a gun, walking "from shot to shot," capturing images with her analytical, appropriative gaze. She dissects these images with a microscope until the microscope's mirror shatters, and then she paws at the world with a steel knife, clawing at it desperately to break through its surface and isolate its parts, like a hunter cleaning his kill. She assumes a perspective of power and control and readily exercises force to see the object of her desire. Although she is intent on seeing, it is the ego of the "I," not the object of the eye, that is emphasized in the passage through the forceful repetition of the first-person pronoun. She exists more in herself than in the world. She remains a voyeur; her vision is both the cause and the result of her detachment.

But at other times Dillard sees quite differently, by "letting go." "When I see this way," she explains, "I sway transfixed and emptied."[9] Like a meditative shaman, she lets the environment slip through her skin, rather than slitting the skin of the environment to peer into it. She empties herself of all intentions, and passively receives the world through her open, unmediating senses. The world captivates her. When she sees in this mode, she writes, it is like walking without a camera: "my own shutter opens, and the moment's light prints on my own silver gut." She becomes more vulnerable to the world, allowing herself to be probed and moved by the surrounding environment. She approaches the world not armed with a camera, microscope, or knife, and instead of shooting images for her analytic brain, she lets the world impress itself upon her senses and her imagination. The individual ego recedes and the "I" is transformed into a state reminiscent of Emerson's transparent eyeball: "I am nothing; I see all."[10]

Clearly, Dillard's two ways of seeing embody two different modes for living in the world. One is active and aggressive, and the other is passive and contemplative. But both are ways of narrowing the gap between separate entities, of encountering the body of the Other. But the trope of seeing, whether through the hunter's or the shaman's eyes, is perhaps too innocuous to convey adequately the stakes of Dillard's project. To heighten the intensity of her narrative quest, Dillard parallels her two ways of seeing with two ways of stalking. While seeing seems to be a harmless, natural action, stalking implies more serious intentions, and perhaps more subversive means. When Dillard goes down to the creek, she says, "I watch and stalk."[11] By doing both together, and writing about them together as though the two acts are the same kind of thing, she brings the act of seeing further into hunting's realm of life and death, in which one stalks the Other in order to overtake and consume it. At the same time, she pushes the idea of stalking into the world of everyday life, suggesting that we are all stalking or being stalked even as we navigate our daily routines and watch the world around us.

As with seeing, one way of stalking is in the mode of the hunter. Dillard writes, "I forge my own passage seeking the creature . . . like Eskimos hunting the caribou herd. I am Wilson squinting after the traces of electrons in a cloud chamber; I am Jacob at Peniel wrestling with the angel."[12] Here again, the writer actively pursues knowledge of the world outside of herself; she looks deliberately for the object of her desire. Stalking muskrats this way, she runs and crawls along the uneven bank, through bushes and grasses, pushing her way through all kinds of obstacles as she chases the agile creatures. She follows tracks and traces with a vengeance, unwilling to be outwitted by the unknown.

But at other times she stalks without pursuing. She writes, "When I stalk this way I take my stand on a bridge and wait, emptied. I put myself in the way of the creature's passage, like spring Eskimos at a seal's breathing hole. Something might come; something might go. I am Newton under the apple tree, Buddha under the bo."[13] In this mode, she waits rather than seeks, and avails herself of the magic around her. Patiently, unobtrusively, she watches fish in the river and she is spellbound by their mysterious movement before her eyes: "They disappear and reappear as if by spontaneous generation: sleight of fish."[14] She is not intent on catching fish; rather, she, herself, is caught in the moment—a true Thoreauvian fisherman. Indeed, a very similar fishing scene is described in *Walden*. Thoreau writes: "It was very queer, especially in dark nights, when your thoughts had wandered to vast and cosmogonal themes in other spheres, to feel this faint jerk, which came to interrupt your dreams and to link you to Nature again. It seems as if I might next cast my line upward into the air, as well as downward into this element, which was scarcely more dense. Thus I caught two fish as it were with one hook."[15]

As Thoreau becomes immersed in the present moment, the external, sensual world merges with the ethereal world of his thoughts, making the two separate realms more continuous. Similarly, Dillard achieves an epiphany, the object of her vision quest, when she relinquishes her own will and allows herself to be captivated by the moment. And it is significant that she achieves this vision, and other similar ones, while standing on a bridge—a symbolic intermediary space, one connecting two separate pieces of land while extending over the fluid presence of the creek. By stalking the world from this perspective, she finds herself suspended in the mystery and power of the present, like a shaman enveloped by a magic that transcends the human realm. It is this form of stalking that serves her best when she wants to see muskrats. While stalking muskrats, she is not a hunter whose body is tense, poised, and ready to take a shot at the first sight of her prey. Adrenaline does not course through her veins and her pulse does not race with anticipation. Her "only weapon [is] stillness."[16] Like Thoreau metaphorically casting his fishing line both up in the air and down in the waters, Dillard casts herself out into the world around her, and waits for a bite to her senses.

Each mode of seeing and stalking involves different ways of approaching the world. One is confrontational and aggressive, and the other is peaceful and receptive, but both lead to an encounter with the Other, and the encounter, Dillard discovers, always involves appropriation and consumption, regardless of the nature of the approach. Throughout the narrative, Dillard witnesses many horrific acts of physical consumption between various animals, insects, and other creatures, and she becomes fascinated with parasite-host and predator-prey relations. The world, it seems, is being eaten out from under her. Amazed and exasperated, she writes, "Is this what it's like . . . a little blood here, a chomp there, and still we live, trampling the grass? Must everything whole be nibbled?"[17] At this moment, the hunting trope seems to be her only recourse for representing encounters with Others, because the "flow of nourishment" between entities appears to be completely unbalanced: contact between entities always involves consumption; one body always dominates the other.

And Dillard is not so presumptuous as to assume that she, herself, a writer about nature, is immune from the cycle of consumption that she sees around her. Although she, as the narrator, never actually kills and eats an animal through the course of her narrative hunt, she recognizes that her own act of seeing is also a matter of consumption. The animals she seeks "have several senses and free will," and just as they do not want to be eaten, "they do not wish to be seen."[18] She calls the objects of her vision "eye food," suggesting that these visions provide nourishment for her even if they are not physically devoured. Similarly, she describes the fish she sees as "spirit food;" they nourish her with their rich symbolic meaning and their magical presence, and she concludes that fishing is an exercise in revelation: spiritual hunting. She writes, "[The fish] are there, they are certainly there, free, food, and wholly fleeting. You can see them if you want to; catch them if you can."[19] Whether we "lure them, net them, troll for them, club them," or simply wait for them to multiply miraculously before our eyes—whether we eat them with our eyes or with our mouths—the fish provide "food"—physical or spiritual fulfillment—for those who diligently seek them.

Because Dillard figures her consumption in terms of spirituality, she is able to push the hunting trope into the realm of the shaman. Dillard's narrative constantly slides between the notions of physical sustenance and spiritual nourishment. For example, she cites an Eskimo shaman who said, "Life's greatest danger lies in the fact that men's food consists entirely of souls."[20] The similarity between *preying* and *praying* thus begins to extend beyond the initial pun, because both have to do with relying on Others for one's survival. Eating the Other becomes a spiritual ritual or sacrament, a form of the Eucharist in which one body is consumed by another. The other side of hunting, therefore, is revealed to be sacrifice, just as the counterpart of the hunter figure is the shaman. Although this suggests that even the shaman's world is not immune from violence, it also suggests that hunting is not devoid of reciprocity. A more sacred, balanced relationship embraces both ethics, acknowledging both the beauty and the violence inherent in all contact between separate entities. Both the hunter's world and the shaman's world encompass parasitism and predation, as well as generosity and sacrifice. As a result, the conventional power dynamic of the hunting trope is revised to suggest a more holistic and honestly reciprocal relationship between separate entities.

The hunt and the spiritual quest are connected even more intimately because of their common origins: Both the hunt for meat and the quest for "eye food" or "spirit food" are born out of hunger and desire. At the end of *Pilgrim at Tinker Creek,* Dillard relates a description of Algonquian Indians in the midst of a winter famine: "[They] ate broth made of smoke, snow, and buckskin, and the rash of pellagra appeared like tattooed flowers on their emaciated bodies—the roses of starvation in a French physician's description; and those who starved died covered with roses."[21] The blood roses, the tracks of the cat over the writer's body, are now revealed to be symptoms of malnutrition. The narrative hunt, this shamanic journey, is a search for fulfillment in the hope of remedying the "flow of nourishment" between the writer's human world and the world of the Others. The writer herself is starving to death, aching for the fullness that comes from connection: "I wonder how many bites I have taken, parasite and predator, from family and friends; I wonder how long I will be permitted the luxury of this relative solitude. Out here on the rocks the people don't mean to grapple, to crush and starve and betray, but with all the good will in the world, we do, there's no other way. We want it; we take it out of each other's hides; we chew the bitter skins the rest of our lives."[22]

Something is missing from everything, Dillard tells us. We grapple and crush and "chew the bitter skins" because we want to break through the surface to get to the meat of it all; we are starving for fulfillment. We are being eaten alive by a lack, and so we voraciously suck out the guts of the world around us, like so many giant water bugs, in order to fill ourselves with something. We *prey* and we *pray* because we are disconnected and thus incomplete.

It is this hunger, this incompleteness, that motivates the hunter's hunt and the shaman's meditation, as well as the writer's narrative. The hunter prowls for meat, the shaman prays for a vision, the writer searches for words, but we are all engaged in the quest for manna. We are constantly hunting for meaning, which is only obtained in the epiphanic moment of the encounter with the Other—and this encounter occurs through the very bodies that both define us as separate, individual entities, and that allow us to touch all that is not us. We need to encounter the flesh of something outside of ourselves in order to substantiate our own being and to feel fuller and more present in the world, for "[w]e are human only in contact, and conviviality, with what is not human."[23]

Our lack, then, also provides us with a space for survival—a breathing hole. It is this gap between Others that facilitates connection. Like the synapses between nerve endings, it offers a point of stimulating contact. At the end of Dillard's narrative hunt, we encounter this in-between space directly. She tells us:

The gaps are the thing. The gaps are the spirit's one home, the altitudes and latitudes so dazzlingly spare and clean that the spirit can discover itself for the first time like a once-blind man unbound. The gaps are the clifts in the rock where you cower to see the back parts of God; they are the fissures between mountains and cells the wind lances through, the icy narrowing fiords splitting the cliffs of mystery. Go up into the gaps. If you can find them; they shift and vanish too. Stalk the gaps. Squeak into a gap in the soil, turn, and unlock—more than a maple—a universe.[24]

For Dillard, stalking muskrats and stalking fish are a means of stalking the gaps. Her goal is not to complete the hunt and physically obtain the creature for herself, although she acknowledges the violence inherent in her own creative quest; rather, she embarks on her hunt in the hope of crossing into the intermediary space of the shaman—the space between worlds, the space between Others, and the space of story.

Dillard's hunt, then, is finally a form of mediation. Even though it involves consumption, it is not a one-way dynamic between predator and prey. Rather, it is part of a more balanced "flow of nourishment" within an extravagant economy. Dillard writes, "Nature is, above all, profligate. Don't believe them when they tell you how economical and thrifty nature is . . . This is a spendthrift economy; though nothing is lost, all is spent."[25] Later, she explains that even as half the population is devoured by the other half, balance is maintained, and "nothing is lost" after all: Of course any predator that decimates its prey will go hungry, as will any parasite that kills its host species. Predator and prey offense and defense . . . usually operate in such a way that both populations are fairly balanced, stable in the middle as it were, and frayed and nibbled at the edges, like a bitten apple that still bears its seeds."[26] Once again, it is the bridge that matters. Although the edges are nibbled away, the space in the middle—a common, connective ground—is maintained. Within the intricacies of an extravagant economy, in the gaps between violence and sacrifice, there is balance: "Waste and extravagance go together up and down the banks, all along the intricate fringe of spirit's free incursions into time."[27]

And so, like the giant water bug that completely devours the frog, Dillard eats the apple core and all of the world, "All of it. All of it intricate, speckled, gnawed, fringed, and free."[28] Starving for connection, covered with blood roses, she both sharpens her claws like a hunter and pummels for milk like the shaman as she reaches to make contact with the world around her. In return, she accepts both the violence and the beauty—the claws and the milk—of the world, because no creature is complete in itself. She, like each of us, is not only the arrow shaft, but also the wounded animal in need of the healing touch of contact. She follows the "straying trail of blood" and her own blood runs into it, in a confluence between herself and Others that flows like the waters of Pilgrim Creek itself, offering a moment of presence like "the wave that explodes over my head, flinging the air with particles at the height of its breathless unroll; it is the live water and light that bears from undisclosed sources the freshest news, renewed and renewing, world without end."[29]

So Dillard does not choose between violence and beauty, between confrontation and consummation; and she does not choose between tropes: She embraces both the mythic figure of the hunter and that of the shaman, because no trope is complete in itself, either. Metaphors can easily become reductive if overused or forced, and they can enforce over-simplistic, dangerous binary oppositions if they remain stagnant. As Dillard explains, "It is the fixed that horrifies us, the fixed that assails us with the tremendous forces of its mindlessness . . . The fixed is the world without fire—dead flint, dead tinder, and nowhere a spark."[30] Language sends sparks into the space between beings, and it stimulates connections that might otherwise remain unseen. "Make Connections," she urges, "let rip; and dance where you can," and her ecstatic language and shifting metaphors keep her own narrative dancing.[31]

We must be willing to move between meanings—to dance in the gaps—if we want to be moved by environmental literature at all. The hunting narrative risks reducing the human relationship of nature to one of conquest and consumption; the shaman narrative risks ignoring the violence that is inherent in many encounters with and in the natural world. Standing alone, each trope is a fiction that closes our eyes to the complexities of the world. Together, they approach truth. As we follow Dillard's "straying trail of blood," the text may serve as a bridge, bringing us closer to the nonhuman world in a form of communion by opening our eyes and our minds. Or, the text may further alienate us from the world if we allow tropes to become traps that reduce the mystery of the Other to a single, definitive representation.

As we read these bloody traces, perhaps we cannot help but be hunters, hunting for meaning in the text, stalking the Other, following the trail of blood to a world we want to make our own. But to prevent our reading from becoming only another form of conquest, we also need to read with the eyes of the shaman, receptively, and without forcing nature into our tired prescribed categories. As Dillard says, "We must somehow take a wider view, look at the whole landscape, really see it, and describe what's going on here. Then we can at least wail the right question into the swaddling band of darkness, or, if it comes to that, choir the proper praise."[32] And she again urges in a later book, *Holy the Firm,* "Let us by all means extend the scope of our charts."[33] As readers and writers of environmental literature, we must analyze and pry and stand between worlds on the narrow bridge of the text, searching—waiting—for meaning. And if we look with the eyes of both the hunter and the shaman, we may, perhaps, see through the lens of language and catch a fleeting glimpse of the world that is always beyond us.

Annie Merrill Ingram

Telling News of the Tainted Land
Environmental Justice Fiction by Women

> The extensive national and international network of community/environmental
> organizations referred to as the environmental justice movement challenges
> dominant meanings of environmentalism and produces new forms of
> environmental theory and action. The term "environmental justice," which
> appeared in the United States sometime in the mid-1980s, questions popular
> notions of "environment" and "nature" and attempts to produce something
> different . . . From the start, the gender, race, and class composition of the
> movement distinguishes it from that of the mainstream environmental
> movement, whose constituents have historically been white and middle class
> and whose leadership has been predominantly male.
> —Giovanna Di Chiro, "Nature as Community: The Convergence of
> Environment and Social Justice"[1]

The phenomenon of environmental racism in the United States originated with
the arrival of European settlers in Native American lands.[2] More recently, key
events in the 1980s sparked the first use of the terms "environmental racism" and
"environmental justice."[3] Environmental racism refers to any kind of environmental degradation directed at or located in communities of color; environmental
justice is the movement that has responded to such discrimination. However, as
cultural critic and environmental historian Giovanna Di Chiro notes in the epigraph above, environmental justice goes beyond redressing environmental racism and toward fundamental redefinitions. One part of the environmental justice
movement's attempt "to produce something different" is the redefinition of "environment"; as Di Chiro further explains, "environment" is not just "wild" or "natural" areas, a place separate from the daily experiences of most humans, but rather
"the place you work, the place you live, the place you play."[4] The central importance of places of employment, residence, and recreation in human culture further
renegotiates the term "environment" within environmental justice literature by asserting that human culture is inextricable from physical environment. As such,

threats to culture are threats to the environment, and environmental activism saves not only landscapes but the entire ecosystem, including humans' livelihoods and cultural memory.

Environmental justice's redefinition of "nature" has by extension redefined "nature writing." In the same way that the environmental justice movement has critiqued and expanded the mainstream environmental movement, environmental justice literature implicitly critiques and explicitly expands the mainstream canon of environmental literature. Environmental justice literature is multicultural, emphasizes social justice, identifies "environment" as the inextricable combination of human culture and natural setting, and attends to issues of race, class, and gender as often as it acknowledges the influence of place. The three novels discussed here all exhibit fundamental attributes of environmental justice literature. Barbara Kingsolver is white, but her novel *Animal Dreams* (1990) focuses on a Native American and Hispanic community. Ana Castillo's *So Far From God* (1993) deals with a Chicano community in the Southwest, and Chickasaw writer Linda Hogan addresses Native American (Cree) culture in *Solar Storms* (1995).[5] Individually and collectively, they represent environmental justice literature's being "multicultural by definition."[6] The plot lines chronicle the social justice work of local activists responding to environmental crises such as a poisoned water supply, toxic waste exposure, and hydroelectric dam construction. In each novel, multivocal narrative strategies not only emphasize diversity but also reinforce the significance of collective memory and cultural pluralism, highlighting the primacy of both human culture and natural setting within the environmental justice themes. Furthermore, all three writers have chosen as their protagonists non-white, working-class women, indicating that the race, class, and gender of marginalized groups are their central concerns. As with all environmental literature, these novels provide rich descriptions of place, and significantly, each novel is set in a borderland: the multicultural Southwest of mythical Grace and actual Tome, Arizona (in *Animal Dreams* and *So Far From God,* respectively), and the Boundary Waters region of the northern United States and southern Canada in *Solar Storms*. These borderland locations function metaphorically as a kind of ecotone, the place where two ecosystems overlap and in which biodiversity is especially rich. Significant ecological processes also occur in the ecotone, and in these novels set in geographical, generational, and cultural borderlands, the possibilities for transformative activism are similarly abundant.[7]

Criticism of environmental justice literature builds on the work of feminist literary critics and ecocritics (environmentalist literary critics), but it also goes a step beyond, linking literary criticism to social activism such that even the act of analyzing a work of literature can be seen as part of a larger activist project. For example, ecocritic Joni Adamson asserts that "issues of race and human rights must be brought into any satisfactory ecocritical discussion of 'nature' and/or 'nature writing,'" a process Adamson labels "transformative ecocriticism."[8] Adamson elaborates on the transformative nature of this ecocritical practice by suggesting that "when we incorporate multicultural perspectives . . . into our discussions, we do not simply broaden or redefine the genre of 'nature writing.' Rather, we open our discussions to diverse perspectives which allow our understanding of environmental literature and ecocriticism to become richer, better, and more complex."[9] In the same way that environmental activists seek to protect the biodiversity of an ecosystem by preserving its richness and complexity, environmental justice writers and critics work toward creating and supporting a literary ecosystem that prizes human diversity and recognizes the transformative potential of literature.

In American literary history, environmental literature has evolved out of a tradition of natural history writing, exemplified in the works of white male authors such as William Bartram, John James Audubon, Henry David Thoreau, John Burroughs, John Muir, and—in our own time—John Elder, Barry Lopez, Peter Matthiessen, and Scott Russell Sanders.[10] The works originally comprising a canon of environmental literature generally addressed only one of the central characteristics of environmental justice literature: the influence of place. This volume of essays publicizes the rich tradition of American women's nature writing, one that rivals that of a male-dominated canon in its scope, variety, and literary merit. But whereas critical attention to American women's nature writing is more than a century old (for example, in contemporary reviews of Susan Fenimore Cooper, Celia Thaxter, and Sarah Orne Jewett), responses to environmental justice literature are far more recent. Most published criticism on environmental justice texts addresses literature written in the past decade or so; ecocritical readings of earlier works with environmental justice concerns comprise a still largely untouched field. Marginalized by a combination of race, gender, class, and the social-justice themes of their work, current American women environmentalist writers have only very recently received serious critical discussion.

The work of ecofeminist theorists and literary critics has inspired and directed much of the recent literary criticism on environmental justice literature—and indeed, has influenced many of the literary works themselves. The alignment of the women's liberation and environmental movements of the 1970s created the field of ecofeminism and found voice in Rosemary Radford Ruether's *New Woman/New Earth: Sexist Ideologies and Human Liberation* (1975), Françoise d'Eaubonne's *Ecologie féminisme: Révolution ou mutation?* (1978), and Carolyn Merchant's *The Death of Nature: Women, Ecology, and the Scientific Revolution* (1980), among other significant early examples. Ecofeminists see the oppression of women and the degradation of natural environments as interconnected, both arising from patriarchal systems of domination. Environmental justice activists (including writers and critics) similarly link environmental racism to other forms of cultural imperialism and environmental destruction. One hopeful sign that environmental justice literature is reaching a wider audience is the recent publication of several important collections of ecofeminist literary criticism and compilations of ecocriticism that include essays on environmental justice literature.[11]

The diversity of perspectives represented in contemporary literature addressing environmental justice issues comes not only from racial, ethnic, geographical, and linguistic variety but also from a multiplicity of genres. Although this essay focuses on three novels published in the past decade, the spectrum of environmental justice literature also includes short fiction, poetry, drama, personal narrative, and performance art, as well as multigeneric compilations. For example, Kingsolver's commitment to grassroots activism also appears in her nonfiction work, *Holding the Line: Women in the Great Arizona Mine Strike of 1983* (1989). Augmenting Castillo's novel is the work of other Chicana writers who are concerned with environmental justice, such as that of Helena María Viramontes (*Under the Feet of Jesus* [1995], a novel; and *The Moths and Other Stories* [1995]), and Cherríe Moraga (*Heroes and Saints* [1994], a play). The environmentalist thrust of Native American writers such as Linda Hogan, Winona LaDuke, and Joy Harjo is generally well known, yet Asian-American and African-American environmental justice authors have also contributed significantly to the genre. Karen Tei Yamashita's *Through the Arc of the Rainforest* (1990), *Brazil-Maru* (1992), and *Tropic of Orange* (1997) all deal with cross-cultural and environmentalist encounters; Octavia Butler, an African-American novelist, best known for her science fiction writing, addresses environmental justice themes in *Wild Seed* (1980), *Parable of the Sower* (1993), and other works. Although the American public quickly became familiar with the results of class-based environmental discrimination when the working-class Love Canal neighborhood gained national media attention for a toxic waste disposal disaster, and the recent book (1995) and film (1998) versions of Jonathan Harr's *A Civil Action* have kept the issue current, cases of environmental racism tend to receive less national recognition or remain long in the public awareness. As more critics and reviewers address literature with environmental justice themes, perhaps this public blind spot will disappear.

Barbara Kingsolver's *Animal Dreams*

The earliest of the texts under consideration here, *Animal Dreams,* tells its story in two voices: the first-person narration of protagonist Cosima "Codi" Noline and the third-person narration of her father, Homero (Homer) Noline, a small-town physician struggling against the onset of Alzheimer's disease. Codi returns to her hometown of Grace ostensibly to take care of her father but also to get some perspective on her own life. Several plot lines interweave in the novel: Homer's failing memory and difficulties in continuing his medical practice, his other daughter Hallie's work in Nicaragua as an agronomist and peace activist, Codi's romantic and professional involvements upon returning to Grace, and the discovery of a local environmental crisis that functions as a metaphorical point of connection for the other plot lines.

Grace's residents are mostly Hispanic and Native American, and Codi realizes early on that the town has become the target of environmental racism. After she takes her high school biology students on a field trip to the local river and they examine samples of its water under microscopes, she discovers that the river water is "dead" (*AD* 110). A mining company has an operation just outside of Grace that leaches gold and molybdenum mining tailings, a process that pours sulfuric acid waste into the Grace river, the water supply that residents use to irrigate their orchards. Already suspicious because of her neighbors' complaints about "fruit drop" and "poison ground," Codi initiates an investigation into the mining company's culpability (*AD* 63). When she asks who owns the water rights to the river, her friend Viola tells her that "nobody around here's got water rights. All these families sold the water rights to the [mining] company in 1939, for twenty-five cents an acre. We all thought we were getting money for nothing" (*AD* 111). Ironically, that is precisely what they get paid for, nothing: no rights to the water that irrigates their orchards, and therefore no apparent legal recourse now that the water is poisoning the soil and killing the fruit. Symbolically representing the hopes of the town for its future, these orchards are "the single self-sufficient aspect of the town's economy" and some of the last evidence of its former agrarian culture.[12]

Not content to let either the mining company or the Environmental Protection Agency ignore the situation, Codi enlists the help of the local "Stitch and Bitch Club" for a fundraising and publicity campaign: The women of the town make peacock piñatas, accompanied by "a written history of Grace and its heroic struggle against the Black Mountain Mining Company," and sell them on the streets of Tucson. These piñatas are one example of what literary critic Krista Comer calls the novel's "Southwestern kitsch": an appropriation of Mexican culture as well as a nostalgic emblem of the upper class now absent from predominantly working-class Grace (in the town's early history, the nine Gracela sisters—brought from Spain to Arizona to wed bachelor miners—kept peacocks).[13] The peacocks can also be seen as the inverse of the "dead birds" that result from cockfighting, a hobby that Codi's lover Loyd Peregrina practices until Codi convinces him to give it up. Yet, whereas Loyd's cockfighting is based in the authentic traditions of several generations of his family and in the culture of his Apache and Pueblo relatives, the folk-art peacocks are crafted out of recycled materials from the mainstream Anglo culture that pervades Grace—for example, "Mrs. Nuñez made peacock wings out of the indigo-colored flyleaves of all twelve volumes of the *Compton's Children's Encyclopedia*" (*AD* 205). Ultimately, the popular folk-art piñatas save the town: As suddenly desirable commodities, they bring the necessary media attention that eventually forces Black Mountain to shut down the leaching operation. In the end, however, although the women's grassroots activism celebrates the power of female ingenuity and solidarity, not to mention the triumph of domestic craft over corporate greed, the cultural dissonance of this solution controverts the otherwise positive efforts against environmental racism.

Other resolutions in the narrative signal more culturally appropriate and sustainable solutions. Codi realizes that her part in initiating and spearheading the effort to save the town is a personal as well as a collective victory. Her work as a temporary (and untrained) science teacher results in her winning "something like teacher of the year," awarded for her "'innovative presentation' and 'spirited development of a relevant curriculum'" (*AD* 290). Among other lessons that she has learned and passed on to her students, she notes, "I'm teaching them how to have a cultural memory . . . I want them to be custodians of the earth" (*AD* 332). Codi has taught her students not only biology but also conservation, activism, and cultural pride. With her "innovative" pedagogy—including experiential learning, field work, and discussion-based classes—she dismantles the hierarchy and domination that originally allowed Black Mountain to gain control of the town.

The novel's ending is not universally happy, however. Memory is a fraught concept in the novel, for not only is Homer losing his memory, but Hallie is killed by *contras* and thus becomes only a memory. Hallie's presence in the novel is primarily textual: Whereas Codi's voice and thoughts are available to readers in much of the narrative, Hallie's voice appears only in Codi's flashbacks or in the letters she writes to her sister. Homer's voice occasionally interrupts the narrative in brief, interspersed chapters; his distant memories (the only ones that remain clear to him) ultimately fill in gaps that are missing from Codi's recollections of her early life. The connections among various forms of memory in the novel—personal, cultural, collective, environmental—create an ecological structure overall, one that connects these disparate narrative strands into an integrated system. Comer's charge that "Kingsolver does not struggle with . . . more serious environmental questions" is well-directed, but where the political thrust and cultural sensitivity of her novel might fail, the narrative structure succeeds, showing the value of experiential learning, grassroots activism, and a multiplicity of voices.[14] In *Animal Dreams,* these transformative practices provide the solutions to environmental problems and give readers some practical exampl es of sustainable activism.

Ana Castillo's *So Far From God*

Cultural memory and environmental stewardship are also central concerns for Ana Castillo, who offers this bilingual dedication to her novel *So Far From God:* "To all the trees that gave their life to the telling of these stories," and *"A m'jito, Marcel, y a las siguientes siete generaciones"* [to my son Marcel, and to the next seven generations]. Castillo's work weaves English and Spanish, reality and magic, in a story about Sofia and her four daughters—Esperanza (Hope), Caridad (Charity), Fe (Faith), and La Loca (the Crazy One)—and their lives in the small town of Tome, Arizona. As in *Animal Dreams,* several different plot lines comprise the novel. Ecocritic Kamala Platt states that "Castillo interconnects a sisterhood of social ills (environmental racism, war, AIDS, and patriarchal domestic violence) through the four sisters, whose experiences with these social ills contribute to their deaths."[15] With long, descriptive chapter headings reminiscent of *Don Quixote* and apparently magical events presented as realism—the resurrection of La Loca from her death as a three-year old, Esperanza's posthumous reappearance in the arroyo that a figure like La Llorona also inhabits, and Caridad's psychic powers—the novel converses with several traditions of Spanish-language fiction. However, Platt identifies *So Far From God* as not so much magical realism as "'virtual realism': a realism that virtually encompasses lived experience and propels it into postmodern fiction."[16] As postmodern, bilingual, virtual realism, *So Far From God* certainly challenges the traditional parameters of American environmental literature. Until chapter 11, in which Fe suffers from exposure to toxic and radioactive chemicals, the novel deals more with complex family interactions and miraculous occurrences than with the detailed place descriptions and environmental concerns more typical of mainstream environmental literature. Yet Castillo's presentation of Tome's cultural ecosystem reveals that here is a complex ecotone that combines languages, blurs the distinctions between reality and magic, aligns formal religion with folk medicine and psychic communication, and still remains grounded in the actual: town politics, the Gulf War, and the economic prosperity brought by the electronics industry.

The promise of high wages lures Fe, the third of Sofia's four daughters, away from a steady and reasonably lucrative job at the local bank to an even more lucrative job at Acme International, a new electronics company in town. She is initially quite happy with her job at Acme, mostly because of the very high pay. Recently married, she buys "the long-dreamed-of automatic dishwasher, microwave, Cuisinart, and the VCR," all "with her own hard-earned money from all the bonuses she earned at her new job" (SFFG 171). When she miscarries a few weeks after starting work, she is concerned, but not enough to quit. Because of her dedication and efficiency, Fe soon moves on to more challenging tasks at Acme: At one point, she is "working in a dark cubicle . . . with a chemical that actually glowed in the dark," which results in her getting "this red ring around her nose and breath that smelled suspiciously of glue" (SFFG 181). Then she starts cleaning machine parts with "ether," and her only protection from the noxious fumes and direct contact is a pair of gloves; however, "not only did the chemical eat them up, it dissolved her manicure, not only the lacquer but the nails themselves!" (SFFG 183). Although the tone here (and in much of the novel) seems playful, like the high camp of telenovelas (Spanish-language soap operas), the message is deadly serious. Standard protective gear is useless, and the corrosive chemical does more than just ruin evidence of vanity (or patriarchal expectations for beauty), it actually dissolves human flesh. Before long, Fe is diagnosed with cancer, and after further "torture" from both the medical establishment (which poisons her already toxic body with more chemicals, as part of a chemotherapy routine) and legal officials (who subpoena her while investigating Acme's use of illegal and toxic chemicals), she demands an explanation from Acme. A classified document explains that "the chemical she more than once dumped down the drain at the end of her day, which went into the sewage system and worked its way to people's septic tanks, vegetable gardens, kitchen taps, and sun-made tea . . . must always be sealed, 'reclaimed,' . . . not left to evaporate" (SFFG 188). The SouthWest Organizing Project (SWOP), which Castillo thanks in her acknowledgments, provides this information about the electronics industry in the Southwest: "Glycol ethers [the chemicals Fe handled] were once considered 'safe' for use as solvents in semiconductor processing. Then studies showed that they caused serious health problems for pregnant women and women of child-bearing age—but only after they had been in use for several years."17 Fe's death from toxic exposure reveals a pernicious example of another kind of borderland, for she is caught between economic prosperity and excessive consumerism, between alleged workplace safety and actual contact with deadly carcinogens.

Moreover, her death implicitly contrasts those of her sisters. La Loca "dies" at age three, only to be mysteriously "resurrected" during her funeral (doctors speculate that she may just have suffered some kind of seizure, but her family and the town prefer to believe that God resurrected her). Esperanza, a TV reporter, is taken hostage and tortured to death during the Gulf War, but her ghost later returns to Tome and converses with various family members. Like Hallie Noline in *Animal Dreams,* Esperanza dies in a foreign country, killed by political insurgents. In both cases, the absence of any "remains," of a body to bury, signals a symbolic form of environmental injustice: There is no corporeal form to return to the earth, no closing of a natural cycle that ends in the organic decomposition of flesh that will in turn enrich the soil. Another sister, Caridad, dies when she leaps off the roof of a pueblo with her lover Esmerelda, as they are pursued by Tsichtinako, a kachina spirit. Nevertheless, the narrator "will refrain from calling [their deaths] tragic," perhaps because Caridad finally encounters true love with Esmeralda (*SFFG* 190). Fe, on the other hand, "just died. And when someone dies that plain dead, it is hard to talk about" (*SFFG* 186). Echoing the well-known slogan of AIDS activists, "Silence = Death," Fe's death is equated with silence. The finality of her death—one so horrific and final that Castillo as author cannot bear to resurrect her through the infinite magical means available—creates a lacuna of silence, a conspicuous gap among the narrative's miraculous revivals and posthumous presences.

By the end of the novel, all four sisters' deaths become contextualized within a wider framework of social justice activism. Just before La Loca dies the second and final time, she participates in the annual Holy Friday procession, one that differs markedly from its predecessors: "No brother was elected to carry a life-size cross on his naked back. There was no "Mary" to meet her son. Instead, some, like Sofi, who held a picture of la Fe as a bride, carried photographs of their loved ones who died due to toxic exposure hung around the necks like scapulars; and at each station along their route, the crowd stopped and prayed and people spoke on the so many things that were killing their land and turning the people of those lands into an endangered species" (*SFFG* 241–42). The stations of the cross inspire recitations such as: "When Jesus was condemned to death, the spokesperson for the committee working to protest dumping radioactive waste in the sewer addressed the crowd"; "Veronica wiped the blood and sweat from Jesus' face. Livestock drank and swam in contaminated canals"; and "Deadly pesticides were sprayed directly and from helicopters above on the vegetables and fruits and on the people who picked them for large ranchers at subsistence wages and their babies died in their bellies from the poisoning. Ayyy! Jesus died on the cross." (*SFFG* 242–43)

In this procession, as well as in the organization that Sofia establishes after all her daughters have died, people speak out against the forces that are destroying their land, water, communities, and individuals. Sofia becomes the first president of M.O.M.A.S., Mothers of Martyrs and Saints—an organization recognizing that the dead still have the power to have an impact on the living. The fictitious M.O.M.A.S. recalls similar organizations that do exist: MELA (Mothers of East Los Angeles/Madres Del Este De Los Angeles), PODER (People Organized in Defense of the Earth and her Resources), and SWOP (SouthWest Organizing Project).[18] By honoring "their loved ones who died as the result of toxic exposure," the participants in the Holy Friday procession and the members of M.O.M.A.S. preserve cultural memory as well as individual memories. Whereas the grassroots response in *Animal Dreams* is initiated (and perpetuated) within an educational context—Codi's innovative pedagogy and the informational campaign that accompanies the sale of the peacock piñatas—in *So Far From God,* religion provides the cultural context for transformative activism. Although M.O.M.A.S. memorializes martyrs and saints, the real force in the organization is the mothers: women without traditional sources of power who nonetheless effect real change as a result of their dedication to a local cause.

Linda Hogan's *Solar Storms*

The main characters of Linda Hogan's *Solar Storms* also fight back when their land and culture are threatened. Four generations of women—seventeen-year-old narrator Angel, her grandmother Bush, great-grandmother Agnes, and great-great grandmother Dora-Rouge—embark on a journey of cultural and environmental reclamation. Residents of a small Native American community called Adam's Rib, located somewhere in the Boundary Waters area between Canada and Minnesota, they learn that a series of dams that are being built to power a hydroelectric plant will also flood millions of acres of land and water, permanently disrupting the indigenous cultures that have thrived on these islands for millennia:

In the first flooding [resulting from the dam construction], . . . they'd killed many thousands of caribou and flooded the land the people lived on and revered. Agents of the government insisted the people had no legal right to the land. No agreement had ever been signed, he said, no compensation offered. Even if it had been offered, the people would not have sold their lives. Not one of them.

Overnight many of the old ones were forced to move. Dams were already going in. The caribou and geese were affected, as well as the healing plants the people needed.

These men's people, my own people, too, had lived there for ever, for more than ten thousand years, and had been sustained by these lands that were now being called empty and useless. If the dam project continued, the lives of the people who lived there would cease to be, a way of life would end in yet another act of displacement and betrayal. *(SS* 57–58)

The dam construction project is linked to centuries of Native American removal, yet unlike some earlier removals, which appropriated Indian territories rich with resources desired by whites, this particular instance of displacement results from an official designation of the lands as "empty and useless," terms that ignore the cultural heritage and human habitation of this sparsely populated area, not to mention its useful sources of food, medicinal plants, and other resources valued by the native inhabitants. As in *Animal Dreams,* questions of ownership seem to justify not only appropriation of the lands but also their permanent alteration.

As the novel's narrator, Angel provides the prospective of a recent arrival to the traditional home of her people. Abandoned by her mother when she was young, Angel has been raised in a series of foster homes far away from Adam's Rib. She returns to the village when she finds evidence of her great-grandmother in some court records. Just as the traditional tribal areas are threatened with cultural extinction and actual fragmentation and division because of the dam-building, so does Angel struggle against the internal fragmentation resulting from her own lack of cultural and familial heritage: "there were no snapshots of me, nothing to say I'd been born, had kin, been loved" *(SS* 26). Throughout the novel, references to broken mirrors, unraveled fabric, unfinished stories, animal skeletons in pieces, broken pacts, and other forms of fragmentation reinforce both Angel's initial anomie and the tribes' impending dislocation. However, in an attempt to counteract the forces behind the dam construction, Angel, Bush, Agnes, and Dora-Rouge decide to join the protest against the dam project, traveling by canoe to a central activist meeting place farther north. Many factors contribute to the danger and difficulty of such a journey: Because of the naturally shifting boundaries between land and water in this area, maps are not always accurate, rendering their route impossible to plan exactly; winter is approaching, making for very cold weather and fewer food sources along the way; the older participants might not have the stamina, and Angel worries about her lack of experience; and the oldest member, Dora-Rouge (perhaps 90 or even 100), cannot walk and therefore must be carried over all of the portages. Yet these women are not to be dissuaded. Although they have different primary motivations—Dora-Rouge is returning to the land of her birth, going home "to die"; Agnes is accompanying her mother; Bush is going to protest the dams; and Angel is searching for her birth mother, reputed to be living somewhere up north—they share the common purpose of refusing to let their familial, communal, and tribal cultures disappear along with the threat of permanent change to the landscape they love.

In keeping with the other two novels' concern with cultural preservation, *Solar Storms* also emphasizes collective memory. For example, Hogan implicitly invokes racial memory when Angel begins dreaming about plants. Although she has never seen them before, she can draw detailed pictures of them, and when she describes them to Dora-Rouge, she discovers that they have important medicinal properties. For Angel, who has never known her family past—let alone that of her entire tribe—these plant dreams are a revelation: "'I knew there'd be another plant dreamer in my family someday,' Dora-Rouge said . . . I got it from blood, she said. I came by it legitimately" (*SS* 171). Angel later reflects on the power of her gift:

I'd never have thought there might be people who found their ways by dreaming. What was real in those land-broken waters, real even to me, were things others might call the superstitions of primitive people. How could it be, I wondered, that all people who came from their own earth, who lived there for tens of thousands of years, could talk with spirits, could hear land speak, and animals? Northern hunters were brilliant hunters. Even now they dream the location of their prey and find it. Could they all have been wrong? I didn't think so. (*SS* 189)

Once Angel experiences the reality of dream messages (a concept she previously rejected when Dora-Rouge insists that her dream-maps are more accurate than the paper ones they have brought with them), she realizes the strength of her culture's beliefs. Like the "redroot" that might heal Agnes, the roots of Angel's Native culture keep her firmly connected to her present as well as to her collective past (*SS* 188). The emphasis on the "real" quality of the dreams in *Solar Storms* echoes the "virtual realism" of Castillo's miracles; in both works, cultural survival depends on a willingness to suspend disbelief in what the dominant culture has labeled "primitive" or "superstition."

As in *Animal Dreams* and *So Far From God,* the ending of *Solar Storms* is bittersweet. Agnes dies unexpectedly before they reach their intended destination, and when Angel finally finds her mother, she is literally on her deathbed, unable to recognize Angel. The tribes' efforts to halt the first phase of the dam project fail, but the protest continues and eventually succeeds more than a year later, after many places have already been flooded, animals have died, and valuable medicinal plant species have been lost forever. However, despite all the sadness, grief, and disappointment, a few details give hopeful closure to the various plot lines of the novel. Angel has undergone a powerful series of transformations as a result of her return to Adam's Rib and her participation in the canoe trip with her older female relatives, finally finding home, family, and her place in both. She leaves Dora-Rouge behind in the land of their ancestors and goes back to Adam's Rib with a newborn half-sister named Aurora, who signifies the dawn of yet another generation of women and the continuation of their struggle against injustice.

In their focus on social justice, multicultural perspectives, and a diversity of voices, these three environmental justice novels all contribute to the redefinition of nature writing and the expansion of the environmental literature canon. Traditionally, canonical environmentalist texts such as John Muir's writings, Aldo Leopold's *A Sand County Almanac,* and Edward Abbey's *Desert Solitaire* have strong messages and promote activism, but they do so for environments that are separate and distinct from the everyday experiences of their audiences. Indeed, part of the appeal of these familiar works is that they provide vicarious escape to the Sierra Nevada mountains or Leopold's weekend farm or the red rock country of present-day Arches National Park. Environmental justice literature, on the other hand, deliberately offers no such escape and does not present its settings as *out there;* instead, it deals with the places where we live, work, and play every day. By focusing on families, local communities, and grassroots activism, the environmental justice novels discussed here offer an alternative view of "nature," one that includes humans and their cultures in a genuinely ecological setting: an environment that survives through interrelation and interconnectedness.

Furthermore, because these novels take place in borderland locations and their plot lines suggest the overlapping qualities of the ecotone, they reinforce our understanding of the power of the margins. As both literary theory and activist movements of the past few decades have shown, revolutions begin among the marginalized. When Kingsolver, Castillo, Hogan, and other women environmental justice writers address the issues that literally hit home, they oppose the structural and conceptual forms of domination that have already significantly destroyed our natures and our environments. As Adamson has argued: "Incorporating multicultural perspectives and literatures into their study and practice would encourage critics and teachers concerned with ecology to join allegiance with those seeking to end oppression by race and class. Only by recognizing the connections between the oppressions of certain peoples and places can literary critics begin to develop ecological theories and practice that can work transformatively toward an ecology of justice."[19] Environmental justice literature already recognizes these connections, and ecocritics such as Adamson, Gaard, and Platt have already begun to "work transformatively toward an ecology of justice." By focusing on American *women* nature writers, *Such News of the Land* promotes one redefinition of nature writing, that not only men go to the woods to "live deliberately," as Thoreau so famously did. This essay encourages not only redefinition of the genre but also a fundamental reconceptualization of the practice of ecocriticism: to end oppression, to uphold justice, and to transform our practices so that all the many environments discussed in this volume not only will endure but also will thrive and remain sustainable for many more generations to come.

Notes

Introduction (pages 1–5)

1. Mary Austin, *The Land of Little Rain* (New York: Dover, 1996), 2.
2. Henry David Thoreau, *Walden,* ed. J. Lyndon Shanley (Princeton, N.J.: Princeton University Press, 1989), 91, 97. Emerson's essay "Nature" contains this distinction: "Nature, in the common sense, refers to essences unchanged by man; space, the air, the river, the leaf. Art is applied to the mixture of his will with the same things, as in a house, a canal, a statue, a picture." *Selections from Ralph Waldo Emerson,* ed. Stephen E. Whicher (Boston: Houghton-Mifflin, 1960), 22.
3. *Selections from Ralph Waldo Emerson,* 24.
4. *Walden,* 175.
5. See here the discussion in Lazar Ziff, *Literary Democracy: The Declaration of Cultural Independence in America* (New York: Viking, 1981), 217.
6. See Mary Rowlandson's *A Narrative of the Captivity and Restoration of Mrs. Mary Rowlandson* found in *Women's Indian Captivity Narratives,* ed. Kathryn Zabelle Derounian-Stodola (New York: Penguin, 1998). An interesting example of the efforts to clarify the notion that gentility and feminity can coexist with wilderness exploration can be found in Isabella Bird's *A Lady's Life in the Rocky Mountains* (London 1879). Bird goes to great length to demonstrate that despite the fact that she finds herself in the wilderness west, she is still a lady—and a civilized English one at that.
7. One can trace this development in a number of key works in American literature, but the idea can easily be seen in a direct line that connects Twain's *The Adventures of Huckleberry Finn* to Kerouac's *On the Road,* Steinbeck's *Travels with Charley,* and William Least Heat Moon's *Blue Highways.* See also Richard Slotkin's *Regeneration Though Violence: The Mythology of the American Frontier, 1600–1860* (Middleton, Conn.: Wesleyan University Press, 1973).
8. *This Incomperable Lande: A Book of American Nature Writing,* ed. Thomas J. Lyon (New York: Penguin, 1991), xiv.
9. On women's struggle to write and publish in a male-dominated arena see Nina Baym, *Women's Fiction: A Guide to Novels by and about Women in America, 1820–1870* (Ithaca, N.Y.: Cornell University Press, 1978); Susan Coultrap-McQuin, *Doing Literary Business: American Women Writers in the Nineteenth Century* (Chapel Hill: University of North Carolina Press, 1990); Mary Kelley, *Private Woman, Public Stage: Literary Domesticity in Nineteenth-Century America* (New York: Oxford University Press, 1984); and Michael Newbury, *Figuring Authorship in Antebellum America* (Stanford: Stanford University Press, 1997).
10. Lawrence Buell, *The Environmental Imagination: Thoreau, Nature Writing, and the Formation of American Culture* (Harvard: Harvard University Press, 1995), 2.
11. The Association for the Study of Literature and the Environment provides an extensive Web site that includes membership information, teaching resources, calls for papers, conference announcements, bibliographies of recent scholarship, and links to additional sites of interest and relevance <http://www.asle.umn.edu>. Current scholarship

on nature writings is found in ISLE: *Interdisciplinary Studies in Literature and the Environment* and in the *American Nature Writing Newsletter* (ASLE news).

The rapidly growing literature and scholarship of nature writing includes Michael Branch et al., *Reading the Earth: New Directions in the Study of Literature and the Environment* (Moscow, Idaho: University of Idaho Press, 1998); Paul Brooks, *Speaking for Nature: How Literary Naturalists from Henry Thoreau to Rachel Carson Shaped America* (Boston: Houghton-Mifflin Co., 1980); Peter Fritzell, *Nature Writing and America: Essays Upon a Cultural Type* (Ames: Iowa State University Press, 1990); Buckner Hollingsworth, *Her Garden Was Her Delight* (New York: Macmillan, 1962); Alfred Kazin, *A Writer's America: Landscape in Literature* (New York: A.A. Knopf, 1988); Ralph Lutts, *The Nature Fakers: Wildlife, Science and Sentiment* (Golden, Colo.: Fulcrum Press, 1990); Thomas Lyon, *This Incomperable Land: A Book of American Nature Writing* (Boston: Houghton Mifflin, 1989); Carolyn Merchant, *The Death of Nature: Women, Ecology, and the Scientific Revolution* (San Francisco: Harper & Row, 1980); Patrick Murphy, *Literature of Nature: An International Sourcebook* (Chicago: Fitzroy Dearborn 1998); Roderick Nash, *Wilderness and the American Mind* (New Haven: Yale University Press, 1967); Daniel Payne, *Voices in the Wilderness: American Nature Writing and Environmental Politics* (Hanover, N.H.: University Press of New England, 1996); Randall Roorda, *Dramas of Solitude: Narratives of Retreat in American Nature Writing* (Albany: State University of New York Press, 1998); Don Scheese, *Nature Writing: The Pastoral Impulse in America* (New York: Twayne Publishers, 1996); Richard Slotkin, *Regeneration Through Violence: The Mythology of the American Frontier, 1600–1860* (Middleton, Conn.: Wesleyan University Press, 1973); Scott Slovic, *Seeking Awareness in American Nature Writing: Henry Thoreau, Annie Dillard, Edward Abbey, Wendell Berry, Barry Lopez* (Salt Lake City: University of Utah Press, 1992); Frank Stewart, *A Natural History of Nature Writing* (Olathe, Kans.: Island Press, 1995), and John Elder, *American Nature Writers*, 2 vols. (New York: Charles Scribner's Sons, 1996).

Anthologies of nature writing include Lorraine Anderson, ed., *Sisters of the Earth: Women's Prose and Poetry about Nature* (New York: Vintage Books, 1991); Lorraine Anderson, Scott Slovic, and John O'Grady, *Literature and the Environment: A Reader on Nature and Culture* (New York: Longman Publishing, 1998); Robert Baron and Elizabeth Darby Junkins, eds., *Of Discovery and Destiny: An Anthology of American Writers and the American Land* (Golden, Colo.: Fulcrum Publishing, 1986); Robert Begiebing and Owen Grumbling, eds., *The Literature of Nature: The British and American Traditions* (Medford, N.J.: Plexus Publishing, 1990); Marcia Myers Bonta, ed., *American Women Afield: Writings by Pioneering Naturalists* (College Station: Texas A. & M. University, 1995); Robert Finch and John Elder, *The Norton Book of Nature Writing* (New York: W. W. Norton, 1990); Sharon Harris, ed., *American Women Writers to 1800* (New York: Oxford University Press, 1996); Karen Kilcup, ed., *Nineteenth-Century American Women Writers* (Malden, Mass.: Blackwell Publishers, 1998); Karen Knowles, ed., *Celebrating the Land; Women's Nature Writings, 1850–1991* (Flagstaff, Ariz.: Northland, 1992); Richard Mabey, ed., *The Oxford Book of Nature Writing* (New York: Oxford University Press, 1995); John Murray, ed., *American Nature Writing 1994* (Corvallis: Oregon State University Press, 1994), and *American Nature Writing 1999* (Corvallis: Oregon State University Press, 1999); Peter Sauer, *Finding Home: Writing on Nature and Culture from Orion Magazine*

(Boston: Beacon Press, 1992); Deborah Strom, *Birdwatching with American Women: A Selection of Nature Writings* (New York: Norton, 1986); *American Nature Writing*, published annually by the Sierra Club; and Walter Levy and Christopher Hallowell, eds., *Green Perspectives, Thinking and Writing about Nature and the Environment* (New York: HarperCollins, 1994).

A few of the recent (and growing) collections of regional nature writing include Pamela Banting, ed., *Fresh Tracks: Writing the Western Landscape* (Victoria, B.C.: Orca Book Publishing, 1998); Michael Branch, Daniel Philippon and John Elder, *The Height of Our Mountains: Nature Writing from Virginia's Blue Ridge Mountains and Shenandoah Valley* (Baltimore: Johns Hopkins University Press, 1998); Steven Gilbar and David Brower, eds., *Natural State: A Literary Anthology of California Nature Writing* (Berkeley: University of California Press, 1998); Andrea Lebowitz, ed., *Living in Harmony: Nature Writing by Women in Canada* (Victoria, B.C.: Orca Book Publishing, 1996); Lucinda MacKethan, *The Dream of Arcady: Place and Time in Southern Literature* (Baton Rouge: Louisiana State University Press, 1980); John Murray, ed., *A Republic of Rivers: Three Centuries of Nature Writing From Alaska and the Yukon* (New York: Oxford University Press, 1990); Vera Norwood and Janice Monk, eds., *The Desert Is No Lady: Southwestern Landscapes in Women's Writing and Art* (Tucson: University of Arizona Press, 1997); Richard Rankin, ed., *North Carolina Nature Writing: Four Centuries of Personal Narratives and Descriptions* (Winston-Salem, N.C.: John F. Blair, 1996); and Gerald Thurmond et al., eds., *The Woods Stretched for Miles: New Nature Writing from the South* (Athens: University of Georgia Press, 1999).

12. Resources for the study of American women's nature writing are growing at a rapid pace. Several periodicals have offered special issues, including *Women's Studies* (vol. 25, no. 5, 1996) on "Women and Nature"; *Critical Matrix: The Princeton Journal of Women, Gender, and Culture* (vol. 10, nos. 1–2, fall 1996) on "M(O)ther Nature"; and PMLA (vol. 114, no. 5, October 1999) offers a "Forum on Literatures of the Environment." Key scholarly works focusing on American women's nature writing include: Marcia Bonta, *Women in the Field: America's Pioneering Women Naturalists* (College Station: Texas A. & M., 1991); Sherrie Innes and Diana Royer, *Breaking Boundaries: New Perspectives on Women's Regional Writing* (Iowa City: University of Iowa Press, 1997); Annette Kolodny, *The Lay of the Land: Metaphor as Experience and History in American Life and Letters* (Chapel Hill: University of North Carolina Press, 1975), and *The Land Before Her: Fantasy and Experience of the American Frontiers, 1630–1860* (Chapel Hill: University of North Carolina Press, 1984); Vera Norwood, *Made from this Earth* (Chapel Hill: University of North Carolina, 1993); and Rachel Stein, *Shifting the Ground: American Women Writers' Revisions of Nature, Gender, and Race* (Charlottesville: University Press of Virginia, 1997). On ecocriticism and ecofeminist writing, see Salleh Ariel, *Ecofeminism as Politics: Nature, Marx, and the Postmodern* (New York: Zed Books, 1997); Greta Gaard, ed., *Ecofeminism: Women, Animals, Nature* (Philadelphia: Temple University Press, 1993); Greta Gaard and Patrick Murphy, eds., *Ecofeminist Literary Criticism: Theory, Interpretation, Pedagogy* (Urbana: University of Illinois Press, 1998); Cheryll Glotfelty and Harold Fromm, eds., *The Ecocriticism Reader: Landmarks in Literary Ecology* (Athens: University of Georgia Press, 1996); Richard Kerridge and Neil Sammells, eds., *Writing the Environment: Ecocriticsm and Literature* (New York: Zed Books, 1998);

Carolyn Merchant, *Radical Ecology: The Search for a Livable World* (New York: Routledge, 1992); Patrick Murphy, *Literature, Nature, and Other: Ecofeminist Critiques* (Albany: State University of New York Press, 1995); and Karen Warren, ed., *Ecofeminism: Women, Culture, Nature* (Bloomington: Indiana University Press, 1997). See also the EcoFeminism Web site at <www.ecofem.org>.

13. *Made from This Earth*, xiv–xv.

14. *Sisters of the Earth: Women's Prose and Poetry about Nature* (New York: Vintage Books, 1991), xvii.

15. See Ulrich, *A Midwife's Tale: The Life of Martha Ballard, Based on Her Diary, 1785–1812* (New York: Knopf, 1990). For an early example in this vein, see Mary R. Beard, *America Through Women's Eyes 1933* (New York: Greenwood Press, 1969); See also Nina Baym, *American Women Writers and the Work of History, 1790–1860* (Piscataway: Rutgers University Press, 1995); Joan Jacobs Brumberg, *The Body Project: An Intimate History of American Girls* (New York: Random House, 1997); Thomas Dublin, *Farm to Factory: Woman's Letters, 1830–1860* (New York: Columbia University Press, 1981); Carol Fairbanks and Sarah Brooks Sundberg, *Farmwomen on the Prairie Frontier: A Sourcebook for Canada and the United States* (Metuchen, N.J.: Scarecrow Press, 1983); Elizabeth Hampstead, *Read This Only to Yourself: The Private Writings of Midwestern Women, 1880–1910* (Bloomington: Indiana University Press, 1982); Julie Roy Jeffrey, *Frontier Women: The Trans-Mississippi West: 1840–1880* (New York: Hill & Wang, 1979); Annette Kolodny, *The Land Before Her: Fantasy and Experience of the American Frontiers, 1630–1860* (Chapel Hill: University of North Carolina Press, 1984); Etta Madden, *Bodies of Life: Shaker Literature and Literacies* (Westport, Conn.: Greenwood Press, 1998); Lillian Schlissel, *Women's Diaries of the Westward Journey* (New York: Schocken Books, 1982); Joanna Stratton, *Pioneer Women: Voices from the Kansas Frontier* (New York: Simon and Schuster, 1981); Ann Taves, *Religion and Domestic Violence in Early New England: The Memoirs of Abigail Abbot Bailey* (Bloomington: Indiana University Press, 1989); and Margo Culley, who provides an extensive bibliography of women's published and unpublished diaries in her edited collection, *A Day at a Time: The Diary Literature of American Women from 1764 to the Present* (New York: Feminist Press at The City University of New York, 1985). See also the following authors for examples of the use of popular culture and popular fiction: Richard Brodhead, *Cultures of Letters: Scenes of Reading and Writing in Nineteenth-Century America* (Chicago: University of Chicago Press, 1994); and the essays in Cathy Davidson, *Reading in America: Literature and Social History* (Baltimore: Johns Hopkins University Press, 1989). See also Michael Denning, *Mechanic Accents: Dime Novels and Working Class Culture in America* (New York: Verso, 1987); Karen Halttunen, *Murder Most Foul: The Killer and the American Gothic Imagination* (Cambridge, Mass.: Harvard University Press, 1998); Karen L. Kilcup, ed., *Nineteenth-Century American Women Writers: A Critical Reader* (Malden, Mass. and Oxford: Blackwell, 1998), and *Soft Canons: American Women Writers and Masculine Tradition* (Iowa City: University of Iowa Press, 1999); Lawrence Levine, *Highbrow/Lowbrow: The Emergence of Cultural Hierarchy in America* (Cambridge, Mass.: Harvard University Press, 1988); David Reynolds, *Beneath the American Renaissance* (New York: Knopf, 1988); Jane Tompkins, *Sensational Designs: The Cultural Work of American Fiction, 1790–1860* (New York: Oxford University Press, 1985).

16. Sarah Orne Jewett, *The Country of the Pointed Firs and Other Stories,* preface by Willa Cather (1956; reprint, New York: Doubleday, 1989), 6.

Gender, Genus, and Genre (pages 9–25)

1. Lorraine Anderson, ed., *Sisters of the Earth: Women's Prose and Poetry About Nature* (New York: Vintage, 1991), xvi.
2. Henry D. Thoreau, *Walden,* ed. J. Lyndon Shanley (1854; Princeton, N.J.: Princeton University Press, 1971); Susan Fenimore Cooper, *Rural Hours,* ed. Rochelle Johnson and Daniel Patterson (1850; Athens: University of Georgia Press, 1998).
3. For an early and influential articulation of Thoreau's distinctiveness from the natural history essayists, see the discussion of "Thoreau and the Thoreauists" in Joseph Wood Krutch, *Great American Nature Writing* (New York: William Sloane Associates, 1950), 71–78. A more recent proponent of this view is Paul Brooks. "Though many fine works in American natural history were published before his time," Brooks argues, "Henry David Thoreau is generally considered the father of the nature essay as a literary form." Paul Brooks, *Speaking for Nature: How Literary Naturalists from Henry Thoreau to Rachel Carson Have Shaped America* (Boston: Houghton Mifflin, 1980), ix.
4. Such texts deserve our attention, according to Sharon M. Harris, because they "challenge our conceptions of what we mean by 'literature,' by the recording of history, and by women's roles in Native, colonial, revolutionary, and federal America. Cumulatively, these texts also begin to define early women's poetics and therein offer many opportunities for reconceptualizing studies of later women's writings and early American studies in general." See Sharon M. Harris, ed., *American Women Writers to 1800* (New York: Oxford University Press, 1996), 161.
5. For an example of the insights gained when seemingly untraditional subjects receive renewed critical attention, see Laurel Thatcher Ulrich, *A Midwife's Tale: The Life of Martha Ballard, Based on Her Diary, 1785–1812* (New York: Knopf, 1990).
6. My choice of topic and the constraints of space prevent me from discussing writing about nature by many other early American women. Much of the work of the three most prominent women writers of the period—Mary Rowlandson, Anne Bradstreet, and Phillis Wheatley—refers to the natural world. Fortunately, their writings have already been studied at length. The writings of other early American women—such as Mary Wright Cooper, Mary Coburn Dewees, and Mary Bartlett—have received far less attention. The songs and stories by and about early Native American women also pose substantial textual and cultural issues that I cannot adequately address here. A comprehensive analysis of early American women's nature writing would obviously need to consider all of these texts and more.
7. Biographical sketches of all three women appear in John A. Garraty and Mark C. Carnes, gen. eds., *American National Biography* (New York: Oxford University Press, 1999); Edward T. James, ed., *Notable American Women, 1607–1950: A Biographical Dictionary* (Cambridge, Mass.: Harvard University Press, 1971); Carla A. Mulford, ed., *American Women Prose Writers to 1820* (Detroit: Gale Research, 1999); and Buckner Hollingsworth, *Her Garden Was Her Delight* (New York: Macmillan, 1962). Additional profiles of Pinckney appear in Harris, *American Women*

Writers to 1800, 111–12; Edward Nicholas, *The Hours and the Ages: A Sequence of Americans* (New York: William Sloan Associates, 1949), 7–38; Eliza Lucas Pinckney, *The Letterbook of Eliza Lucas Pinckney, 1739–1762*, ed. Elise Pinckney (Chapel Hill: University of North Carolina Press, 1972), hereafter cited in text as *LEP;* Harriott Horry Ravenel, *Eliza Pinckney* (New York: Scribner's, 1896); Constance B. Schulz, "Eliza Lucas Pinckney (1722–1793)," in *Portraits of American Women: From Settlement to the Present*, ed. G. J. Barker-Benfield and Catherine Clinton (New York: St. Martin's, 1991), 65–81; Frances Leigh Williams, *Plantation Patriot: A Biography of Eliza Lucas Pinckney* (New York: Harcourt, Brace, 1967); and Nancy Woloch, *Women and the American Experience* (New York: Knopf, 1984), 51–64. Additional profiles of Colden appear in Marcia Myers Bonta, *Women in the Field: America's Pioneering Women Naturalists* (College Station: Texas A&M University Press, 1991), 5–8; James Britten, "Jane Colden and the Flora of New York," *Journal of Botany* 33 (1895): 12–15; Jane Colden, *Botanic Manuscript,* ed. H. W. Rickett and Elizabeth C. Hall (New York: Chanticleer Press, 1963), hereafter cited in text as *BM;* Harris, *American Women Writers to 1800*, 105; Brooke Hindle, "A Colonial Governor's Family: The Coldens of Coldengham," *New York Historical Society Quarterly* 45 (1961): 233–50; Beatrice Scheer Smith, "Jane Colden (1724–1766) and Her Botanic Manuscript," *American Journal of Botany* 75 (1988): 1090–96; and Anna Murray Vail, "Jane Colden, An Early New York Botanist," *Torreya* 7 (1907): 21–34. Unless otherwise noted, biographical details were compiled from these sources. No visual images of these women are known to exist.

8. Joan Hoff Wilson, "Dancing Dogs of the Colonial Period: Women Scientists," *Early American Literature* 7 (1973): 230.

9. C. P. Snow, *The Two Cultures: and a Second Look* (New York: Cambridge University Press, 1964).

10. John Elder, ed., *American Nature Writers* (New York: Scribner's, 1996), xiii.

11. Howard Mumford Jones, *O Strange New World: American Culture: The Formative Years* (New York: Viking, 1964), 345. Three good surveys of science in eighteenth-century America are John C. Greene, *American Science in the Age of Jefferson* (Ames: Iowa State University Press, 1984); Brooke Hindle, *The Pursuit of Science in Revolutionary America, 1735–1789* (Chapel Hill: University of North Carolina Press, 1956); and Raymond Phineas Stearns, *Science in the British Colonies of America* (Urbana: University of Illinois Press, 1970).

12. Pinckney's letters about the natural world were written mostly during the 1740s, Logan's gardener's calendar first appeared in 1752, and Colden completed her botanic manuscript in the 1750s. Crèvecoeur's *Letters from an American Farmer* was published in 1782; White's *Natural History of Selborne* in 1789; and Bartram's *Travels* in 1791.

13. Ann B. Shteir has also observed this pattern: "During the Linnaean years, many women came to botany through their families and became helpmates and 'fair associates' to fathers, husbands, or brothers. A pattern of family apprenticeship is not unique to botanical or scientific culture but mirrors instead the connection through which women came into art or music." See Ann B. Shteir, *Cultivating Women, Cultivating Science: Flora's Daughters and Botany in England, 1760–1860* (Baltimore: Johns Hopkins University Press, 1996), 50.

14. Interestingly, all three women also corresponded with the Charleston naturalist Alexander Garden.

15. Christoph Irmscher makes a related claim about the pursuit of pre-Darwinian natural history by men. "American natural history," he claims, . . . "only superficially avoids what it very often becomes—a form of autobiography." See Christoph Irmscher, *The Poetics of Natural History: From John Bartram to William James* (New Brunswick, N.J.: Rutgers University Press, 1999), 8.

16. The maiden name of her mother, Anne, is unknown.

17. Eliza's two brothers, George, Jr., and Thomas, remained at school in London.

18. Charles Town became Charleston in 1783, after the Revolution.

19. "Lucerne" is alfalfa; "Casada" is probably cassava, whose root is a food in the tropics.

20. David Ramsay, *The History of South Carolina*, 2 vols. (1809), vol. 2, 212. The loss of the bounty, competition from the East Indies, and the rise of cotton eventually drove Carolina indigo out of production by the end of the century. For more on indigo production in the South, see Lewis Cecil Gray, *History of Agriculture in the Southern United States to 1860* (Washington, D.C.: Carnegie Institution, 1933), 290–97.

21. A 1740 letter from Eliza to her father, in which she declines two potential suitors, suggests that she may have had an "Inclination" toward Charles Pinckney since at least that time. She writes: "I hope heaven will always direct me that I may never disappoint you; and what indeed could induce me to make a secret of my Inclination to my best friend, as I am well aware you would not disapprove it to make me a Sacrifice to Wealth, and I am as certain I would indulge no passion that had not your aprobation." See Pinckney, *Letterbook*, 6.

22. For Eliza's later letters, see Elise Pinckney, ed., "Letters of Eliza Luca Pinckney, 1768–1782," *South Carolina Historical Magazine* 76 (1975), 143–70.

23. For more on science in colonial Charleston, see "The Charles Town Scientific Community" in Stearns, *Science*, 593–619.

24. Richard Beale Davis, *Intellectual Life in the Colonial South, 1585–1763* (Knoxville: University of Tennessee Press, 1978), 359.

25. Hennig Cohen, *The South Carolina Gazette, 1732–1775* (Columbia: University of South Carolina Press, 1953), 31.

26. Ann Manigault, "Extracts from the Journal of Mrs. Ann Manigault, 1754–1781," with notes by Mabel L. Webber, *South Carolina Historical and Genealogical Magazine* 20 (1919), 205n9.

27. See Cohen, *South Carolina Gazette,* 55; and Edmund Berkeley and Dorothy Smith Berkeley, *Dr. Alexander Garden of Charles Town* (Chapel Hill: University of North Carolina Press, 1969), 154n4.

28. Manigault, "Extracts," 205n9.

29. Manigault, "Extracts," 205; Julia Cherry Spruill, *Women's Life and Work in the Southern Colonies,* introduction by Anne Firor Scott (1938; New York: Norton, 1998), 278.

30. John Bartram, *The Correspondence of John Bartram, 1734–1777,* ed. Edmund Berkeley and Dorothy Smith Berkeley (Gainesville: University Press of Florida, 1992), 517; hereafter cited in text as *CJB*. Bartram frequently failed to punctuate the end of a sentence and capitalize the first word of the next sentence. To increase readability, I have added initial capital letters and closing punctuation marks where needed. Why Bartram called Logan "an elderly widow Lady" in 1761 is unclear, since her husband did not die until 1 July 1764. Logan's offhand remark, in a 14 September 1764 letter, that she would send Bartram some plants in the spring of 1765, "if I live till then," suggests that her husband's death may have affected her own physical

or psychological health, although she makes no explicit mention of his passing. See Bartram, *Correspondence,* 636–37. Logan herself died on 28 June 1779.

31. A notice in the *South Carolina Gazette* of 6 December 1751 announces: "Just Published, and to be sold by the Printer hereof, The South Carolina Almanack for the year 1752," which includes a "Gardners Kalander, done by a Lady of this Province and esteemed a very good one." See Mabel L. Webber, comp., "South Carolina Almanacs to 1800," *South Carolina Historical and Genealogical Magazine* 15 (1914): 73.

32. A reprint of the 1756 calendar appears in Ann Leighton, *American Gardens in the Eighteenth Century: "For Use or For Delight"* (Boston: Houghton Mifflin, 1976), 211–15. A booklet reprinting the monthly entries of the 1756 and 1798 calendars next to one another appeared in 1976. See Alice Logan Wright, ed., *A Gardener's Kalendar: Done by a Colonial Lady* (Charleston: National Society of the Colonial Dames of America in the State of South Carolina, 1976).

33. Ramsay, *History of South Carolina,* vol. 2, 227.

34. Sarah Pattee Stetson, "American Garden Books Transplanted and Native, Before 1807," *William and Mary Quarterly* (ser. 3) 3 (1946): 353.

35. For more on the use of lunar astrology in eighteenth-century agriculture, see Herbert Leventhal, *In the Shadow of the Enlightenment: Occultism and Renaissance Science in Eighteenth-Century America* (New York: New York University Press, 1976), 39–44; and Agnes Arber, "The Doctrine of Signatures, and Astrological Botany," in *Herbals: Their Origin and Evolution: A Chapter in the History of Botany, 1470–1670,* 3rd ed. (1938; New York: Cambridge University Press, 1986), 247–63.

36. I quote from the unpaginated 1796 edition of the calendar, the first to be published under Logan's name.

37. Alexander Garden, "The Description of a New Plant," *Essays and Observations, Physical and Literary,* 2nd ser., 3 vols. (Edinburgh: Philosophical Society of Edinburgh, 1754–71), vol. 2 (1756), 1–7.

38. See Berkeley and Berkeley, *Dr. Alexander Garden,* 47–48. Garden eventually got the honor for which he had hoped when the botanist John Ellis later named the cape jasmine, *Gardenia jasminoides,* for him. See Berkeley and Berkeley, *Dr. Alexander Garden,* 158–62. For more on the naming of the *Gardenia,* see Margaret Denny, "Naming the Gardenia," *Scientific Monthly* 67 (1948), 17–22.

39. David Rains Wallace, "The Nature of Nature Writing," *New York Times Book Review,* 22 July 1984, 1.

40. Pamela Regis, *Describing Early America: Bartram, Jefferson, Crèvecoeur, and the Rhetoric of Natural History* (DeKalb: Northern Illinois University Press, 1992).

41. Robert Finch and John Elder, eds., *The Norton Book of Nature Writing* (New York: Norton, 1990), 19. See also Peter A. Fritzell, "The Mosquito in My Garden: Early American Selves and Their Nonhuman Environments," *Nature Writing and America: Essays Upon a Cultural Type* (Ames: Iowa State University Press, 1990), 109–52.

42. Cadwallader Colden, *The Letters and Papers of Cadwallader Colden,* vol. 5: 1755–1760, *Collections of the New York Historical Society* 54 (1921): 203; hereafter cited in text as *LPCC.*

43. Cadwallader Colden, "Plantae Coldenghamiae in provincia Noveboracensi Americes sponte crescentes," *Acta Societatis Regiae Scientiarum Upsaliensis,* vol. 4 (1743), 81–136; vol. 5 (1744–50), 47–82.

44. For more on Colden's scientific pursuits, see Louis Leonard Gitin, "Cadwallader

Colden as Scientist and Philosopher," *New York History* 16 (1935): 169–77; and Stearns, *Science,* 559–75. See also Asa Gray, ed., "Selections from the Scientific Correspondence of Cadwallader Colden with Gronovius, Linnaeus, Collinson, and other Naturalists," *American Journal of Science and Arts* 44 (1843): 85–133.

45. Like John Bartram, Cadwallader Colden frequently failed to punctuate the end of a sentence. To increase readability, I have added closing punctuation marks where needed. Although Colden's tone may seem condescending to modern readers, Colden was in fact quite progressive for his time. More than ten years earlier, Colden had made similar remarks about women and botany in a 13 November 1742 letter to Peter Collinson: "The Ladies are at least as well fitted for this Study as the men by their natural curiosity & the accuracy & quickness of their Sensations. It would give them means of imploying many idle hours both usefully & agreably." See Colden, *Letters and Papers,* vol. 2, *Collections* 51 (1918), 282. He was also quite proud of Jane's accomplishments, as he noted in a 15 February 1758 letter, probably to Robert Whytt: "she is more curious & accurate than I could have been her descriptions are more perfect & I believe few or none exceed them." See Colden, *Letters and Papers,* vol. 5, *Collections* 54 (1921), 217.

46. James Edward Smith, ed., *A Selection of the Correspondence of Linnaeus and Other Naturalists* (1821; New York: Arno Press, 1978), vol. 1, 39; hereafter cited in text as *SCL.*

47. Colden was not afraid, however, to disagree with Linnaeus (at least in private) if the evidence demanded it. Of the Seneca snakeroot (*Polygala senega*), she wrote in her botanical journal: "Linnaeus describes this as being a Papilionatious Flower, and calls the two largest Leaves of the Cup *Alae,* but as they continue, till the Seed is ripe and the two flower Leaves, and its appendage fold together. I must beg Leave to differ from him." See Colden, *Botanic Manuscript,* 53. In *The Social Ladder,* Mrs. John King Van Rensselaer states that the Swedish botanist Peter Kalm also paid tribute to Jane by naming a wild flower after her. According to Van Rensselaer, "His delicate compliment was not received cordially by the Colden family. To them, publicity of this kind seemed shocking, if not disgraceful. 'What!' exclaimed an aunt of Jane's, 'name a weed after a Christian woman!'" Although Kalm did meet Jane during his travels in America from 1748 to 1751, I have found no evidence to support Van Rensselaer's claim. See Mrs. John King Van Rensselaer, *The Social Ladder* (New York: Henry Holt, 1924), 116–17. For more on Kalm's visit to America, see Martti Kerkkonen, *Peter Kalm's North American Journey: Its Ideological Background and Results* (Helsinki: Finnish Historical Society, 1959). See also the correspondence of Cadwallader Colden with Kalm in Colden, *Letters and Papers,* vol. 4, *Collections* 53 (1920).

48. Raymond Phineas Stearns has argued that the work of Jane Colden "excited considerable comment in the natural history circle, more, it would appear, because of her sex than because of her abilities . . . [N]o one appears to have accepted her work on the same level as that of her father and of other male collectors in the colonies. However able Jane Colden may have become as a botanist, her unconventional activities as a woman relegated her to the outer rim of the natural history circle." Joan Hoff Wilson similarly uses Colden's reception as an example of her argument that women scientists were "dancing dogs of the colonial period." In both cases, my own interpretation of the evidence suggests otherwise. See Stearns, *Science,* 566–67; and Wilson, "Dancing Dogs," 225–28.

49. See John Brett Langstaff, *Doctor Bard of Hyde Park,* introduction by Nicholas Murray Butler (New York: E. P. Dutton, 1942), 47. See also John McVickar, *A Domestic Narrative of the Life of Samuel Bard, M. D.* (New York: The Literary Rooms, 1822).

50. And a successful experiment at that. Walter Rutherfurd, who visited Coldengham in the mid-1750s, wrote to a friend in Scotland about her efforts: "She makes the best cheese I ever ate in America." See Vail, "Jane Colden," 32.

51. For a vivid example of her affection for Farquhar, see her 16 May 1759 letter to her sister Katharine in Colden, *Letters and Papers,* vol. 9, *Collections* 68 (1935), 175–76.

52. Titled "Flora Nov.-Eboracensis," it can be found in the British Museum (Natural History) Catalogue, No. 26.e.19. In 1963, the Garden Club of Orange and Dutchess Counties, New York, published fifty-seven of her drawings and descriptions as Jane Colden's *Botanic Manuscript,* and it is to this publication that I will refer. See Colden, *Botanic Manuscript.*

53. Arber, *Herbals,* 268. For more on herbals, medicine, and botany in the English Renaissance, see A. C. Crombie, *Medieval and Early Modern Science,* vol. 2 (Cambridge, Mass.: Harvard University Press, 1967); Allen G. Debus, *Man and Nature in the Renaissance* (New York: Cambridge University Press, 1978); and Karen Meier Reeds, "Renaissance Humanism and Botany," *Annals of Science* 33 (1977): 519–42.

54. Collinson refers to John Martyn's translations of Joseph Pitton de Tournefort's *Compleat Herbal: or, The botanical institutions of Mr. Tournefort* (London, 1716–30) and *History of plants growing about Paris, with their uses in physick; and a mechanical account of the operation of medicines* (London, 1732). Her father had also requested a copy of Robert Morison, *Plantarum historiae universalis Oxoniensis* (Oxford, 1680–99), which Collinson did not send.

55. According to Joseph Kastner in *A Species of Eternity,* William Bartram had written to Colden, asking for help with his drawing, but soon received the reply: "I can teach you no more. You have already surpassed me in skill and imagination" (p. 80). Unfortunately, this quotation appears to be fabricated, part of Marjory Bartlett Sanger's "interpretive" biography, *Billy Bartram and His Green World* (New York: Farrar, Straus, and Giroux, 1972), 31.

56. See Vail, "Jane Colden," 32.

57. See Stearns, *Science,* 567n142.

58. See Britten, "Jane Colden," 14; Hollingsworth, *Her Garden Was Her Delight,* 30; Joseph Ewan, ed., *Botanical and Zoological Drawings, 1756–1788,* by William Bartram (Philadelphia: American Philosophical Society, 1968), 4; and Colden, *Botanic Manuscript,* 25.

59. Vera Norwood, *Made from this Earth: American Women and Nature* (Chapel Hill: University of North Carolina Press, 1993), 60.

60. For more on the illustration of herbals, see Frank J. Anderson, *An Illustrated History of the Herbals* (New York: Columbia University Press, 1977); Arber, *Herbals,* 185–246; Wilfred Blunt and Sandra Raphael, *The Illustrated Herbal* (New York: Thames and Hudson, 1979); and Gill Saunders, *Picturing Plants: An Analytical History of Botanical Illustration* (Berkeley: University of California Press, 1995).

61. The roots of *Polygala senega* and *Aristolochia serpentaria* were reputed to be antidotes for snake venom.

62. John Clayton and John Frederic Gronovius, *Flora Virginica,* 2nd ed. (Lugduni Batavorum, 1762).

63. Apparently, Garden had been rebuked once before for his overzealousness in praising Jane's abilities. In a 23 May 1755 letter, Garden wrote to Colden:

> By your second letter I find that I have very innocently offended Both you & Miss Colden by some expressions that insensibly dropt from my pen as archetypes of what my heart dictated in was on sincerity. This gives me real concern & give me leave to assure you I shall endeavour as far as in my power to amend any thing in my conduct or manner of writing that you are kind enough to point out as wrong. I trust that Both you & your Daughter will forgive me for once, I shall be more sparing in saying what I think is due to such merit for the future.

See Colden, *Letters and Papers,* vol. 5, *Collections* 54 (1921), 11.

64. William Martin Smallwood and Mabel Sarah Coon Smallwood, *Natural History and the American Mind* (New York: Columbia University Press, 1941), 92–93. The letter also contains a postscript describing the best ways of conveying correspondence between Coldengham and Edinburgh.

65. Greg Dening, "Introduction: In Search of a Metaphor," *Through a Glass Darkly: Reflections on Personal Identity in Early America,* ed. Ronald Hoffman, Mechal Sobel, and Fredrika J. Teute (Chapel Hill: University of North Carolina Press, 1997), 2.

66. Elder, *American Nature Writers,* xiii.

Sentimental Ecology (pages 27–36)

1. Susan Fenimore Cooper, *Rural Hours,* eds. Rochelle Johnson and Daniel Patterson (1850; reprint, Athens: University of Georgia Press, 1998). First published in the summer of 1850, the book was so much a success that her father negotiated with Putnam, the publisher, at the end of November 1850 to obtain a better bargain and higher percentage per copy. See *Rural Hours,* ed. David Jones (1887; reprint, Syracuse: Syracuse University Press, 1968), xxiv. The book eventually went through five more American and one more English edition, including a "fine" volume complete with color plates of local birds. Susan Cooper produced a new edition in 1868 with a new preface, and a "new and revised" edition in 1887, which condensed a great deal of the original.

2. Susan Fenimore Cooper edited the 1853 American edition of John Leonard Knapp's *Country Rambles in England; or Journal of a Naturalist* (Buffalo, N.Y.: Phinney And Co., 1853).

3. Lawrence Buell, *The Environmental Imagination: Thoreau, Nature Writing, and the Formation of American Culture* (Cambridge, Mass.: Harvard University Press, 1995), 266.

4. Sandra Zagarell, "Narrative of Community: The Identification of Genre," *Signs: Journal of Women in Culture and Society* 13, no. 3 (1988): 499. My emphasis.

5. Max Oelschlaeger, *The Idea of Wilderness* (New Haven: Yale University Press), 104.

6. Buell, *Environmental Imagination,* 406.

7. Alan Taylor, *William Cooper's Town: Power and Persuasion on the Frontier of the Early American Republic* (New York: Vintage, 1995), 119. William Cooper, Susan's

grandfather and the founder of Cooperstown, began heavily promoting Otsego County as a major source of maple sugar in June of 1789. In *The Pioneers* (1823; reprint New York: Signet, 1964, 1980), James Cooper's most environmentally informed novel, Judge Marmaduke Temple forbids the burning of maple wood in his estate's fireplaces both because maple trees are "precious gifts of nature," and because they are "mines of comfort and wealth" (101). Thus we can see maple sugar in terms of an important economic commodity, as a product of the Otsego region, and as an ecological treasure.

8. Buell, *Environmental Imaginations,* 407.

9. Ibid., 220.

10. *Rural Hours,* 4. Most references to *Rural Hours* are from the 1850 edition, reprinted in 1998 and edited by Rochelle Johnson and Daniel Patterson (Athens: University of Georgia Press, 1998). References to Cooper's later additions such as "Later Hours" are from the 1887 edition, reprinted in 1968 and edited by David Jones (Syracuse: Syracuse University Press, 1968). References are cited parenthetically in the text.

11. Thoreau also "questions seasonal categorization rigorously" by seeking to "free himself 'from the tyranny of chronological time,' to redefine November for example from a 'calendrical unit' to a 'phenomenological category of thought.'" Buell, *Environmental Imagination,* 228.

12. Buell, *Environmental Imagination,* 223.

13. *Rural Hours,* 98.

14. The importance of the almanac in creating the domestic pastoral also appears in Nathaniel Hawthorne's short story "Roger Malvin's Burial," in *Mosses From an Old Manse: The Centenary Edition of the Works of Nathaniel Hawthorne,* vol. X [n.p.] (Ohio State University Press, 1974), 337–60. When the family leaves the settlement to travel into the unknown wilderness, they carry with them "the current year's Massachusetts Almanac, which, with the exception of an old black-letter Bible, comprised all the literary wealth of the family" (354). These two literary works help the family in attempting to bring domestic comfort into the wilds, and thus also mediate between civilization and wild nature.

15. *Rural Hours,* 96.

16. The spotless dairy the mistress keeps rivals that of Aunt Fortune in Susan Warner's domestic sentimental novel, *The Wide Wide World,* and the industry in the dairy is mirrored by the other household duties.

17. *Rural Hours,* 100.

18. Vera Norwood, *Made from This Earth: American Women and Nature* (Chapel Hill: University of North Carolina Press, 1993), 29–30.

19. *Rural Hours,* 194.

20. Ibid.

21. Norwood, *Made from This Earth,* 28.

22. Ibid., 12.

23. Many of the pseudo-scientific works for women were condescending at best. One French author (L. F. Raban) avoids scientific terms because of their difficulty. Norwood, *Made from This Earth,* 13. Other books reduced their scientific goals even more, completely giving up any pretense of being scientific and provided "simplified codes for love, flattery, jealousy, motherhood, and nationalism." Norwood, *Made from This Earth,* 18.

24. *Rural Hours,* 83.
25. The relationship between stewardship and naming is clear in Genesis, when Adam is given the task of naming. By naming the plants and animals, Adam not only gains sovereignty but also undertakes a strong measure of responsibility at the same.
26. *Rural Hours,* 83.
27. Annie Finch, "The Sentimental Poetess in the World: Metaphor and Subjectivity in Lydia Sigourney's Nature Poetry," *Legacy* 5, no. 2 (1988): 5.
28. *Rural Hours,* 86. Cooper's italics.
29. Ann Bermingham, in *Landscape and Ideology: The English Rustic Tradition, 1740–1860* (Berkeley: University of California Press, 1986), proposes that "there is an ideology of landscape" that "embodied a set of socially and, finally, economically determined values" (3). The ideology of landscape as represented by both painters and poets reflects national concerns because the landscape is the face of the country. In developing a set of expressions and techniques to understand and manipulate landscape representations, artists and writers participate in the creation of a national identity.
30. James McKusick, "'A language that is ever green': The Ecological Vision of John Clare," *The University of Toronto Quarterly* 61, no. 2 (1991 Winter): 242. McKusick's term is perhaps more apt in this instance. His neologism, "ecolect," is used to describe the poetry of the British rustic poet John Clare, and refers to an idiolect that is specifically rooted in the local landscape. Because of the intimate connection between nature or landscape poetry and the natural environment, this term has more resonance.
31. *Rural Hours,* 202. (Cooper's italics.)
32. The Coopers lived in Paris from 1826 until 1833 (Jones xiv).
33. *Rural Hours,* 222–23.
34. Ibid., 209.
35. Ibid.
36. Ibid., 210.
37. Ralph Waldo Emerson, "The Poet," *Selections from Ralph Waldo Emerson,* ed. Stephen E. Whicher (Boston: Houghton, 1957), 238.
38. The novel centers around a young woman, Elinor, who lives with her doting grandfather and aunt in a rural village, probably a fictionalized Cooperstown, about "fifteen years since," or in the late 1820s and early 1830s. Elinor, who is often described as "plain," or, worse yet, ugly, is pleasant, kind, and popular among the upper classes in Longbridge. As the action begins, Elinor accepts a proposal of marriage from her long-time friend, Harry Hazlehurst, and seems to be on the path toward domestic happiness. Harry, though, soon falls in love with another family friend, a beautiful but shallow woman whose morals, sense, and intelligence pale beside Elinor's. This shallow woman, Jane, is not interested in Hazlehurst but marries a young rake who leads her to the brink of financial ruin before he dies. In the meantime, a presumably long-lost heir arrives in Longbridge to challenge Harry's right to his fortune. Harry eventually triumphs over the impostor, regains his inheritance, and finally marries Elinor by the end of the novel. While these events are transpiring, myriad subplots are unraveling. Another young friend of the Wyllys family, Charlie Hubbard, determinedly pursues his passion for landscape painting, becoming an important practitioner and advocate of American art before his untimely death in a boating accident. Other minor

characters marry, make fortunes, lose fortunes, or fall into scandal and disrepute. Although the number of characters and plot developments can be a distraction, each event and person has a distinct effect on Elinor, and Elinor's family often influences other characters in important ways.

39. Rosaly Torna Kurth, "Susan Fenimore Cooper: A Study of Her Life and Works" (Ph.D. diss., Fordham University, 1974), 124, 112.

40. In the "Preface to the Leatherstocking Tales," which opens *The Deerslayer,* James Cooper addresses critics who claim that he has presented a "more favorable picture of the redman than he deserves." He goes on to say, "The critic is understood to have been a very distinguished agent of the government, one very familiar with Indians, as they are seen at the councils to treat for the sale of their lands, where little or none of their domestic qualities come in play, and where indeed, their evil passions are known to have fullest scope" (x). Here Cooper seems to understand the Indians, and is able to make excuses for their "evil passions," but he fails to see that the Indians have been wronged in any way. Instead, they are at a disadvantage in the councils, so their lesser nature comes in play.

41. *Rural Hours,* 209.

42. Ibid., 108.

43. Susan M. Levin, "Romantic Prose and Feminine Romanticism" *Prose Studies* 10, no. 2 (Sept. 1987): 178–95. Levin points out that "romantic women writers find the forms of conventional language at odds in some way with the realities their writing presents," which in turn creates this Kristevan discourse (185). As Levin notes, this discourse contradicts standard language by its infantile search for a maternal figure and appears "without discernible contours" (186). I propose taking this further by saying that Susan Cooper does not want to fuse with the mother in this example, but wants to become the nurturing figure.

44. *Rural Hours,* 112.

45. Ibid.

46. Walter Levy and Christopher Hallowell, the editors of the eco-conscious textbook reader, *Green Perspectives,* have chosen this passage as the first excerpt.

47. *Rural Hours,* 128.

48. Ibid., 133.

49. Ibid., 132.

50. Ibid., 135.

51. Ibid.

"Appropriated Waters" (pages 37–46)

1. Henry David Thoreau, *Walden* (New York: Norton, 1992), 129.

2. Mary Austin, *The Land of Little Rain* (Albuquerque: University of New Mexico Press, 1974), 139.

3. Ibid.

4. In an introductory chapter to his anthology of American nature writing, for example, Thomas Lyon writes that "Thoreau placed science and fact as helpful tools within the transcending philosophical context of experience, and presented this view of the world in accomplished, artistic prose. It is thus fair to say that the possibilities of the nature

essay as a modern literary form were first outlined in Thoreau's first essay, published in July, 1842" (Thomas J. Lyon, *This Incomperable Lande* [New York: Penguin, 1991], 52). In Lawrence Buell's *The Environmental Imagination* (Cambridge: Harvard University Press, 1995), Buell notes of Thoreau that "no writer in the literary history of America's dominant subculture comes closer than he to standing for nature in both the scholarly and popular mind," even though "Environmental nonfiction predates Thoreau by a longer span than he predates us" (2). Buell continues, "a mere generation after Thoreau it began to seem as if nothing had preceded him. Late nineteenth-century readers began to forget that Audubon, Wilson, and the Bartrams had existed, at least as literary figures; they derived Thoreau from Emerson, then marked Thoreau as the effective start of the nature essay. They exaggerated both Thoreau's originality and Emerson's influence on the genre" (397–98). More recently, in Randall Roorda's *Dramas of Solitude* (Albany: SUNY Press, 1998), Roorda observes that "'Thoreau' refers to the model text and a model experience, in both of which nature writing is generically constituted" (27).

5. Mikhail Bakhtin, *The Dialogic Imagination,* trans. Caryl Emerson and Michael Holquist (Austin: University of Texas Press, 1981), 272.

6. Ibid., 280.

7. Aside from admiring its eloquence and acumen, my citation of Bakhtin's passage is meant to serve a specific purpose. In my understanding, Bakhtinian theory can be situated as a sort of turning point between a theory of historical influence that one might judge as being myopic in scope, and a theory of intertextuality in which an author becomes a function of language, the text itself then necessarily unable to offer an intended "political" critique. Though the effectivity of my readings of particular passages lies in their correspondence in *Walden* and *Land,* the two texts can be profitably read against each other regardless of authorial intention. In citing Bakhtin, my own intention is to acknowledge, yet avoid, the pitfalls of the two types of "influence criticism" that precede and follow him: that is, to have my interpretive cake and eat it, too. For a brief but informative overview of the history of theories of literary influence, see Jay Clayton and Eric Rothstein's "Figures in the Corpus: Theories of Influence and Intertextuality," in *Influence and Intertextuality in Literary History,* ed. Jay Clayton and Eric Rothstein (Madison: University of Wisconsin Press, 1991).

8. Mary Austin, *Earth Horizon* (Boston: Houghton Mifflin, 1932), 165.

9. Ibid., n. 371.

10. Ibid., 112.

11. Ibid.

12. Austin, *The Land of Little Rain,* 103.

13. Austin, *Earth Horizon,* 371.

14. Ibid., 5.

15. Thoreau, *Walden,* 131.

16. Ibid., 132.

17. Austin, *The Land of Little Rain,* xv. See also Annie Dillard, *Pilgrim at Tinker Creek* (N.Y.: Harper & Row, 1985), 54–56.

18. Austin, *The Land of Little Rain,* 81.

19. Vera Norwood, "Heroines of Nature," in *The Ecocriticism Reader* ed. Cheryll Glotfelty and Harold Fromm (Athens: University of Georgia Press, 1996), 334; Elizabeth Ammons, *Conflicting Stories* (New York: Oxford, 1991), 92; Buell, *The Environmental*

Imagination, 177 and 175. Critics such as Lois Rudnick have noted the significance of this gesture, not only in *Land* but in *California* and *The Land of Journey's Ending.* In *Re-Naming the Land,* she states that Austin "rejects Anglo names for geographic landmarks and uses the original Indian and Hispanic designations because they express the land's natural characteristics rather than the individual discoverer's ego" ("Re-Naming the Land: Anglo Expatriate Women in the Southwest," in *The Desert Is No Lady,* ed. Vera Norwood and Janice Monk [New Haven: Yale University Press, 1987] 16). Rudnick also emphasizes Austin's belief that an "authentic American culture must be rooted in the land where artists lived and must build upon the available indigenous expressions" (24).

20. Austin, *The Land of Little Rain,* 82.
21. Thoreau, *Walden,* 129.
22. Ibid., 130.
23. Austin, *The Land of Little Rain,* 84, 40.
24. Thoreau, *Walden,* 154.
25. Austin, *The Land of Little Rain,* 22.
26. Thoreau, *Walden,* 140, 146.
27. Ibid., 143.
28. Austin, *The Land of Little Rain,* 3.
29. Ibid., 9.
30. Ibid., 72.
31. Ibid., 46.
32. I am reminded of a remark made by Herman Melville upon reading some of Emerson's writing, as noted by F. O. Mathiessen: "To one who has weathered Cape Horn as a common sailor what stuff all this is" (*American Renaissance* [New York: Oxford University Press, 1968], 158).
33. For example, see "Sex Emancipation Through War," originally published in 1918, in Reuben J. Ellis's collection of a portion of Austin's nonfiction:

> The world is really a very feminine place, a mother's place, conceptive, brooding, nourishing: a place of infinite patience and infinite elusiveness. It needs to be lived in more or less femininely, and the chief reason why we have never succeeded in being quite at home in it is that our method has been almost exclusively masculine. We have assaulted the earth, ripped out the treasure of its mines, cut down its forests, deflowered its fields and left them sterile for a thousand years. We have lived precisely on the same terms with our fellows, combatively, competitively, geocentrically. Nations have not struggled to make the world a better place, but only to make a more advantageous place in it for themselves. Man invented the State in the key of maleness, with combat for its major occupation, profit the spur and power the prize. This is the pattern of our politics, our economic and our international life, a pattern built not on common human traits of human kind, but on dominant sex traits of the male half of society. It is even marked, in certain quarters of the earth, with intrinsic male weaknesses, the strut, the flourish, the chip-on-its-shoulder. The greed of exclusive possessions, the mastery of the seas, the control of world finance. (*Beyond Bor-*

ders: The Selected Essays of Mary Austin, ed. Reuben J. Ellis [Carbondale: Southern Illinois University Press 1996], 45).

34. See Michael T. Gilmore, "Walden and the 'Curse of Trade,'" in *Ideology and Classic American Literature,* ed. Sacvan Bercovitch and Myra Jehlen (Cambridge: Cambridge University Press, 1986), 293–310.
35. Thoreau, *Walden,* 12.
36. See Gilmore, "Walden and the 'Curse of Trade.'"
37. Ester Stineman, *Mary Austin: Song of a Maverick* (New Haven: Yale University Press, 1989), 60.
38. Ibid., 53.
39. Austin, *The Land of Little Rain,* 106.
40. Ibid.
41. Thoreau, *Walden,* 76.
42. Austin, *The Land of Little Rain,* 110.
43. Ibid.
44. Thoreau, *Walden,* 3.
45. Austin, *The Land of Little Rain,* 110.
46. Ibid., 139.
47. Ibid., 141.

"Recounting" the Land (pages 47–54)

1. Mary Austin, Letter to Carey McWilliams, 12 July 1930 (unpublished manuscript in the Mary Austin Collection, The Huntington Library, Pasadena, Calif.).
2. While Austin and her work are becoming more well known, a brief sketch of her life remains helpful. Born Mary Hunter in 1868, Austin grew up in Carlinville, Illinois, and graduated from Blackburn College in 1888. That same year, she moved west with her mother and brother, who had reserved a homestead claim in southern California, near the area of General Edward Beale's Tejon Ranch and Fort Tejon. After marrying Wallace Austin in 1891, Mary Austin moved to the arid Owens Valley region, east of the Sierras and west of Death Valley and the Mojave Desert. Here, while raising her mentally handicapped daughter and relying on her husband's inconsistent income, she began in earnest to write for publication. By the mid-1900s, she was splitting her time among the Owens Valley, San Francisco, and the writers colony at Carmel.

   Eventually, Austin divorced Wallace and transferred custody of her daughter to an institution. Around 1910, she left California, moving first to Europe and then to New York. In 1924, she returned to the West, spending the rest of her life in Santa Fe, but she never again lived in California. Although Austin remained extremely prolific throughout her life (publishing some thirty-five novels and book-length works and some two hundred articles and stories), and although she often returned to her earlier concerns of region, gender and race, her later work never addressed the Western land as directly and intimately as her California writings had. Austin died in Santa Fe in 1934.

   See also Mary Austin, *Earth Horizon* (New York: The Library Guild, 1932); T. M. Pearce, *The Beloved House* (Caldwell, Idaho: The Caxton Printers, Ltd., 1940) and *Mary Hunter Austin* (N.Y.: Twayne Publishers, 1965); Augusta Fink, *I-Mary: A*

Biography of Mary Austin (Tucson: University of Arizona, 1983); Esther Stineman, *Mary Austin: Song of a Maverick* (New Haven: Yale University Press, 1989); and Peggy Pond Church, *Wind's Trail: The Early Life of Mary Austin* (Santa Fe: Museum of New Mexico Press, 1990).

3. The notion of "nature for its own sake" echoes the title of John Van Dyke's 1898 meditation on nature and aesthetics, a volume that anticipated his 1901 book of nature writing, *The Desert* (N.Y.: Scribner's Sons, 1901).

4. Ernest Thompson Seton, *The Book of Woodcraft and Indian Lore* (1912; reprint Garden City, N.Y.: Doubleday, 1926), 3; Roderick Nash, *Wilderness and the American Mind* (New Haven: Yale University Press, 1967).

5. Ann H. Zwinger, "Introduction," in *Writing the Western Landscape,* by Mary Austin and John Muir, ed. Ann H. Zwinger (Boston: Beacon Press, 1994), x.

6. Austin herself describes being compelled by Muir and draws similarities between their visions of wild nature. Austin was familiar with Muir's writings and had known him when they were both associated with the circle of writers at Carmel. Of these writers, Austin states in her autobiography:

> I recall John Muir the most distinctly, a tall lean man with the habit of talking much, the habit of soliloquizing. He told stories of his life in the wild, and of angels; angels that saved him; that lifted and carried him; that showed him where to put his feet; he believed them. I told him one of mine; except that I didn't see mine. I had been lifted and carried; I had been carried out of the way of danger; and he believed me. I remember them still.

Austin connected with Muir's mystical vision of wilderness, feeling that the wild had also spiritually "saved" her.

Tellingly, however, Austin continues her story of meeting Muir by explaining that she could not completely identify with his image of angels; for her, "the pietistic characteristics of the angels she had heard of prevented such identification." Instead, "[t]he human experience, which in the general mind could be most easily made to illustrate what she felt in the desert wild, was . . . the One Woman." Austin clearly departs from Muir in her distinct awareness of the gender-specificity of her experience (Mary Austin, *Earth Horizon,* 298, 188).

7. Reviews of *The Land of Little Rain,* by Mary Austin: *The Dial* (December 1903): 422; *The New York Times,* November 1903: 862; *The Nation* (May 1904): 392; and *Out West* (July 1903): 679.

8. Mary Austin, *The Land of Little Rain* (1903; reprinted in *Stories from the Country of Lost Borders,* ed. Marjorie Pryse [New Brunswick; Rutgers University Press, 1987]), 15.

9. Austin, *The Land of Little Rain,* 12.

10. Ibid., 57.

11. James Work similarly sees Austin as writing "in deep yet self-forgetting sympathy with the desert life-struggles" (James C. Work, "The Moral in Austin's *The Land of Little Rain,*" in *Women and Western American Literature,* ed. Helen Winter Stauffer and Susan J. Rosowski [Troy, N.Y.: Whitston, 1982], 299). And Lawrence Buell claims that "it is a mark of the maturity of [Austin's] environmental vision . . . that her protagonist is the land" and that she "allows the book to be taken over by other

peoples' stories and her speaker to imagine the desert as it might look through the eyes of birds and animals" (Lawrence Buell, *The Environmental Imagination* [Cambridge: Harvard University Press, 1995]), 80, 176.

12. Austin, *The Lane of Little Rain*, 3.

13. Ibid., 4.

14. Ibid., 9.

15. William Cronon, "A Place for Stories: Nature, History, and Narrative," *The Journal of American History* 78, no. 4 (1992): 1367.

16. Austin, *The Land of Little Rain*, 4.

17. Buell, *The Environmental Imagination,* 176; William J. Scheik, "Mary Austin's Disfigurement of the Southwest in *The Land of Little Rain*," *Western American Literature* 27, no. 1 (1992): 43.

18. Austin, *The Land of Little Rain*, 9.

19. Ibid., 3.

20. Ibid., 46.

21. Mary Austin, "Tejon Diary, 1888–1910" (unpublished manuscript in the Mary Austin Collection, The Huntington Library, Pasadena, Calif.), 26.

22. Scheik, "Disfigurement," 37.

23. Patrick Murphy, "Voicing Another Nature," in *A Dialogue of Voices: Feminist Literary Theory and Bakhtin,* ed. Karen Hohne and Helen Wussow (Minneapolis: University of Minnesota Press, 1994), 62.

24. Despite widespread praise of *The Land of Little Rain,* like most of Austin's books, it did not sell widely. Personally and financially, this became a very serious issue for Austin who, for most of her life, had to earn a living from her writing. "I think I have it in me to do bigger novels than any body in the west is doing," she wrote to W. S. Booth at Houghton Mifflin in 1907, "but I can not do them and work at other things for a living . . . I do not know how to make you understand any better than I have done, how you miss getting at [the people my books are written for]" (Mary Austin, Letter to W. S. Booth, June 1907 [unpublished mss in the Mary Austin Collection, The Huntington Library, Pasadena, Calif.]). She laid the blame for his lack of success on the problematic marketing of the Eastern publishing establishment, who, according to Austin, neither knew nor cared about Western audiences well enough to market her work appropriately. As Karen Langlois discusses, while Austin "believed that her regional books could be successfully marketed in the West" and that "her fellow Californians [were] a 'natural' clientele for her western work," Houghton Mifflin, her primary publisher, "was marketing her work to intellectually and culturally elite 'hardcore' book buyers in the East" (Karen Langlois, "Mary Austin and Houghton Mifflin Company: A Case Study in the Marketing of a Western Writer," *Western American Literature* 23, no. 1 [1988]: 34–35).

25. Patrick Murphy, "Prolegomenon for an Ecofeminist Dialogics," in *Feminism, Bakhtin, and the Dialogic,* ed. Dale M. Bauer and S. Jaret McKinstry (Albany: State University of New York Press, 1991), 49.

26. Murphy, "Prolegomenon," 53.

27. Murphy, "Voicing Another Nature," 16.

28. Susan Sniader Lanser, *Fictions of Authority: Women Writers and Narrative Voice* (Ithaca: Cornell University Press, 1992), 21–22.

29. Austin, *The Land of Little Rain*, 4.

30. Ibid., 3.
31. Cronon, "A Place for Stories," 1375.
32. *Earth Horizon,* 197.
33. Ibid., 194.

Gender and Genre (pages 59–67)

1. Marcia Myers Bonta, ed., *American Women Afield: Writings by Pionering Women Naturalists* (College Station: Texas A. & M. University Press, 1995), xii.
2. Richard H. Brodhead, *Cultures of Letters: Scenes of Reading and Writing in Nineteenth-Century America* (Chicago: University of Chicago Press, 1993).
3. See for example, Annette Kolodny, *The Lay of the Land: Metaphor as Experience and History in American Life and Letters* (Chapel Hill: University of North Carolina Press, 1975), 3–9.
4. Ralph H. Lutts, *The Nature Fakers: Wildlife, Science and Sentiment* (Golden, Colo.: Fulcrum, 1990), ix–x.
5. Brodhead describes Jewett's *Country of the Pointed Firs* specifically in terms of her use of the "topoi of vacationing" to a "restorative place," 145–50.
6. Lawrence Buell, *The Environmental Imagination: Thoreau, Nature Writing and the Formation of American Culture* (Cambridge: Harvard University Press, 1995), 44–45.
7. Olive Thorne Miller wrote more than twenty nature books in the period from 1870 to 1915. Her *First Book of Birds* (1899) was one of the most popular and useful bird books not only for children but also for amateur birding enthusiasts. Other popular works by Miller include *Little Brothers of the Air* (Boston, 1892) and *Bird-Ways* (Boston, 1885). In addition to Olive Thorne Miller, other popular bird books for children were written by Mabel Osgood Wright: *Birdcraft: A Field Book of Two Hundred Song, Game and Water Birds* (1895); Neltje Blanchan: *Bird Neighbors* (1897); and Florence Merriam (later Florence Merriam Bailey), *A-Birding on a Bronco* (1896), *Birds of Village and Field* (Boston, 1898), and *Birds through an Opera Glass* (1900). Wright's *Birdcraft* and Miller's *The Second Book of Birds* also contained superlative illustrations by Louis Agassiz Fuertes. Mabel Wright also was the editor of the children's section of *Bird-Lore.*
8. On the domination of male institutions, see for example Carolyn Merchant, *The Death of Nature: Women, Ecology and the Scientific Revolution* (San Francisco: Harper & Row, 1980), and Margaret W. Rossiter, *Women Scientists in America: Struggles and Strategies to 1940* (Baltimore: Johns Hopkins University Press, 1982). Exceptions to a local focus by women nature writers include the later work of Florence Merriam Bailey, *A-Birding on a Bronco,* and her last book, *Among the Birds in the Grand Canyon Country* (1939), which are closer to the wilderness studies of John Muir and others naturalist adventurers.
9. See Bonta, *American Women Afield,* xii.
10. Florence Merriam Bailey, for example, includes the following whimsical comment on the affinity of squirrels and partridges for music in her *Birds Through an Opera Glass:* "One winter they seemed to show a fondness for music, often coming close to the house (although 'painfully shy') as I was playing the piano . . . the squirrels not

only nibble their corn with complacent satisfaction when the music box is wound for them, but have even let themselves be stroked when a peculiarly pathetic air was whistled! Who dare say what forest concerts the pretty creatures may get up on the long winter evenings when they are tired frolicking on the moonlit snow!" (Boston: Houghton Mifflin, 1889) reprinted in Bonta, *American Women Afield,* 99.

11. See Rossiter, *Women Scientists in America,* 78; see also Paul H. Oehser, "In Memoriam: Florence Merriam Bailey," *Auk* 69 (1952): 26.

12. John Burroughs, "The Literary Treatment of Nature," *Atlantic Monthly* 94 (1904): 38. All quotations from Burroughs in this paragraph come from this article.

13. Burroughs, Preface to *Boy and Man,* reprinted in *John Burroughs' America: Selections from the Writings of the Naturalist,* ed. Farida A. Wiley (Mineola, N.Y.: Dover, 1997), 3.

14. Miller, *First Book of Birds,* quoted in Joseph Kastner, *A World of Watchers* (San Francisco, Sierra Club, 1986), 165.

15. Florence Merriam Bailey, *Birds through an Opera Glass* (Boston: Houghton Mifflin, 1889).

16. See for example, Karen Kilcup, *Nineteenth-Century American Women Writers* (Cambridge, Mass. and Oxford, U.K.: Blackwell, 1997).

17. See Randall Roorda, *Dramas of Solitude: Narratives of Retreat in American Nature Writing* (Albany: State University of New York Press, 1998).

18. Buell, *The Environmental Imagination,* 397.

19. It appears, both from my survey of material published in the *Atlantic Monthly* in the period from 1870–1890 and from my examination of critical commentary from this period, that there were no definitive guidelines for the "nature essay" or objective criteria for distinguishing it from the "rural sketch." Some account of the actual physical environment was clearly required, but just how much description, how scientific it had to be, how much freedom the individual writer had to express her feelings or to move, as Thoreau himself characteristically did, between close observation and idealized abstraction, appeared to be a matter of choice or personal style.

20. See William M. Tanner, *Essays and Essay Writing Based on "Atlantic Monthly" Models* (Boston: Atlantic Monthly Press, 1920).

21. Buell, *The Environmental Imagination,* 405–408.

22. Buell, *The Environmental Imagination,* 406.

23. Frank Norris attacked realism (by women in particular) in "A Plea for Romantic Fiction," *Boston Evening Transcript* 18 (December 1901); reprinted in *Literary Criticism of Frank Norris,* 76. Ann Douglas Wood, "The Literature of Impoverishment: The Women Local Colorists in America 1865–1914" (*Women's Studies* 1 [1972]:14) has been frequently cited and critiqued by more recent studies of women's regional writing.

24. Quoted in Bonta, *American Women Afield,* 20.

25. Buell notes that "the field of environmental nonfiction [in Thoreau's time] was itself a patchwork, in some ways more 'polyphonic,' more 'heteroglossic,' than that of the novel," 397.

26. Brodhead, *Cultures of Letters,* 122.

27. Rose Terry Cooke, *Root-Bound and Other Sketches* (Ridgewood, N.J.: Gregg Press, 1968).

28. Willa Cather's essay in *Not Under Forty* (N.Y.: A. Knopf, 1922) helped shape later

understandings of Jewett's fiction and the aim and method of regionalist fiction. Cather writes: "Miss Jewett wrote of everyday people who grew out of the soil, not about exceptional people at odds with their environment" (82). Cather called Jewett's sympathetic intuition "a gift from heart to heart." The regionalists differentiated "sympathy" from "sentimentalism," which by the end of the century had the negative connotation of "false emotionalism."

29. James Lane Allen, "Local Color," *Critic* 9, January 1886, as cited in Donna M. Campbell, *Resisting Regionalism: Gender and Naturalism in American Fiction, 1885–1915* (Athens, Ohio: University Press, 1997), 17.

30. Paula Blanchard, *Sarah Orne Jewett: Her World and Her Work* (Reading, Mass.: Addison-Wesley, 1994), 224–25.

The Harvester and the Natural Beauty of Gene Stratton-Porter (pages 68–74)

1. Ralph H. Lutts, *The Nature Fakers: Wildlife, Science and Sentiment* (Golden, Colo.: Fulcrum Publishing, 1990), 8.

2. Bertrand F. Richards, *Gene Stratton-Porter* (Boston: Twayne, 1981), 22.

3. Jeanette Porter Meehan, *The Lady of the Limberlost: The Life and Letters of Gene Stratton-Porter* (New York: Doubleday, 1928), 99–110.

4. Gene Stratton-Porter, *Let Us Highly Resolve* (1927. New York: Aeonian Press, n.d.), 339.

5. Lutts, *The Nature Fakers,* 14.

6. Richards, *Gene Stratton-Porter,* 22.

7. Lutts, *The Nature Fakers,* 30–36.

8. *The Book Review Digest* 7.12 (December 1911): 379–80.

9. Meehan, *The Lady of Limberlost,* 155,160.

10. Ibid., 154.

11. Gene Stratton-Porter, *The Harvester* (1911; reprint, Indianapolis: Indiana University Press, 1987), 24.

12. Ibid., 25.

13. Ibid., 23–44.

14. Ibid., 35.

15. Ibid., 29.

16. Ibid., 87–88.

17. Ibid., 34.

18. Ibid., 171.

19. Ibid., 63.

20. Ibid., 156.

21. Ibid., 117.

22. Ibid., 119.

23. Ibid., 118.

24. Ibid., 29.

25. Lutts, *The Nature Fakers,* 36–42.

26. Richards, *Gene Stratton-Porter,* 131.

27. *Book Review Digest* (1912), 366.

28. Stratton-Porter, *The Harvester,* 50, 116.

29. Ibid., 139.
30. Ibid., 29.
31. Ibid., 505.
32. Richards, *Gene Stratton-Porter,* 53, 79.
33. Stratton-Porter, *The Harvester,* 283.
34. Jane S. Bakerman, "Gene Stratton-Porter," *American Women Writers* 4 (1982): 178–80.
35. Stratton-Porter, *The Harvester,* 195.
36. Ibid., 229–30.
37. Ibid., 511.
38. Gene Stratton-Porter, *Let Us Highly Resolve,* 300.
39. Ibid., 300.
40. Stratton-Porter, *The Harvester,* 306.

Willa Cather as Nature Writer (pages 75–84)

1. Susan J. Rosowski, "Willa Cather's Ecology of Place," *Western American Literature* 30, no. 1 (1995): 47.
2. Willa Cather, *O Pioneers!* (Boston: Houghton Mifflin, 1941), 65.
3. Growing interest in ecological literature as a genre has produced a number of anthologies and reference works of American nature writers, including *The Oxford Book of Nature Writing* (New York: Oxford University Press, 1995) and *The Norton Book of Nature Writing* (New York: W.W. Norton, 1990), neither of which includes an entry on Cather. The two-volume *American Nature Writers,* ed. John Elder (New York: Charles Scribner's Sons, 1996) provides no focused study on Cather, although Nicholas O'Connell makes passing reference to Cather's *O Pioneers!* and *The Professor's House* as "set[ting] the stage for an emerging American tradition of ecofiction" (1042). A notable exception is Lorraine Anderson's 1991 collection *Sisters of the Earth: Women's Prose and Poetry About Nature* (New York: Vintage-Random House) in which "The Ancient People" from *Song of the Lark* (1915) appears. Anderson notes that the land is a "central figure" in Cather's work and that one of her primary themes is "that if treated properly, the land is a source of well-being and . . . a source of deep solace" (159).
4. Willa Cather, *Not Under Forty* (New York: Knopf, 1936), v.
5. Nicholas O'Connell, "Contemporary Ecofiction," in *American Nature Writers,* 2: 1041.
6. E. K. Brown, an early biographer of Cather, does little more than assert that Cather's detailed accounts of the land in all its moods and seasons illustrate her appreciation of the region's beauty and serves as a nostalgic recalling of her childhood impressions (*Willa Cather: A Critical Biography* [New York: Knopf, 1953]). Mildred Bennett's study of Cather's early years (*The World of Willa Cather* [Lincoln: University of Nebraska Press, 1951]) asserts that "Miss Cather's feeling for the earth, the grasslands and the trees of Nebraska was responsible for the flavor of her best novels" (138) and recounts Cather's "particularly sensitive" response to the prairie's odors, light, and flowers (139). In *The Landscape and the Looking Glass,* John H. Randall, III, places Cather within the romantic tradition of living close to nature and argues that the

"virtues of the garden . . . form the core of Willa Cather's system of social values for the rest of her life" ([Boston: Houghton Mifflin, 1960], 74). Randall notes that. for Cather, the fragile and pristine landscape was threatened by human defilement unless, like Antonia Shimerda and Alexandra Bergson, one came to nature with "love and yearning." In the chapter "Antiphony of the Land," Edward and Lillian Bloom focus on Cather's "subtle fusion of the real and the ideal" in the treatment of her materials but caution that "above all, she ever reminds us that the land is the manifestation of a divine, supersensible force" (*Willa Cather's Gift of Sympathy* [Carbondale: Southern Illinois University Press, 1964], 24, 27). David Stouck, in *Willa Cather's Imagination,* places her early stories and novels (through *A Lost Lady,* 1923) within a shifting pastoral mode. While Stouck notes Cather's use of the pastoral ideal of the green world as a retreat, a sanctuary of innocence, he sees the pastoral mode shifting to interior landscapes of the artist's withdrawal into self in the later works (Lincoln: University of Nebraska Press, 1975).

7. Glen A. Love, "Revaluing Nature: Toward an Ecological Criticism," in *Old West–New West: Centennial Essays,* ed. Barbara Howard Meldrum (Moscow: University of Idaho Press, 1993), 290–93.

8. Kathleen Butterly Nigro, "Compassing Landscapes: Nature and Art as Sacred Space in Jewett, Cather, Glasgow, and Austin" (Ph.D. diss., Saint Louis University, 1997), abstract in *Dissertation Abstracts International* 58 (1998): 3134A.

9. Rosowski, "Willa Cather's Ecology of Place," 42–47.

10. Cheryl and John N. Swift, "Natives and Transplants: Cultivating the Wild in Cather's Remembered Gardens" (paper presented at the Seventh International Willa Cather Seminar, Winchester, Virginia, June 1997), n.p.

11. Only Doo-ho Shin in "Ecological Feminist Dialogics and Ecofeminist Approach to Willa Cather" explicitly connects Cather to the ecofeminist movement. Using *Song of the Lark* as an example, Shin argues that Thea Kronborg's heightened awareness of the natural world in Panther Canyon fails to aid her in moving beyond a "self-assertive and self-gaining consciousness" that would see nature as "things-in-themselves" rather than "things-for-us." Shin thus concludes that Thea's "feminist and ecological vision fails to meet the ecofeminist perspective" (*American Canadian Studies 3* [1994], 102).

12. Cather, *O Pioneers!,* 65.

13. Greta Gaard, "Living Interconnections with Animals and Nature," in *Ecofeminism: Women, Animals, Nature,* ed. Greta Gaard (Philadelphia: Temple University Press, 1993), 2.

14. Gretchen T. Legler, "Ecofeminist Literary Criticism," in *Ecofeminism: Women, Culture, Nature,* ed. Karen J. Warren (Bloomington: Indiana University Press, 1997), 228.

15. "Lure of Nebraska, Irresistible, Says Noted Authoress," *Omaha Daily Bee,* 29 October 1921, quoted in Mildred R. Bennett, *The World of Willa Cather,* 140.

16. Eleanor Hinman, "Willa Cather, Famous Nebraska Novelist," *Lincoln Sunday Star,* 6 November 1921, quoted in Brent L. Bohlke, ed., *Willa Cather in Person* (Lincoln: University of Nebraska Press, 1986), 47.

17. Rosowski, "Willa Cather's Ecology of Place," 37.

18. Hinman, "Willa Cather," 47.

19. "Woman's Club," *Hastings Daily Tribune*, 22 October 1921, quoted in *Willa Cather in Person*, 146.
20. "Miss Cather in Lincoln," *Lincoln State Journal*, 2 November 1921, quoted in *Willa Cather in Person*, 40.
21. Bennett, *The World of Willa Cather*, 143.
22. Willa Cather, "Prairie Dawn," in *April Twilights* (1903), rev. ed., ed. Bernice Slote (Lincoln: University of Nebraska Press, 1968), 38.
23. Willa Cather, *The World and the Parish: Willa Cather's Articles and Reviews, 1893–1902*, ed. William M. Curtin (Lincoln: University of Nebraska Press, 1970), 1: 111–12.
24. Willa Cather, "El Dorado: A Kansas Recessional," *Willa Cather's Short Fiction, 1892–1912*, ed. Virginia Faulkner (Lincoln: University of Nebraska Press, 1970), 294.
25. Ibid., 308.
26. Willa Cather, "The Enchanted Bluff," in *Willa Cather's Short Fiction*, 69.
27. Ibid., 72.
28. Willa Cather, *My Ántonia* (1918; reprint, Sentry ed., Boston: Houghton Mifflin, 1954), 233.
29. Randall, *The Landscape and the Looking Glass*, 19.
30. Hinman, "Willa Cather," 46.
31. Randall, *"The Landscape and the Looking Glass*, 151.
32. Ibid., 150–51.
33. "State Laws Are Cramping," *Lincoln Evening State Journal*, 31 October 1921, quoted in *Willa Cather in Person*, 148.
34. Lorraine Anderson, introduction to *Sisters of the Earth: Women's Prose and Poetry About Nature* (New York: Random House, 1991), xvii.
35. Quotations from Willa Cather's novels are cited in the text with the following abbreviations: *OO: One of Ours* (1922; reprint, New York: Vintage Random House, 1991); and *LL: A Lost Lady* (1925; reprint, Scholarly ed., Lincoln: University of Nebraska Press, 1997).
36. Nancy Chodorow, *Reproduction of Mothering* (Berkeley: University of California Press, 1978); and Carol Gilligan, *In a Different Voice* (Cambridge: Harvard University Press, 1982), quoted in Gaard, "Living Interconnections," 2.
37. Legler, "Ecofeminist Literary Criticism," 229.
38. Nina Schwartz, "History and the Invention of Innocence in *A Lost Lady*," *Arizona Quarterly* 56, no. 2 (1990): 49.
39. Demaree C. Peck, *The Imaginative Claims of the Artist in Willa Cather's Fiction* (Selingsgrove, Penn.: Susquehanna University Press, 1996), 174.
40. Ibid.
41. Annette Kolodny, *The Lay of the Land: Metaphor as Experience and History in American Life and Letters* (Chapel Hill: University of North Carolina Press, 1975), 4.
42. Morton D. Zabel, "The Tone of Time," in *Modern Critical Views: Willa Cather*, ed. Harold Bloom (New York: Chelsea House, 1995), 43.
43. Schwartz, "History and the Invention of Innocence," 47.
44. Peck, *The Imaginative Claims of the Artist*, 181.
45. Cather, *O Pioneers!*, 308.

"That Florida Flavor" (pages 85–94)

1. Zora Neale Hurston, "Go Gator and Muddy the Water," reprinted in *Go Gator and Muddy the Water: Writings by Zora Neale Hurston from the Federal Writers' Project,* ed. Pamela Bordelon (New York & London: W. W. Norton & Company, 1999), 69–70. All subsequent citations of Hurston's FWP work are taken from this edition and are cited according to title of essay/sketch and page number.

2. Hurston, "How It Feels To Be Colored Me," reprinted in *Zora Neale Hurston: Folklore, Memoirs, and Other Writings,* ed. Cheryl A. Wall (New York: The Library of America, 1995), 827. Hurston's Florida pride is well known. Ann R. Morris and Margaret M. Dunn observe that, in his biography on Hurston, Robert Hemenway describes her as having "the map of Florida on her tongue" (90; quoted in Morris and Dunn, "Flora and Fauna in Hurston's Florida Novels," in *Zora in Florida,* ed. Steve Glassman and Kathryn Lee Seidel [Orlando: University of Central Florida Press, 1991], 1–12). See also Robert Hemenway, *Zora Neale Hurston: A Literary Biography* (Urbana: University of Illinois Press, 1977).

3. Hurston was hired by the editorial staff of the Florida FWP on 25 April 1938 and worked for them until the spring of 1939. For a biography on Hurston specifically emphasizing her time with the FWP, see Pamela Bordelon's comprehensive and insightful piece, "Zora Neale Hurston: A Biographical Essay," in *Go Gator and Muddy the Water,* 3–59. Christopher D. Felker observes that Hurston's job entailed promoting Florida culture and being among a special group of writers who were "to see themselves as symbols and repositories 'of new national values: strenuousity, political activism, the outdoor life.'" In this way, Hurston was part of "the nation's first vast and cohesive 'culture industry,'" 146–47. For the full discussion of how Hurston's involvement with the WPA was a contribution to "intrinsic history" as well as to "extrinsic history," see Christopher D. Felker, "'Adaptation of the Source': Ethnocentricity and 'The Florida Negro'," in *Zora in Florida,* 146–58. General information on the Progressive Era and the work that came out of it can be found in Christopher Wilson, *The Labor of Words: Literary Professionalism in the Progressive Era* (Athens: University of Georgia Press, 1985).

4. Hurston, *Their Eyes Were Watching God* (New York: Harper Perennial, 1990), 183. All subsequent citations from *Their Eyes* are taken from this edition.

5. Bordelon, "Zora Neale Hurston," 25.

6. Hurston, "Other Negro Folklore Influences," 93.

7. Lawrence Buell, *The Environmental Imagination: Thoreau, Nature Writing, and The Formation of American Culture* (Cambridge, Mass. and London: Harvard University Press, 1995), 7. I have cited only Buell's first definition of an environmental text because it is most applicable to Hurston's FWP work. Buell also lists three other definitions of environmental writing that might also be applied to Hurston but are not my focus in this chapter. Buell's other definitions hold that, in environmental texts, "[t]he human interest is not understood to be the only legitimate interest," "[h]uman accountability to the environment is part of the text's ethical orientation," and "[s]ome sense of the environment as a process rather than as a constant or a given is at least implicit in the text" 6–8.

Although many critics discuss the presence of nature in Hurston's writing, only

Morris and Dunn, in "Flora and Fauna in Hurston's Florida Novels," discuss her role as a nature writer. They argue that "[i]n *Jonah's Gourd Vine* (1934), *Their Eyes Were Watching God* (1937), and *Seraph on the Suwanee* (1948), Hurston uses her knowledge of the plants and animals of Florida to provide realistic settings, to compose authentically natural speech for her characters, and to create central symbols," 1. What is especially provocative about their study is their implication that Hurston defies the traditional definition of nature writer—which, as Paul Bryant avers, "is a specific genre of non-fiction prose about nature . . ." ("Nature Writing and the American Frontier," in *The Frontier Experience and the American Dream: Essays on American Literature,* eds. David Mogen, Mark Busby, and Paul Bryant [College Station: Texas A & M University Press, 1989], 205). In my own study, I would like to redirect attention to Hurston's nonfiction essays, though I have observed that her FWP writing features a mixture of styles in the sense that Hurston's folklore is riddled with fact and her nonfiction is riddled with fictive elements.

8. Hurston, quoted in Cheryl Wall, "Zora Neale Hurston's Traveling Blues," *Women of the Harlem Renaissance* (Bloomington and Indianapolis: Indiana University Press, 1995), 179.

9. Hurston, "Proposed Recording Expedition into the Floridas," 67.

10. Rachel Stein, *Shifting the Ground: American Women Writers' Revisions of Nature, Gender, and Race* (Charlottesville and London: University Press of Virginia, 1997), 6. Stein cites cultural historians such as Perry Miller and Henry Nash Smith, who note that "Americans conceived of the uniqueness of their national character as a result of their domestication of the vast wilds of the New World." In her own work, Stein demonstrates how women writers such as Emily Dickinson, Zora Neale Hurston, Alice Walker, and Leslie Marmon Silko voice "cultural perspectives counter to the prevailing mythos," 15. In her chapter on Hurston, Stein specifically argues that Hurston underscores "the harsh effects of [the] negative identification of black women with lower nature" and presents "Afro-Caribbean Voodoo spirituality as offering a more unbounded concept of gender and race" 20. Indeed, as Cheryl Wall indicates in her discussion of Hurston, hoodoo supported Hurston's ideals for gender equality and for the significance of life. Historian Lawrence Levine notes that, in hoodoo, people were considered "part of, not alien to, the Natural Order of things, attached to the Oneness that bound together all matter, animate and inanimate, all spirits, visible or not" (quoted in Wall, *Women of the Harlem Renaissance,* 172; see Lawrence W. Levine, ed., *Black Culture and Black Consciousness: Afro-American Folk Thought from Slavery to Freedom* [Oxford: Oxford University Press, 1977], 58–59).

For some of the more recent views on the ways in which the American landscape and peoples have been conceived and represented historically and in literature, see Annette Kolodny, *Lay of the Land: Metaphor as Experience and History in American Life and Letters* (Chapel Hill: University of North Carolina Press, 1975) and *The Land before Her: Fantasy and Experience of the American Frontiers, 1630–1860* (Chapel Hill: University of North Carolina Press, 1984); Roy Harvey Pearce, *Savagism and Civilization: A Study of the Indian and the American Mind* (Berkeley: University of California Press, 1988); Frederick Jackson Turner, *Beyond Geography: The Western Spirit against the Wilderness* (New Brunswick: Rutgers University Press, 1990); and Vera Norwood, *Made from This Earth: American Women and Nature* [Chapel Hill: University of North Carolina Press, 1993], 1).

11. Hurston, "Go Gator and Muddy the Water," 69.

12. Hurston, "Proposed Recording Expedition into the Floridas," 63.

13. Hurston, "Other Negro Folklore Influences," 93.

14. Hurston, "Go Gator and Muddy the Water," 69, 78.

15. Hurston, "Other Negro Folklore Influences," 90.

16. Hurston, "Proposed Recording Expedition into the Floridas," 63.

17. Ibid., 66.

18. Hurston, "Proposed Recording Expedition into the Floridas," 65.

19. Ibid., 63–64.

20. Morris and Dunn, "Flora and Fauna," 4–5.

21. Bordelon, *Go Gator and Muddy the Water,* 128

22. Hurston, "Turpentine," 129. "Often," Steve Glassman and Kathryn Lee Seidel observe, "the black labor in [the sawmill and turpentine camps] was forced in a system known as debt peonage, embarrassingly similar to antebellum slavery, made acceptable by a thin legal sugar coating" (Introduction, *Zora in Florida,* x).

23. Hurston, "The Jacksonville Recordings," 167.

24. Hurston, *Seraph on the Suwanee,* quoted in "Flora and Fauna," 4.

25. Hurston, "The Citrus Industry," 132–33.

26. Ibid., 132, 135.

27. Wall, "Zora Neale Hurston's Traveling Blues," 140.

28. Hurston, "The Jacksonville Recordings,"160. These songs and Hurston's interviews are part of the WPA's Jacksonville recording session in June 1939; see Bordelon, *Go Gator and Muddy the Water,* 157–58 and Hurston, 158–77.

29. Hurston, "Proposed Recording Expedition into the Floridas," 66.

30. Hurston, "The Fire Dance," 155. Bordelon states that before Hurston wrote "Performance Pieces" as the last of her FWP assignments in 1939, "The Fire Dance" was performed throughout Florida in 1933 and 1934 as "the midnight scene" from Hurston's Broadway play. See Bordelon, Foreword, *Go Gator and Muddy the Water,* xiii.

31. Hurston, quoted in "Flora and Fauna," 5.

32. Much of Tea Cake's charm, in fact, derives from his ties to nature and his independence from white society. As Cheryl Wall comments, Tea Cake is "a veritable man of nature, or natural man, [who] seems at ease being who and what he is" ("Zora Neale Hurston's Traveling Blues," 188). Janie and Tea Cake's life in the muck revolves around the seasons and the soil, and their relationship to the earth is an egalitarian one as is their relationship to one another. Through the black folk culture that Tea Cake and the muck embody, Janie experiences self-fulfillment. Nature, as Michael Awkward comments, serves as Janie's "instructor." See "'The Inaudible Voice of it All': Silence, Voice and Action in *Their Eyes Were Watching God,*" in *Inspiriting Influences: Tradition, Revision, and Afro-American Women's Novels* (New York: Columbia University Press, 1989), 18.

33. Hurston, "Other Negro Folklore Influences," 91.

34. Hurston, "Proposed Recording Expedition into the Floridas," 65.

35. Hurston, "The Sanctified Church," 97.

36. See Stein's Introduction in *Shifting the Ground* for a comprehensive view of this topic, 1–22.

37. Bordelon, *Go Gator and Muddy the Water,* 94. For more on Hurston's distinction

between traditional spirituals and "neo-spirituals," see her 1934 essay "Spirituals and Neo-Spirituals," reprinted in *Zora Neale Hurston: Folklore, Memoirs, and Other Writings,* 869–74; Wall, *Women of the Harlem Renaissance,* 158; Felker, "Adaptation of the Source," 150–52; and Paul Gilroy, "'Jewels Brought from Bondage': Black Music and the Politics of Authenticity," in *The Black Atlantic: Modernity and Double Consciousness* (Cambridge: Harvard University Press, 1993), 72–110.

38. Hurston, "Other Florida Guidebook Folktales," 112–13.

39. Hurston, "Go Gator and Muddy the Water," 76.

40. Ibid.

41. Hyperbole is popular in African-American oral tradition; note the "contest in hyperbole" in chapter 6 of *Their Eyes.* For a critical analysis of this "mule talk" and of Hurston's use of hyperbole, see Wall, *Women of the Harlem Renaissance,* 186.

42. Hurston, "Go Gator and Muddy the Water," 70–71.

43. Mary Jane Lupton provides an in-depth analysis of the hurricane and flood in the novel. Lupton argues that Janie is an heroic figure who survives both social and natural animosity and is, as such, "a woman whose actions are larger, even, than heroic" ("Zora Neale Hurston and the Survival of the Female," *The Southern Literary Journal* 15 no. 1 [Fall 1982]: 53). See also Missy Dehn Kubitschek, "'Tuh de Horizon and Back': The Female Quest in *Their Eyes Were Watching God,*" reprinted in *Modern Critical Interpretations of Zora Neale Hurston's Their Eyes Were Watching God,* ed. Harold Bloom (New York: Chelsea House Publishers, 1987), 19–33.

44. Cornwell, "Searching for Zora in Alice's Garden: Rites of Passage in Hurston's *Their Eyes Were Watching God* and Walker's *The Third Life of Grange Copeland,*" in *Alice Walker and Zora Neale Hurston: The Common Bond,* ed. Lillie P. Howard (Westport, Conn.: Greenwood Press, 1993), 100.

45. Hurston, "Go Gator and Muddy the Water," 76.

46. For their example, Morris and Dunn look to the association of An' Dangie Devoe, the conjure woman in *Jonah's Gourd Vine,* with the poisonous Palma Christi plant ("Flora and Fauna," 4).

47. Hurston, "Other Florida Guidebook Folktales," 116.

48. Hurston, "Go Gator and Muddy the Water," 70.

49. Ibid.

50. Henry David Thoreau, *Walden and Other Writings by Henry David Thoreau,* ed. Joseph Wood Krutch (New York: Bantam Books, 1981), 221.

51. Henry Louis Gates, Jr., "Zora Neale Hurston: 'A Negro Way of Saying,'" Afterword, *Their Eyes Were Watching God* (New York: Harper Perennial, 1990), 193.

52. Hurston, *Their Eyes,* 6.

"A Happy Rural Seat of Various Views" (pages 95–109)

I would like to thank Karen Kilcup, Annie Ingram, Katie Ryan, Dottie Burkhart, Steve Brandon, and Laura Shearer for critiquing earlier versions of this essay.

1. Sarah Orne Jewett, *Deephaven* (Boston: Houghton Mifflin, 1894), 225.

2. Jewett, "An October Ride," in *Sarah Orne Jewett: Novels and Stories* (New York: Library of America, 1994), 603.

3. For an overview of Jewett criticism, see the introduction to *Jewett and Her Contem-*

poraries: Reshaping the Canon, ed. Karen Kilcup and Thomas S. Edwards (Gaines-ville: University Press of Florida, 1999), 1–27.

4. Richard Cary, *Sarah Orne Jewett* (New York: Twayne, 1962), 54–55.
5. Josephine Donovan, *Sarah Orne Jewett* (New York: Frederick Ungar, 1980), 11–12.
6. Ibid., 12. Donovan writes: "In an early sketch Sarah lamented the destruction of a nearby forest: 'The woods I loved best had all been cut down the winter before. I had played under the great pines when I was a child, and I had spent many a long after-noon under them since. There never will be such trees for me any more in the world. I know where the flowers grew under them, and where the ferns were greenest, and it was as much home to me as my own house.'" Donovan also notes Jewett's response to the "plan to subdivide the beautiful Piscataqua riverbank": "such developments gave her 'a hunted feeling like the last wild thing that is left in the fields.'"
7. Marcia B. Littenberg, "Transcendentalism to Ecofeminism: Celia Thaxter and Sarah Orne Jewett's Island Views Revisited," in *Jewett and Her Contemporaries,* 138. Com-paring Jewett and Thaxter to Thoreau, Littenberg observes that the main "difference lies in representing oneself as connected to nature, not simply as a careful observer. It is also marked by the narrator's sensitivity to and acceptance of points of view differ-ent than her own" (145–46).
8. I am using the 1997 edition of the text, edited by Sarah Way Sherman, which "repro-duces the 1896 first edition of Sarah Orne Jewett's *The Country of the Pointed Firs* (Hanover, N.H.: University Press of New England, 1997), vii–lviii. Subsequent quota-tions in the text will be abbreviated as *CPF,* followed by a page number. Sherman in-cludes the additional Dunnett Landing stories: "A Dunnet Shepherdess," "The Foreigner," "The Queen's Twin," and "William's Wedding" as separate sketches, in-stead of incorporating them, thus "correcting" the order imposed by such editors as Willa Cather. I am using the additional four stories that deal with Dunnett Landing primarily because they all concern one community and help to give more insight into Jewett's treatment of nature. See Sherman's introduction for a further discussion of this publishing history (xlii–xliii).
9. For example, in *Nature Writing: The Pastoral Impulse in America* (New York: Twayne, 1996), Don Scheese defines nature writing in this way: "The typical form of nature writing is a first-person, nonfiction account of an exploration, both physical (outward) and mental (inward), of a predominantly nonhuman environment, as the protagonist follows the spatial movement of pastoralism from civilization to nature. In its key emphases, nature writing is a descendent of other forms of written dis-course: natural history, for its scientific bent (the attempt to explain the workings of the physical universe over time); spiritual autobiography, for its account of the growth and maturation of the self in interaction with the forces of the world; and travel writ-ing (including the literature of exploration and discovery), for its tracing of a physical movement from place to place and recording of observations of both new and familiar phenomena" (6). See also Michael Cohen's article "Literary Theory and Nature Writ-ing" in *American Nature Writers,* ed. John Elder, 2 vols. (New York: Charles Scribner's Sons, 1996), 1099–1113, and John Elder's introduction to *American Nature Writers* for further discussions of the definition of nature writing.
10. See, for example, *New Literary History* 30, no. 3 (1999) issue, entirely devoted to ec-ocriticism as well as the Forum in the October 1999 PMLA. For discussion of the def-inition and purpose of ecocriticism, see Cheryll Glotfelty, "Introduction: Literary

Studies in an Age of Environmental Crisis," in *The Ecocriticism Reader: Landmarks in Literary Ecology,* ed. Cheryll Glotfelty and Harold Fromm (Athens, Ga.: University of Georgia Press, 1996), xviii.

11. Glotfelty explains further: "Despite the broad scope of inquiry and disparate levels of sophistication, all ecological criticism shares the fundamental premise that human culture is connected to the physical world, affecting it and affected by it. Ecocriticism takes as its subject the interconnections between nature and culture, specifically the cultural artifacts of language and literature. As a critical stance, it has one foot in literature and the other on land; as a theoretical discourse, it negotiates between the human and the nonhuman (*The Ecocriticism Reader,* xix). For a further discussion of holism, see Carolyn Merchant's *Earthcare: Women and the Environment* (New York: Routledge, 1995), 88.

12. For an overview of key events, see Scheese's chronology, *Nature Writing,* xviii–xix.

13. Jewett, "A White Heron," in *Novels and Stories.*

14. Sherman, *Sarah Orne Jewett: An American Persephone* (Hanover, N.H.: University Press of New England, 1989), 161.

15. Jewett was deeply interested in educating "city" people about the rural environments. For example, in her introduction to *Deephaven,* she says, "There is a noble saying of Plato that the best thing that can be done for the people of a state is to make them acquainted with one another" (3), and that "Human nature is the same the world over, provincial and rustic influences must ever produce much the same effects upon character, and town life will ever have in its gift the spirit of the present, while it may take again from the quiet of hills and fields and the conservatism of country hearts a gift from the spirit of the past" (6).

16. Sherman, *American Persephone,* 108; Marilyn E. Mobley, "Rituals of Flight and Return: The Ironic Journeys of Sarah Orne Jewett's Female Characters," *Colby Library Quarterly* 22, no. 1 (1986): 40.

17. Gwen Nagel, "This Prim Corner of the Land Where She Was Queen': Sarah Orne Jewett's New England Gardens," *Colby Library Quarterly* 22, no. 1 (1986): 61.

18. Laurie Crumpacker, "The Art of the Healer: Women in the Fiction of Sarah Orne Jewett," *Colby Library Quarterly* 19, no. 3 (1983): 157.

19. Karen Oakes attributes this secrecy to the perhaps abortive properties of her Indian remedy (Karen Oakes, "'Colossal in Sheet-Lead': The Native American and Piscataqua-Region Writers," in *The Literature of the Colonial Revival, 1860–1930,* ed. Sarah L. Giffen and Kevin D. Murphy (York, Maine: Old York Historical Society, 1992), 172.

20. See Sandra Zagarell's "*Country's* Portrayal of Community and the Exclusion of Difference," in *New Essays on The Country of the Pointed Firs,* ed. June Howard (New York: Cambridge University Press, 1994), 39–60. Also see Elizabeth Ammons' discussion in *New Essays on the Country of the Pointed Firs,* 81–99.

21. Sherman observes in her introduction in *Country of the Pointed Firs* that "if Mrs. Blackett represents an idealized vision of the pastoral world's moral potential, Mrs. Todd represents its more troubled struggle to overcome personal and cultural limitations" (xxxii–xxxiii).

22. Phillip Terrie, "Local Color and a Mythologized Past: The Rituals of Memory in *The Country of the Pointed Firs,*" *Colby Library Quarterly* 23, no. 1 (1987): 16–25. Terrie, for example, attributes this inability to an over-dependence on memory. He

explains that "those who mythologize the past to the point of losing their ability to function in the present isolate themselves from the community; they also represent that tendency of some regional fiction to idealize the past and thus retreat from the realities of the present" (18).

23. Sherman, *American Persephone*, 215.

24. See Marcia McClintock Folsom's article for more discussion on this quality: "Tact is a Kind of Mind-Reading': Empathetic Style in Sarah Orne Jewett's *The Country of the Pointed Firs*," in *Critical Essays on Sarah Orne Jewett*, ed. Gwen Nagel (Boston: G.K. Hall, 1984), 76–89.

25. For a more developed discussion of how Jewett complicates gender, see Karen Oakes's "'All That Lay Deepest in Her Heart': Reflections on Jewett, Gender, and Genre," as well as Margaret Roman's critical study *Sarah Orne Jewett: Reconstructing Gender* (Tuscaloosa: University of Alabama Press, 1992).

26. Karen Oakes has explored the connections between Native Americans and women in her "Colossal" essay, "'All That Lay Deepest in Her Heart': Reflections on Jewett, Gender, and Genre," *Colby Library Quarterly*, 26 no. 3 (1990): 142–160.

27. Barbara Johns calls Joanna "a living suicide" ("'Mateless and Apealing': Growing into Spinsterhood in Sarah Orne Jewett," in *Critical Essays on Sarah Orne Jewett*, 161); while Richard Carson discusses how "Nature here is represented by the sterility of the accumulated shells around her and the woven mats made from dried reeds . . . Joanna's life is hardened into a kind of spiritual death" ("Nature and Circles of Initiation in *The Country of the Pointed Firs*," *Colby Library Quarterly* 21, no. 3 [1985]: 157).

28. Paula Blanchard, *Sarah Orne Jewett: Her World and Her Work* (Reading, Mass.: Addison-Wesley, 1994), 98–99.

29. Robin Magowan, "Pastoral and the Art of Landscape in *The Country of the Pointed Firs*," in *Appreciation of Sarah Orne Jewett: 29 Interpretive Essays*, ed. Richard Cary (Waterville, Maine: Colby College Press, 1973), 188.

30. Lynn White, for example, explains criticism of the Christian viewpoint in her article "The Historical Roots of Our Ecologic Crisis," in *The Ecocriticism Reader*, 8–14.

31. Mark Stoll, *Protestantism, Capitalism, and Nature in America* (Albuquerque: University of New Mexico Press, 1997), 3.

32. This inclusion of different spiritual views may have to do with her own varied religious background. Cary notes how "she embraces components of a mystical affinity with nature which is transcendentalism, a scaled biological destiny which is Darwinism, and a belief in the absolute sovereignty of an anthropomorphic God which is Christian dogma" (*Sarah Orne Jewett*, 73). Donovan notes her belief in a kind of ESP because of her vision of Celia Thaxter after Thaxter's death and also points out that although she "was brought up a Congregationalist and was confirmed an Episcopalian," she experienced dissatisfaction with conventional religion because of its strong judgments (*Sarah* 17, 18). Blanchard, likewise, argues that Jewett's spirituality in *Deephaven* "envisions no cold Oversoul but a highly personal and attentive Deity, whose love for humanity is expressed not only through nature and the workings of Providence but through human affection and moral responsibility" (*Her World*, 102). Sherman, in her introduction to *The Country of the Pointed Firs*, refers to Jewett's "The Decay of Churches," which expresses her feelings that "old-time ministers lost their congregations because they ignored parishioners' human needs, failed to offer compassion and comfort, instead of theory and judgment" (xxxvii).

33. Marilyn M. Fisher, "Community and Earthly Salvation: Christian Intimations Within the Setting of Jewett's Pointed Firs," *Literature and Belief* 10 (1990): 74. Critic Marilyn Fisher examines many of Jewett's natural symbols and characters as emblematic of a Christian worldview. She asserts that "Nature conveys a sense of order, confers upon the community its distinctive character, and constitutes the background and stimulus for salvific activity. Jewett also notes the hand of a Creator in the natural order of things in "An October Ride" (603–604).

34. Josephine Donovan, "A Woman's Vision of Transcendence: A New Interpretation of the Works of Sarah Orne Jewett," *Massachusetts Review* 21 (1980): 377–80; Elizabeth Ammons, "Jewett's Witches," in *Critical Essays on Sarah Orne Jewett* (Boston: G. K. Hall, 1984), 165–84.

35. Sherman, *American Persephone,* 103–108.

36. Josephine Donovan, "Jewett and Swedenborg," *American Literature* 65, no. 4 (1993): 732. Cary sees elements of "An October Ride" as evidence of links to "the mystic ecology of Thoreau at Walden Pond" (*Sarah Orne Jewett,* 56). Sherman also describes in the introduction to *Country of the Pointed Firs,* how in "A Winter Drive" (1881), Jewett "confessed to a faith in the 'old doctrine' of Hylozoism, 'the theory of the soul of the world, of a life residing in nature, and that all matter lives; the doctrine that life and matter are inseparable.' Thus trees and plants also have embodied souls" (xxxvii). A more general view of animism reveals many of the same assumptions. Christopher Manes discusses animism as the following: "Among its characteristics is the belief (1) that all the phenomenal world is alive in the sense of being inspirited—including humans, cultural artifacts, and natural entities, both biological and 'inert' and (2) that not only is the nonhuman world alive, but it is filled with articulate subjects, able to communicate with humans" (Christopher Manes, "Nature and Silence" in *The Ecocriticism Reader: Landmarks in Literary Ecology,* ed. Cheryl Glotfelty and Harold Fromm [Athens: University of Georgia Press, 1996], 17–18).

37. Christopher Manes, "Nature and Silence," 16.

38. Marjorie Pryse, "Sex, Class and 'Category Crisis': Reading Jewett's Transitivity" in *Jewett and Her Contemporaries: Reshaping the Canon,* ed. Karen L. Kilcup and Thomas S. Edwards (Gainesville: University Press of Florida, 1999), 57.

39. See Sarah Way Sherman's introduction (*CPF* ix–xviii) for discussion of these cultural changes and Jewett's response to them. Similar to Jewett, Cooper's ecological interests were founded on "her theory . . . that knowledge of place causes people to approach the land more humble and gratefully and with less greed" (*Rural Hours,* xviii). For a more complete discussion of Cooper, see Rochelle Johnson and Daniel Patterson's introduction in Susan Fenimore Cooper, *Rural Hours,* ed. Rochelle Johnson and Daniel Patterson (Athens: University of Georgia Press, 1998), ix–xxii.

40. Kilcup and Edwards, *Jewett and Her Contemporaries,* 18.

41. Willa Cather, Preface, *The Country of the Pointed Firs,* by Sarah Orne Jewett (Garden City, N.Y.: Doubleday, 1956), 6.

The Best Way Is the Simplest (pages 110–119)

1. Florence A. Merriam, *Birds through an Opera-Glass* (New York: Houghton Mifflin, 1891), 1.

2. For the biographical information about Merriam in this paragraph and those that follow, I am indebted to Harriet Kolfalk's biography, *No Woman Tenderfoot: Florence Merriam Bailey, Pioneer Naturalist* (College Station: Texas A & M University Press, 1989), 37; and *Biographies of Members of the American Ornithologists' Union,* ed. Paul H. Oehser (Washington, D.C.: [s.n.], 1954).

3. Oehser, *Biographies of Members of the American Ornithologists' Union,* 86.

4. Vera Norwood, *Made from This Earth, American Women and Nature* (Chapel Hill: University of North Carolina Press, 1993), 43.

5. Norwood, *Made from This Earth,* 45.

6. Merriam, *Birds through an Opera-Glass,* 25.

7. Ibid., 145–46.

8. Ibid., 188. Interestingly, the spring 1998 issue of *Environmental Ethics* contains an article by M. M. Van de Pitte, "'The Female Is Somewhat Duller': The Construction of the Sexes in Ornithological Literature" 20: 1, which unknowingly employs Merriam's argument to discuss the "outrageous and unnoticed" sexism in ornithology. Outrageous, perhaps, but not unnoticed, at least not by Merriam in 1883.

9. Merriam, *Birds through an Opera-Glass,* 28.

10. Florence A. Merriam, *Birds of Village and Field* (New York: Houghton Mifflin, 1898), 27.

11. Norwood, *Made from This Earth,* 46. Norwood rightly argues that the late-century shift to narratives of "heroic exploration" effectively silence the quieter, more home-oriented voices in nature writing, including those of males.

12. Evidence of Merriam's commitment to social activism is her lifelong work with the Playground and Recreation Association of America, the National Housing Association, and the National Child Labor Committee; until her death she was on the board of Washington's Working Boys' Home.

13. Evelyn Morantz-Sanchez, "Field Work and Family North American Women Ornithologists, 1900–1950," in *Uneasy Careers and Intimate Lives,* ed. Pnina G. Abir-Am and Dorinda Outram (New Brunswick, N.J.: Rutgers University Press, 1987), 76. Ornithology, Morantz-Sanchez argues, was attractive to women for two reasons: its deemphasis on university education and its flexibility—unlike, say other field sciences like botany and geology—which appealed to those women burdened with domestic responsibilities.

14. Merriam, *Birds of Village and Field,* Prefatory Notes, iii.

15. Merriam, *Birds through an Opera-Glass,* 23.

16. Merrian, *Birds of Village and Field,* 33.

17. Merriam, *Birds through an Opera-Glass,* 3.

18. Merriam, *Birds of Village and Field,* 30.

19. Ibid., 82.

20. Paul Brooks, *Speaking for Nature: How Literary Naturalists from Henry Thoreau to Rachel Carson Shaped America* (Boston: Houghton Mifflin, 1980), 174. Merriam was also active in the anti-feathers campaign directed toward Northampton milliners, and she founded the Smith College Audubon Club on 17 March 1886, only a month after the national Audubon Club was founded by George Bird Grinell.

21. Kofalk, *No Woman Tenderfoot,* 37.

22. Clara Barrus, *The Life and Letters of John Burroughs* (Boston: Houghton Mifflin, 1925), 92.

23. Kofalk, *No Woman Tenderfoot,* 32.
24. Merriam described the work that would later constitute *Birds through an Opera-Glass* to the Smith College class letter as "Appearing in homeopathic quantities (the poor world couldn't bear much of such highly concentrated knowledge at once) in that far-famed journal, *The Audubon Magazine*" (Kofalk, *No Woman Tenderfoot,* 41).
25. Barrus, *Life and Letters of John Burroughs,* 45.
26. Christoph Irmscher, "Violence and Artistic Representation in John James Audubon," *Raritan* 15 (Fall 1995): 2.
27. Merriam, *Birds through an Opera-Glass,* 186–87.
28. Aaron Clark Bagg and Samuel Atkins Eliot, *Birds of the Connecticut Valley in Massachusetts* (Northampton, Mass.: Hampshire Bookshop, 1937), 17; Kofalk, *No Woman Tenderfoot,* 49.
29. Randall Roorda, "Where the Summit Bears: Narrative Logic in Thoreau's 'Ktaadn,'" *Arizona Quarterly* 53 (Autumn 1997): 3.
30. Merriam, *Birds of Village and Field,* 107–108.
31. Merriam, *Birds through an Opera-Glass,* 19.

Writing the Swamp (pages 125–134)

Note: parts of this article also appeared in my own "Our Lady of the Glades: Marjory Stoneman Douglas and *The Everglades: River of Grass*" and "Marjory Stoneman Douglas: 1890–1998," cited below.

1. William Frederick Badé, *The Life and Letters of John Muir,* vol. 1 (Boston: Houghton Mifflin, 1924), 259.
2. For a selection of Douglas's short fiction, see the two collections of her short stories edited by Kevin M. McCarthy, *Nine Florida Stories by Marjory Stoneman Douglas* (Jacksonville: University of North Florida Press, 1990) and *A River in Flood and Other Florida Storie*s (Gainesville: University Press of Florida, 1998). For a critical study of the role of advocacy in Douglas's short fiction, see McCarthy's "How Marjory Stoneman Douglas Crusaded for South Florida in Her Short Works," *Journal of Florida Literature* 3 (1997): 15–22. For discussion of Douglas as a novelist, see Melissa Walker, "*Freedom River:* A Book for the Next Millennium," *Journal of Florida Literature* 3 (1997): 35–44. For bibliographies of Douglas's work, see Rosalie E. Leprosky, "Marjory Stoneman Douglas Bibliography," *Journal of Florida Literature 3* (1997): 55–74, and Michael Branch, "Marjory Stoneman Douglas: 1890–1998," *ISLE: Interdisciplinary Studies in Literature and Environment* 5, no. 2 (Summer 1998): 123–27.
3. Kim Heacox, *Visions of a Wild America: Pioneers of Preservation* (Washington, D.C.: National Geographic Society, 1996), 172.
4. Marjory Stoneman Douglas, "How I Wrote My Everglades Book," in *The Book Lover's Guide to Florida,* ed. Kevin M. McCarthy (Sarasota, Fla.: Pineapple Press, 1992), 319.
5. Marjory Stoneman Douglas and John Rothchild, *Voice of the River* (Englewood, Fla.: Pineapple Press, 1989), 191.
6. Marjory Stoneman Douglas, *The Everglades: River of Grass,* Fiftieth Anniversary ed. (Sarasota, Fla.: Pineapple Press, 1997), 5.

7. Ibid., 33.
8. Ibid., 31.
9. Melissa Walker, "Marjory Stoneman Douglas" in *American Nature Writers,* 2 vol., ed. John Elder (New York: Scribners, 1996), 1: 238.
10. Douglas, *River of Grass,* 61.
11. Ibid., 34.
12. Heacox, *Visions of a Wild America,* 172.
13. Douglas, *River of Grass,* 92.
14. For a contemporary American woman nature writer's treatment of Native American and environmental issues in the Everglades, see Linda Hogan, *Power* (New York: Norton, 1998).
15. Douglas, *River of Grass,* xviii.
16. Ibid., 379.
17. Ibid., 286.
18. Rachel Carson, *Silent Spring,* 1962 (Boston: Houghton Mifflin, 1987), 297.
19. Douglas, *River of Grass,* 323.
20. Ibid., 375.
21. For discussion of Douglas's literary approach to hope and despair, see Betsy S. Hilbert, "Marjory Stoneman Douglas and an Antidote to Despair," *Journal of Florida Literature* 3 (1997): 45–54.
22. Douglas, *River of Grass,* 385.
23. Douglas, *River of Grass,* xvii. These fifty years of service are even more remarkable when we consider that Douglas was already nearly sixty years old when she published *River of Grass* in 1947. In her introduction to Thomas E. Lodge's *The Everglades Handbook: Understanding the Ecosystem* (Boca Raton, Fla.: St. Lucie Press, 1994), she notes, with characteristic aplomb: "Although I have been almost totally blind for some years now, I can still see clearly that the Everglades continues to need help."
24. Cyril Zaneski, "Coming Together," in Marjory Stoneman Douglas's *The Everglades: River of Grass,* Fiftieth Anniversary ed. (Sarasota, Fla.: Pineapple Press, 1997), 453.
25. Following the text of the first edition (1947), subsequent Rinehart (1947–1962) and Hurricane House (1965) editions concluded with the chapter "The Eleventh Hour," a grim but ultimately optimistic look at how the devastating environmental effects of large-scale drainage efforts might be mitigated by then-current plans to have the Corps of Engineers modify water flow in the Glades by an ambitious network of levees, canals, and pumps. The Mockingbird Books edition (1974–present) added an "Author's Afterword" in which Douglas commented upon the regional environmental history of the quarter century since the first edition and forcefully conveyed the discouraging news that the Corps's replumbing of the Glades had only exacerbated the ecological devastation of South Florida. A decade after the Banyan Books revised edition (1978), Pineapple Press published a revised edition (1988) that expanded *River of Grass* to include "Forty More Years of Crisis," a new final chapter written by Randy Lee Loftis with Marjory Stoneman Douglas. The chapter included discussion of species endangerment and extinction; biodiversity loss; water timing, occlusion, and salinity; urban sprawl; degradation of habitat within Everglades National Park; climate change; aquifer depletion; soil loss; phosphorus pollution; and other current environmental problems in and around the Everglades. Pineapple's Fiftieth anniversary edition, published

in 1997 to mark the golden anniversary of *River of Grass*, adds yet another chapter, "Coming Together," written by environmental journalist Cyril Zaneski. This chapter provides additional information on changes in the Glades since the publication of the Loftis/Douglas chapter in 1988 and discusses recent problems such as seagrass die-off in Florida Bay and mercury contamination of wildlife in the Everglades. Ultimately, Zaneski argues that the late 1990s is a time of arrival for efforts to save the Glades, a historical moment in which greater understanding of this complex ecosystem has led to the heightened awareness and cooperation Douglas so long advocated.

26. Paul Anderson, "Clinton Honors 'Grandmother of the Glades,'" *Miami Herald* 1 (December 1993), 1A+; Zaneski, "Coming Together," 433.

The Animal Anthropology of Sally Carrighar (pages 136–145)

1. Sally Carrighar, *Home to the Wilderness: A Personal Journey* (Boston: Houghton Mifflin, 1973), 275.
2. Ibid.
3. See the Introductory Note to *One Day on Beetle Rock*, written by the then-director of the California Academy of Sciences, Robert C. Miller. "These are stories of the adventures of animals, but with a difference—the stories are of actual animals in an actual place, as the author has observed them. She has watched carefully and reported truthfully, always with sensitive understanding and a keen awareness of beauty. The tales are fiction, yes, but fiction closely parallel with fact. This is real natural history." The phrase "fictional natural history" is my own but is clearly suggested, I believe, by Miller's words.
4. In the book proposal she prepared prior to the publication of *One Day on Beetle Rock*, Carrighar writes: "The forest is most interesting the way it is. No animals talking a human language and feeling human emotions can be as appealing as an animal meeting its own problems with its own equipment" (Sally Carrighar Collection, Mugar Library, Boston University, Box 10, typed manuscript, page 1).
5. Carrighar writes: "My method in this book (*One Day on Beetle Rock*) is to try to give the reader a direct sense of the animals' consciousness. It will be legitimate to include sensations, the promptings of instinct, and such other mental traits as the scientists know that animals possess" (Ibid., 2).
6. Carrighar writes: "There is almost nowhere in fiction a realistic picture of a wilderness community. A few realistic stories have been written about single animals, but perhaps none about a group of interrelated species. Yet the subject has infallible appeal: variety of situation, suspense, characters who show courage, tenderness, skill" (Ibid., 1). In her autobiography *Home to the Wilderness*, Carrighar provides a retrospective explanation for her understanding of individual animals in communities in *One Day on Beetle Rock*. She sought "to portray the pattern but devote most attention to individual creatures in it. I wanted to tell how these animals were related to one another but to show chiefly what was interesting to the creatures themselves. They did not see themselves as strands in a net" (278).
7. Carrighar's 1965 book *Wild Heritage* (Boston: Houghton Mifflin), a survey of the recent findings of ethologists, is premised on the evolutionary link between human beings and other animals. The title of the book's first chapter, "Animals and Men: The

Blurred Borderline," indicates clearly that in Carrighar's view, as well as in the view of the ethologists whose work she presents, the wild heritage is our own.

8. Gale Lawrence, "Sally Carrighar," in *American Nature Writers,* vol. 1, ed. John Elder (New York: Charles Scribner's Sons, 1996), 142.

9. Carrighar, *Home to the Wilderness,* 279.

10. Anthropologist Barbara Noske calls for "an anthropology of animals" in her book *Beyond Boundaries: Humans and Animals* (Montreal: Black Rose Books, 1997). Noske argues that as "the science of the Other" anthropology has the potential to expand its field of inquiry to encompass the study of the subjecthood of nonhuman animals in large part because it possesses participant observation as a field method of inquiry and tends to be "holistic in its approach to the Other" (168–69). Such an approach to animals is needed, she argues because "so far little attention has been paid (and least of all by mainstream science) to the possibility that it is not just human subjects who socially and collectively construct their world but that animal subjects do so too. These animal constructs may be markedly different from ours but may be no less real" (157–58). Without having specifically linked her own project to anthropology and the field method of participant observation, Carrighar's work and her stated positions on animal consciousness are nevertheless consistent, I believe, with the premises and goals Noske outlines.

11. For a fuller discussion of this point, see Ralph H. Lutts, "The Wild Animal Story: Animals and Ideas," in *The Wild Animal Story,* ed. Ralph H. Lutts (Philadelphia: Temple University Press, 1998), 9.

12. Ibid., 10.

13. Vera Norwood, *Made from this Earth: American Women and Nature* (Chapel Hill: University of North Carolina Press, 1993), 212.

14. Sally Carrighar Collection, Mugar Library, Boston University, Box 10, typed manuscript, page 1.

15. Carrighar, *Home to the Wilderness,* 264.

16. Ibid., 265.

17. Ibid., 266.

18. Ibid., 266–67.

19. Noske, *Beyond Boundaries,* 169.

20. Philosopher Mary Midgely helpfully approaches this question from within the context of human to human interactions, noting that:

> none of us need suspect that (in spite of constant success) we are constitutionally unable to pronounce on other people's feelings, that we are locked away in a Cartesian solitude. Solipsists apart, most of us would be ready to accept that we can know something about human feelings. Ought we to hesitate about extending this ability to our contacts with animals? Do we really not know what they feel?" (*Beast and Man: The Roots of Human Nature* [1979; reprint, New York: Routledge, 1995], 347)

> She continues: "The species barrier is, in itself, irrelevant. Members of one species do in fact often succeed in understanding members of another well enough for both prediction and a personal bond. Nothing more is necessary" (348).

21. The issue of human sympathy for animals is germane to my argument but beyond the

scope of this paper. A convincing argument for the possibility of sympathy with and for animals across the species boundaries is made by John A. Fisher, "Taking Sympathy Seriously: A Defense of Our Moral Psychology Toward Animals," *Environmental Ethics* 9 (1987): 197–215.

22. Carrighar, *Home to the Wilderness,* 193–96.
23. Ibid., 278.
24. Ibid., 280–83.
25. Ibid., 297.
26. Ibid., 302–303.
27. Ibid., 303–304.
28. Writing of the possibility of extending the discipline of anthropology, particularly the method of participant observation, to animal study, Barbara Noske notes that an anthropologist seeking access to an unfamiliar society "tread[s] upon this unknowable ground with respect rather than with disdain" (*Beyond Boundaries,* 169). Carrighar seems to have intuitively recognized this requirement.
29. Carrighar, *Home to the Wilderness,* 304.
30. Carrighar, *Wild Heritage,* 105. Barbara Noske notes that "a participant-observer does not just work with her mind . . .—she has to immerse herself body, mind and soul in the Other's sphere . . . Good participatory observation is basically an exercise in empathy." (*Beyond Boundaries,* 169).
31. Carrighar, *Wild Heritage,* 174–76.
32. Carl Gans, "The Sensory World of Animals," in *Animal Intelligence: Insights into the Animal Mind,* ed. R. J. Hoage and Larry Goldman (Washington, D.C.: Smithsonian Institute Press, 1986), 89.
33. The concept of the "self-world" or "Umwelt" is taken from the work of the pioneer ethologist Jakob von Uexkull, "A Stroll through the Worlds of Animals and Men: A Picture Book of Invisible Worlds," ed. and trans. Claire H. Schiller, *Instinctive Behavior: The Development of a Modern Concept* (1934; reprint, New York: International Universities Press, 1957), 5–80. This is "the world as it appears to the animals themselves, not as it appears to us" (5). His point is that the environment we perceive surrounding animals we observe is a product of our human "Umwelt" and very likely fundamentally different from "the 'Umwelten' built up by the animals themselves and filled with the objects of their own perception" (64). Barbara Noske describes the concept of "inwardness" as one of the properties that distinguish living beings from machines. It is "a mysterious inner side, a self if you like, which is bound to evade a science which contents itself with registering from without . . . It is this inwardness, the animal's self and integrity which is disenchanted when science attempts to abolish all non-physical forces from the picture of nature, or at least tries to give them the status of epiphenomenon" (*Beyond Boundaries,* 61–62).
34. Carrighar, *Home to the Wilderness,* 279.
35. Carrighar, *One Day on Beetle Rock* (1944; reprint, Lincoln: University of Nebraska Press, 1978), 108.
36. Ibid., 123.
37. Ibid., 184–85.
38. Carrighar, *One Day at Teton Marsh* (1947; reprint, Lincoln: University of Nebraska Press, 1979), 29.
39. Ibid., 30.

40. Carrighar, *One Day on Beetle Rock,* 173.
41. Ibid., 173–74.
42. Ibid., 174.
43. Ibid., 15.
44. Ibid., 101.
45. Ibid., 148–49.
46. Ibid., 91.
47. Interestingly, Carrighar distinguishes (perhaps in idealized terms) the actions and life ways of the native people of the arctic from those of people of European origin. In writing of native people, she includes them as full participants within their ecological communities. For example, in her description of an Eskimo hunter's pursuit of a beluga whale cow and calf in *Icebound Summer,* Carrighar foregrounds the hunter's and the whale's parallel concerns. The hunter is a parent too. Concerned for the health of an ailing child, he hunts not for sport but out of necessity. "[H]ere on this northwest coast of Alaska the food chains of nature still included the human race—this was one of the few places left on earth where that was true" (164).
48. Carrighar, *Icebound Summer,* 180–82.
49. Ibid., 185.
50. Carrighar, *The Twilight Seas* (New York: Weybright and Talley, 1975), 65.
51. Carrighar, *Icebound Summer,* 73–74.
52. Ibid., 74.
53. Ibid., 75.
54. Ibid., 81–82.
55. Ibid., 91–92.
56. Carrighar, *The Twilight Seas,* 150.
57. Ibid., 152.
58. Carrighar, *Wild Voice of the North* (Garden City, N.Y.: Doubleday, 1959), 151–52.
59. Patrick Murphy, *Literature, Nature, and Other: Ecofeminist Critiques* (Albany: State University of New York Press, 1995), 24.
60. Midgley, *Beast and Man,* 347.

A View of Her Own (pages 146–159)

1. Donald W. Meinig, "Introduction," *The Interpretation of Ordinary Landscapes: Geographic Essays,* ed. Donald W. Meinig (New York: Oxford University Press, 1979), 6; Yi-Fu Tuan, "Thought and Landscape: The Eye and the Mind's Eye," in *The Interpretation of Ordinary Landscapes,* 101.
2. The organization, process, and information for this article draw on a much longer version found in Tamara Fritze, "A View of Her Own: A Cultural Study of Rural Western Women's Dooryard Gardens," Ph.D. diss., Washington State University, 1997.
3. Two scholars who have examined women's relationship with the land are Annette Kolodny and Vera Norwood. Kolodny, in *The Land Before Her: Fantasy and Experience of the American Frontiers, 1630–1860* (Chapel Hill: University of North Carolina Press, 1984) suggests that women coming west preferred the landscape of the prairies because they resembled a flower garden and lent themselves to the creation (or

re-creation) of home and community on a cultivated land, expanding the realm of domesticity beyond the interior of the home. Norwood, in "Women's Place: Continuity and Change in Response to Western Landscapes," in *Western Women: Their Land, Their Lives,* ed. Lillian Schlissel, Vicki Ruiz, and Janice Monk (Albuquerque: University of New Mexico Press, 1988), indicates that women further west found value in the land as a place of home; the land became a source of freedom for them, a place where they could gain power and independence through encounters with nature. Some found greater equality and the possibility for a life much less passive and more active than what they had in the urban East. Forced to live a life of greater self-sufficiency, they used their role as producer to express their creativity.

4. Vera Norwood notes many anthropologists suggest that men are responsible for culture and its creations, and women serve as mediators for nature and culture. Norwood disputes this theory, indicating that implied within it is a tension or challenge between nature and culture that is not felt by all cultural groups. She suggests that to use the binary at all is to limit one's study to those members of cultures who perceive it as reality ("Women's Place," 166). Other scholars also point out the falsehood of the nature-culture binary. Gardener and writer Michael Pollan, in the "Afterword" of *Keeping Eden: A History of Gardening in America* ed. Walter T. Punch (Boston: Bullfinch, 1992), suggests that this false binary dominates white American thinking and results in their only two contributions to the field of landscape: the lawn (pure culture) and the wilderness area (pure nature). He suggests that we must relinquish this binary to improve the health of the planet; for Pollan, the place to begin this change in perception is in the garden, a place that requires a blending of nature and culture (261).

5. Meinig, "Reading the Landscape: An Appreciation of W. G. Hoskins and J. B. Jackson," in *The Interpretation of Ordinary Landscapes,* 228.

6. Tillie Olsen, *Yonnondio: From the Thirties* (New York: Delacorte, 1974) 130, 155.

7. Danielle Klaveano, interview by author, Cashup, Washington, 27 June 1996.

8. Yi-Fu Tuan, *Space and Place: The Perspective of Experience* (Minneapolis: University of Minnesota, 1977), 146.

9. Willa Cather, *O Pioneers! Early Novels and Stories* (New York: Viking, 1987), 177–78.

10. Donna Donnelly, interview by author, Dayton, Washington, 14 July 1996.

11. Cather, *Oh Pioneers!,* 203, 281.

12. Yi-Fu Tuan, *Passing Strange and Wonderful: Aesthetics, Nature, and Culture* (Washington, D.C.: Island Press, 1993), 214.

13. Sandy Simmons, interview by author, Walla Walla, Washington, 18 July 1996.

14. Jeanne Wakatsuki Houston, "Rock Garden," in *Dreamers and Desperadoes: Contemporary Short Fiction of the American West,* ed. Craig Lesley and Katheryn Stavrakis (New York: Dell, 1993) 195–202.

15. Mary Maier, interview by author, Lapwai, Idaho, 24 July 1996.

16. Janie Tippett, interview by author, Joseph, Oregon, 22 July 1996.

17. Davonna Fritze, interview by author, Coeur d'Alene, Idaho, 12 October 1996.

18. Mary Cunningham, interview by author, Pasco, Washington, 28 June 1998.

19. Bev Mill, interview by author, Benton City, Washington, 19 June 1996.

20. Pat Mora, *Nepantla: Essays from the Land in the Middle* (Albuquerque: University of New Mexico Press, 1993), 27.

Tending the Southern Vernacular Garden (pages 160–169)

1. Elizabeth Lawrence, *Gardening for Love: The Market Bulletins,* ed. Allen Lacy (Durham: Duke University Press, 1987), 24. At her death, Lawrence left over eight boxes of market bulletins, newspaper clippings, and correspondence related to her study. From this material and an incomplete manuscript, Lacy edited and introduced it as *Gardening for Love* three years after Lawrence's death. The unpublished manuscript and correspondence of Lawrence are housed in the Elizabeth Lawrence Collection, Cammie Henry Research Center, Northwestern State University of Louisiana, Natchitoches.

2. Vera Norwood, *Made from This Earth: American Women and Nature* (Chapel Hill: University of North Carolina Press, 1993), 134.

3. Elizabeth Lawrence, "An Autobiography," *Herbertia* 10 (1943): 13. Valuable biographical accounts appear in the following: Lacy's introduction to *Gardening for Love,* 1–22; Mimi Fuller Foster, "A Gardening Legacy," *The Atlanta Journal/The Atlanta Constitution,* Sunday, 14 January 1994, B1–2, B5; and Emily Herring Wilson, "Elizabeth Lawrence, 1904–1985," in *North Carolina Women Making History,* ed. Emily Herring Wilson and Margaret Supplee Smith (Chapel Hill: University of North Carolina Press, 1998).

4. Lawrence, "An Autobiography," 13.

5. Ibid.

6. Elizabeth Lawrence, "Gardens of the South," *The University of North Carolina Library Extension Publication,* 11, no. 2 (April 1945): 11.

7. Lucinda Hardwick MacKethan, *The Dream of Arcady: Place and Time in Southern Literature* (Baton Rouge: Louisiana State University Press, 1980), 3. Many have explored why the region's most formative myth is Arcadian, a return to or preservation of the old times and simpler ways, marked by a working respect for nature and a practice of neighborliness. Lewis Simpson's study focuses on both historic and literary dimensions in *The Dispossessed Garden: Pastoral and History in Southern Literature* (Athens: University of Georgia Press, 1975). Albert Cowdrey offers a very useful environmental history in *This Land, This South* (Lexington: The University Press of Kentucky, 1983). In short, most agree with Richard Gray's assessment that a sense of place is "one of the structuring principles of Southern myth." *Writing the South: Ideas of an American Region* (Cambridge, Mass.: Harvard University Press, 1980), 173.

8. Henry V. Hubbard and Theodora Kimball, *Landscape Design* (Cambridge, Mass.: Harvard University Press, 1917). Lawrence's copy of this text, with some penciled, marginal notes, is in the Elizabeth Lawrence Library Collection, Cherokee Garden Library, Atlanta History Center, Atlanta, Georgia.

9. See David A. Lockmiller, *History of the North Carolina State College of Agriculture and Engineering* (Raleigh, N.C.: Edwards and Broughton, 1939), chap. VII.

10. Suzanne Turner, *The Gardens of Louisiana: Places of Work and Wonder* (Baton Rouge: Louisiana State University Press, 1997), 17.

11. Lawrence, "Gardens of the South," 5.

12. Cited in Elizabeth Lawrence, *Through the Garden Gate,* ed. Bill Neal (Chapel Hill: University of North Carolina Press, 1990), x.

13. Ibid., 48.

14. Ibid., 51.

15. Ibid., 36.

16. Ibid., 23.

17. Ibid., 25. In her own organization of materials for the market bulletin study, Lawrence often filed together manuscript pages of individual sections (for example, "The Mississippi Market Bulletin") with the letters of correspondents (Kim Kimery). See the index to the Elizabeth Lawrence Collection. In her study, then, the bulletins were usually associated with the gardeners she met through their advertisements.

18. Ruth Dormon to Elizabeth Lawrence, 1942. Caroline Dormon Collection, Cammie Henry Research Center, Eugene Watson Library, Northwestern State University of Louisiana, Natchitoches.

19. Dormon to Lawrence, 30 October 1942; 14 September 1943, Elizabeth Lawrence Collection. Lawrence to Ruth Dormon, 20 November 1942; 22 March 1943; April 1943; 17 May 1943; 17 July 1943, Caroline Dormon Collection. This last letter, in which Lawrence describes to Dormon the hymenocallis bloom, illustrates the passionate details of their exchange: "It is the loveliest by far that I have had. The cup is usually deep, ¼". The tube, 3", the segments 3½" long. The scape 18", 5 fls. To the cape."

20. Lawrence to Dormon, 17 September 1943, Caroline Dormon Collection.

21. Dormon to Lawrence, 24 October 1943, Elizabeth Lawrence Collection.

22. Lawrence, *Gardening for Love,* 41.

23. Lawrence, *A Southern Garden,* xxvii.

24. Lawrence, *Gardening for Love,* 4.

25. Allen Lacy, *Home Ground: A Gardener's Miscellany* (Boston: Houghton Mifflin, 1992); Steve Bender and Felder Rushing, *Passalong Plants* (Chapel Hill: University of North Carolina Press, 1993); Scott Ogden, *Garden Bulbs for the South* (Dallas, Tex.: Taylor Publishing Co., 1994); William C. Welch and Greg Grant, *The Southern Heirloom Garden* (Dallas: Taylor Publishing, 1995).

26. Norwood, chap. 5.

27. May Brawley Hill, *Grandmother's Garden: The Old-Fashioned American Garden, 1865–1915* (New York: Harry N. Abrams, 1995). See especially her chapter on "Grandmother's Garden in the Middle Atlantic and South," 162–77.

28. Lawrence, *Gardening for Love,* 35–46.

29. Richard Westmacott, *African-American Gardens and Yards in the Rural South* (Knoxville: University of Tennessee Press, 1992).

30. John Brinckerhoff Jackson, *Discovering the Vernacular Landscape* (New Haven: Yale University Press, 1984), 47.

31. Ibid., 154–55.

32. Lawrence, *Gardening for Love,* 24.

Linda Hasselstrom (pages 170–177)

1. Interest in Hasselstrom's work is reflected in the following studies: Robin Kacel, "An Interview with Linda Hasselstrom," *Writing!* (September 1988): 11–13; John Murray, "The Rise of Nature Writing: America's Next Great Genre?" [Nature-Writing Symposium], *Manoa* 4, no. 2 (Fall 1992): 73–96; Paul Higbee, "At Home on the Range,"

South Dakota Writers (September/October 1992): 23–26; Geraldine Sanford, "The Dichotomy Pulse: The Beating Heart of Hasselstrom Country," *South Dakota Review* 30, no. 3 (Autumn 1992): 130–55; and Jed Stavick, "A Vegetarian Critical Response to *Dakota Bones:* Becoming a Wrangler of Moral Dilemmas with Linda Hasselstrom," *South Dakota Review* 33, no. 2 (Summer 1995): 120–29.

2. Linda Hasselstrom, *Windbreak: A Woman Rancher on the Northern Plains* (Valecitos, N.M.: Barn Owl Books, 1987); *Going Over East: Reflections of a Woman Rancher* (Golden, Colo.: Fulcrum Publishing, 1987); *Land Circle: Writings Collected from the Land* (Golden, Colo.: Fulcrum Publishing, 1991); *Roadside History of South Dakota* (Missoula, Mont.: Mountain Press Publishing, 1994).

3. Her books of poetry are *Caught by One Wing* (1984; reprint, Granite Falls, Minn.: Spoon River Poetry Press, 1990); *Roadkill* (Peoria, Ill.: Spoon River Poetry Press, 1987); and *Dakota Bones: The Collected Poems of Linda Hasselstrom* (Granite Falls, Minn.: Spoon River Poetry Press, 1993). Her collection of essays, entitled *Feels Like Far: A Rancher's Life on the Great Plains,* was published (New York: The Lyons Press) in 1999.

4. See Hector St. John de Crèvecoeur, *Letters from an American Farmer* (1871; reprint, New York: Dutton, 1962) and Hyppolite Taine, *Lectures on Art,* 2 vols., trans. John Durand (1875; reprint, New York: Henry Holt-AMS Press, 1971).

5. See Terry Tempest Williams, *Refuge: An Unnatural History of Family and Place* (New York: Pantheon Books, 1991) and Mary Clearman Blew, *All but the Waltz: Essays on a Montana Family* (New York: Viking, 1991).

6. Hasselstrom, "On Writing Western," *Roundup Magazine* 4, no. 2 (December 1996): inside back cover.

7. Hasselstrom, *Windbreak,* 144–45.

8. Murray, "The Rise of Nature Writing," 85.

9. Hasselstrom, "The Real Western Brand," *Rocky Mountain News* (26 March 1995): 20A.

10. Hasselstrom, "On Writing Western." See Teresa Jordan, *Riding the White Horse Home: A Western Family Album* (New York: Pantheon Books, 1993); Blew, *All but the Waltz;* and Williams, *Refuge;* and Dayton Hyde, *Yamsi: A Year in the Life of a Wilderness Ranch* (Corvallis: Oregon State University Press, 1996).

11. Hasselstrom, "The Price of Bullets," *The Missouri Review* 12, no. 1 (1989): 7–16.

12. Hasselstrom, "Nighthawks," in *The Soul of Nature: Celebrating the Spirit of the Earth,* ed. Michael Tobias and Georgianne Cowan (New York: Plume-Penguin, 1996), 71.

13. Hasselstrom, *Going Over East,* 13.

14. Hasselstrom, *Windbreak,* 62.

15. Ibid., 193.

16. Ibid., 56.

17. Ibid., 54.

18. Ibid., 90.

19. Ibid., 91.

20. Hasselstrom, "Rancher Roulette," in *Dakota Bones,* 102–103.

21. Hasselstrom, *Windbreak,* 116.

22. Sanford, "The Dichotomy Pulse," 134.

23. Hasselstrom, *Windbreak,* 124–25.

24. Ibid., 125.
25. Ibid., 100–101.
26. Hasselstrom, "Journal of a Woman Rancher," *Life* (July 1989): 94.
27. Kacel, "An Interview," 12.
28. Hasselstrom, *Going Over East,* 36.
29. Ibid.
30. Hasselstrom, *Roadside History of South Dakota,* xi.
31. Ibid.
32. Hasselstrom, *Land Circle,* xix.
33. Hasselstrom, *Going Over East,* 5–6.
34. Higbee, "At Home on the Range," 25.
35. Hasselstrom, *Going Over East,* 53.
36. Ibid.
37. Hasselstrom, *Dakota Bones,* 25.
38. Hasselstrom, *Windbreak,* 104–05.
39. Ibid., 122–23.
40. Hasselstrom, *Going Over East,* 98.
41. Ibid., 77.
42. Ibid., 40–41.
43. Hasselstrom, *Land Circle,* 340.
44. Hasselstrom, "Nighthawks," 71, 65.
45. James Clyman, *Journal of a Mountain Man,* ed. Linda Hasselstrom (Missoula, Mont.: Mountain Press Publishing, 1984).
46. Linda Hasselstrom, ed., *Next-Year Country: One Woman's View* (Hermosa, S.D.: Lame Johnny Press, 1978).
47. Hasselstrom, "How I Became a Broken-In Writer," in *Imagining Home: Writing from the Midwest,* ed. Mark Vinz and Thom Tammaro (Minneapolis: University of Minnesota Press, 1995), 157.
48. Hasselstrom, *Land Circle,* 223–26.
49. Ibid., 266–88.
50. Hasselstrom, "Planting Peas," in *Dakota Bones,* 56; "Digging Potatoes," in *Windbreak,* 23.
51. Hasselstrom, "How I Became," 152.
52. Ibid., 153.
53. See Tillie Olsen, *Silences* (New York: Delacorte Press-Seymour Lawrence, 1978); Meridel Le Sueur, *Ripening: Selected Work, 1927–1980,* ed. Elaine Hedges (Old Westbury, N.Y.: Feminist Press, 1982); and Adrienne Rich, *Adrienne Rich's Poetry and Prose,* ed. Barbara Charlesworth Gelpi and Albert Gelpi (New York: W. W. Norton, 1993).
54. Hasselstrom, "On Writing Western."
55. Hasselstrom, *Windbreak,* 170.
56. Linda Hasselstrom, Gaydell Collier, and Nancy Curtis, eds., *Leaning into the Wind: Women Write from the Heart of the West* (Boston: Houghton Mifflin, 1997), xvi. Hasselstrom is currently preparing a second anthology.
57. Hasselstrom, "How I Became," 157.
58. Harold P. Simonson, *Beyond the Frontier: Writers, Western Regionalism and a Sense of Place* (Fort Worth: Texas Christian Uuniversity Press, 1989), 143.

59. Ibid.
60. Hasselstrom, *Going Over East,* 200.

The Land as Consciousness (pages 178–203)

1. Leslie Marmon Silko, *Storyteller* (New York: Little, Brown [Arcade], 1981), 51. Subsequent references in the text are to this edition.
2. Historians and literary scholars frequently refer to the concept of the American Adam and Eve, as for example in R. W. B. Lewis, *The American Adam: Innocence, Tragedy, and Tradition in the Nineteenth Century* (Chicago: University of Chicago Press, 1955), and Barbara Taylor, *Eve and the New Jerusalem: Socialism and Feminism in the Nineteenth Century* (London: Virago, 1983). Many other writers within the field of American studies use the concepts of the garden and wilderness to detail European visions of the new world; see for example the works of Annette Kolodny, Roderick Nash, and Leo Marx. Frederick Turner applies the concept "ecocritically" in "Cultivation and the American Garden," in *The Ecocriticism Reader: Landmarks in Literary Ecology,* ed. Cheryll Glotfelty and Harold Fromm (Athens: University of Georgia Press, 1996), 40–51.
3. Annette Kolodny details men's and women's visions of the landscape as garden in *The Lay of the Land: Metaphor as Experience in American Life and Letters* (Chapel Hill: University of North Carolina Press, 1975) and *The Land Before Her: Fantasy and Experience of the American Frontiers, 1630–1860* (Chapel Hill: University of North Carolina Press, 1984). Carolyn Merchant and other scholars of the Puritan culture document the numerous references to a "howling wilderness." See Carolyn Merchant, *Ecological Revolutions: Nature, Gender and Science in New England* (Chapel Hill: University of North Carolina Press, 1989).
4. Stephen Trimble, ed., *Words from the Land: Encounters with Natural History Writing,* expanded edition (Reno: University of Nevada Press, 1995), 2 (emphasis added).
5. Leslie Marmon Silko, "Landscape, History, and the Pueblo Imagination." in *The Ecocriticism Reader,* 265.
6. Lee Schweninger, "Writing Nature: Silko and Native Americans as Nature Writers," *MELUS* 18, no. 2 (1993): 49.
7. Helen Jaskowski, *Leslie Marmon Silko: A Study of the Short Fiction* (New York: Twayne, 1998). 36, 17.
8. Glotfelty and Fromm, *The Ecocriticism Reader,* 266.
9. Gregory Salyer, *Leslie Marmon Silko* (New York: Twayne, 1997), 1.
10. Silko, "Landscape, History, and the Pueblo Imagination," 273, 274, 275.
11. Jaskowski, 3 (emphasis added).
12. Silko, "Interior and Exterior Landscapes," in *The Ecocriticism Reader,* 37, 42–43; cited in Jaskowski, *Leslie Marmon Silko: A Study of the Short Fiction,* 3.
13. Silko, interview with Lawrence Evers and Denny Carr, in *The Ecocriticism Reader,* 95.
14. Ibid., 115
15. For criticism of Silko's agendas, see for example Suzanne Ruta's review of "Dances with Ghosts," in *Gardens in the Dunes* in *New York Times Book Review* 18 (April, 1999): 31. Helen Jaskowski details Silko's mode of working with dual audiences in *Leslie Marmon Silko: A Study of the Short Fiction* (New York: Twayne, 1998).

16. See Trimble's *Words from the Land.*

17. Paula Gunn Allen, "The Ceremonial Motion of Indian Time: Long Ago, So Far," in *The Sacred Hoop: Recovering Feminine in American Indian Traditions* (Chicago: University of Chicago Press, 1990), 149.

18. Leslie Marmon Silko, *Ceremony* (1977, reprint; New York: Viking Press, 1986), 4. Subsequent references in the text are to this edition.

19. Jaskowski, *Leslie Marmon Silko: A Study of the Short Fiction,* 36.

20. Ibid., 34.

21. Interview with Silko by Thomas Irmer (Alt-X Berlin/Leipzig correspondent), Internet: <http://www.altx.com/interviews/silko.html>.

22. Silko, *Almanac of the Dead* (New York: Penguin, 1991), 555. Subsequent references in the text to this edition are indicated by *AD,* followed by a page number.

23. See Thomas Irmer's Internet interview.

24. Silko, *Gardens in the Dunes* (New York: Viking Penguin, 1986), 15. Subsequent references in the text to this edition are indicated by *GD,* followed by a page number.

25. Trimble, *Words from the Land,* 2.

26. Norwood, *Made from This Earth: American Women and Nature* (Chapel Hill: University of North Carolina Press, 1993), 180.

A New *Mestiza* Primer (pages 204–216)

Thank you to Kathleen Brown and Harriet Rohmer at Children's Book Press, the premiere publisher of multicultural children's literature in the United States, for granting permission to use illustrations from Gloria Anzaldúa's two children's books. Illustrations reprinted with permission of the publisher, Children's Book Press, San Francisco, California. Illustrations from *Prietita and the Ghost Woman* are copyright 1996 by Christina Gonzalez. Illustrations from *Friends from the Other Side* are copyright 1993 by Consuelo Méndez.

I am extremely grateful to Tom Edwards for his enthusiastic encouragement and limitless patience. The completion of this work was severely impaired by the murder of a family member and all its tragic and complicated circumstances. Phillip Serrato, one of the most brilliant graduate students with whom I've had the fortune to work, provided the foundation for this writing endeavor, helping my return to what most gives me hope. A heartfelt thanks to my partner Charles Hiroshi Garrett and colleague Amy Ongiri for their generous comments on earlier drafts of this essay.

1. Devon G. Peña, "Introduction," in *Chicano Culture, Ecology, Politics: Subversive Kin,* ed. Devon G. Peña (Tucson: University of Arizona Press, 1998), i.

2. For example, neither Robert M. Torrance's *Encompassing Nature:* A Sourcebook (Washington, D.C.: Counterpoint, 1998) or Mary Joy Breton's *Women Pioneers for the Environment* (Boston: Northeastern University Press, 1998) features work by U.S. Latina/o nature writers or environmental activists. An inclusion of these authors undoubtedly would have further enriched these works by adding to their critical discussion of the genre's treatment of culture, identity, and politics.

3. Gloria Anzaldúa, "Preface," in *Borderlands/La Frontera* (San Francisco: Spinsters/Aunt Lute, 1987), i.

4. Donna Perry, *Backtalk: Women Writers Speak Out* (New Brunswick: Rutgers University Press, 1993), 21.

5. Cherríe Moraga, *The Last Generation* (Boston: South End Press, 1993), 174.

6. Anzaldúa, *Borderlands/La Frontera,* 79.

7. Raul A. Fernandez, *The Mexican-American Border: Issues and Trends* (Notre Dame: University of Notre Dame Press, 1989), 1.

8. Oralia Garza de Cortes, "Behind the Golden Door: The Latino Immigrant Child in Literature and Films for Children," in *The New Press Guide to Multicultural Resources for Young Readers,* ed. Daphne Muse (New York: The New Press, 1997), 452.

9. Peter H. Schuck and Rogers M. Smith, *Citizenship without Consent: Illegal Aliens in the American Polity* (New Haven: Yale University Press, 1985), 109.

10. Leo Chavez, *Shadowed Lives: Undocumented Immigrants in American Society* (Fort Worth: Harcourt Brace Publishers, 1992), 18.

11. Anzaldúa, *Borderlands/La Frontera,* 91.

12. Perry, *Backtalk,* 21.

13. As Fernandez indicates in *The Mexican-American Border,* "The Lower Rio Grande Valley of Texas . . . has been known variously as the 'Valley of Tears' because of the poverty of many of its inhabitants or the 'Magic Valley' because of its long season for growing fruits and vegetables" (37). In *Border Visions: Mexican Cultures of the Southwest United States* (Tucson: The University of Arizona Press, 1996), Carlos Velez-Ibañez provides a thorough discussion of the concentration of poverty in the southwestern border region. See especially chapter 2, "The American *Entrada:* 'Barrioization' and the Development of Mexican Commodity Identity"; chapter 5, "The Distribution of Sadness: Poverty, Crime, Drugs, Illness, and War"; and pp. 297–99 n. 11.

14. Anzaldúa, *Borderlands/La Frontera,* 20.

15. Ibid., 11.

16. Ibid., 3, 2.

17. Frances Ann Day, *Latina and Latino Voices in Literature for Children and Teenagers* (Portsmouth: Heinemann, 1997), 57.

18. Anzaldúa, *Borderlands/La Frontera,* 82.

19. Rodolfo Gonzales, *Yo Soy Joaquín/I am Joaquín* (New York: Bantam, 1972).

20. Cherríe Moraga, "Giving Up the Ghost," in *Heroes and Saints & Other Plays* (Albuquerque: West End Press, 1994), 1–35.

21. Anzaldúa, *Borderlands/La Frontera,* 8.

22. Malia Davis, "Philosophy Meets Practice: A Critique of Ecofeminism through the Voices of Three Chicana Activists," in *Chicano Culture, Ecology, Politics,* 201.

23. See, for example, Joe Hayes, *La Llorona: The Weeping Woman* (El Paso: Cinco Punto Press, 1987).

24. For one example of how the *curandera* is portrayed as simultaneously feared and revered, see Louie The Foot González's short story "Doña Toña of Nineteenth Street," in *Growing Up Chicana/o,* ed. Tiffany Ana López (New York: William Morrow, 1993).

25. Gloria Anzaldúa, "La Prieta," in *This Bridge Called My Back: Writings by Radical Women of Color,* ed. Cherríe Moraga and Gloria Anzaldúa (New York: Kitchen Table/Women of Color Press, 1983), 201.

26. Cherríe Moraga, *Loving in the War Years: lo que nunca pasó por sus labios* (Boston: South End Press, 1983), 101–102.

27. The specific examples I have in mind here are the two jointly authored books by Sally

Morgan and Rosie Harlow, *Pollution and Waste* and *Garbage and Recycling* (New York: Kingfisher, 1995); and Susan E. Goodman's *Ultimate Field Trip I: Adventures in the Amazon Rain Forest* (New York: Simon & Schuster Aladdin Paperbacks, 1999).

28. M. Jimmie Killingsworth and Jacqueline S. Porter, "Ecopolitics and the Literature of the Borderlands: The Frontiers of Environmental Justice in Latina and Native American Writing," in *Writing the Environment: Ecocriticism and Literature,* ed. Richard Kerridge and Neil Sammels (London & New York: Zed Books Ltd., 1998), 206.

Stalking a Prayer (pages 217–226)

1. Annie Dillard, *Pilgrim at Tinker Creek* (New York: Harper Perennial, 1988), 1.

    Since the publication of *Pilgrim at Tinker Creek* in 1974, Dillard has said that neither the tomcat scene described here, nor the giant waterbug scene referred to later in the paragraph, actually happened to her. Some writers and readers have criticized Dillard for such fictional embellishment in a work classified as nonfiction (See, for example, C. L. Rawlins, "Dillard's Cat," *Northern Lights* [Summer 1998]: 16–18). However, the importance of the two scenes in framing and representing the major themes of the text remains fundamental. It is not the project of this essay to resolve this controversy, or to analyze writerly decisions that occur outside of the text itself. Rather, this essay considers how shifting rhetorical strategies within the text mediate the boundaries between the human and nonhuman worlds. Because it is awkward to repeatedly distinguish between the author and the narrator in a text that is classified as nonfiction, I will often refer to the narrator as Dillard; but the reader should note that I do not read the text as autobiography, and Dillard herself, in the afterword of the Harper Perennial twenty-fifty anniversary edition of *Pilgrim at Tinker Creek,* describes the "I" of the text as a point of view, and not necessarily as a strict representation of herself.

2. Ibid., 5.

3. Ibid., 226.

4. Ibid., 12.

5. David Abram, *The Spell of the Sensuous: Perception and Language in a More-Than-Human World* (New York: Vintage Books, 1997), 7.

6. Richard Slotkin, *Regeneration through Violence: The Mythology of the American Frontier, 1600–1860* (Middletown, Conn.: Wesleyan University Press, 1973), 49.

7. Dillard, *Pilgrim at Tinker Creek,* 76.

8. Ibid., 31.

9. Ibid.

10. Ralph Waldo Emerson, "Nature" in *The Complete Works of Ralph Waldo Emerson,* vol. 1, Centenary Edition (New York: Houghton Mifflin, 1903), 10.

11. Dillard, *Pilgrim at Tinker Creek,* 182.

12. Ibid., 184.

13. Ibid.

14. Ibid., 185.

15. Henry David Thoreau, *The Illustrated Walden* (Princeton: Princeton University Press, 1973), 5.

16. Dillard, *Pilgrim at Tinker Creek,* 187.

17. Ibid., 227.
18. Ibid., 184.
19. Ibid., 186.
20. Ibid., 239.
21. Ibid., 265.
22. Ibid., 239–40.
23. Abram, *Spell of the Sensuous,* 22.
24. Dillard, *Pilgrim at Tinker Creek,* 269.
25. Ibid., 65.
26. Ibid., 237.
27. Ibid., 266.
28. Ibid., 271.
29. Ibid., 103.
30. Ibid., 68.
31. Ibid., 96.
32. Ibid., 9.
33. Annie Dillard, *Holy the Firm* (New York: Harper & Row, 1988), 50.

Telling News of the Tainted Land (pages 227–238)

1. In William Cronon, ed., *Common Ground: Rethinking the Human Place in Nature* (New York: Norton, 1996), 300.
2. See, for example, Mary Lee Kerr and Charles Lee, " From Conquistadors to Coalitions: After Centuries of Environmental Racism, People of Color Are Forging a New Movement for Environmental Justice," *Southern Exposure* 21, no. 4 (1993): 8–19.
3. Kamala Platt states that "the term *environmental racism* was coined in 1982 during demonstrations against a PCB landfill slotted for Warren County, North Carolina, a predominantly African-American area" ("Ecocritical Chicana Literature: Ana Castillo's 'Virtual Realism,'" in *Ecofeminist Literary Criticism: Theory, Interpretation, Pedagogy,* ed. Greta Gaard and Patrick Murphy [Urbana: University of Illinois Press, 1998], 140). According to the *Environmental Encyclopedia* by William P. Cunningham et al. (Detroit: Gale Research, 1994), "the term environmental racism was coined in a 1987 study conducted by the United Church of Christ that examined the location of hazardous waste dumps and found an 'insidious form of racism'" (303). The definition further explains, "exposure to environmental risks is significantly greater for racial and ethnic minorities than for nonminority populations"; "official response to environmental problems may be racially biased"; and "in response to the concern that traditional environmentalism does not recognize the social and economic components of environmental problems and solutions, a national movement for 'environmental and economic justice' has spread across the country" (303).
4. Di Chiro, "Nature as Community," 301.
5. The editions used in this essay are: Barbara Kingsolver, *Animal Dreams* (New York: HarperPerennial, 1991), hereafter cited parenthetically within the text as *AD,* followed by a page number; Ana Castillo, *So Far From God* (New York: Penguin/Plume, 1994), cited as *SFFG;* and Linda Hogan, *Solar Storms* (New York: Scribner, 1997), cited as *SS.*

6. See Kamala Platt, "Environmental Justice: Multicultural by Definition," *Phoebe: An Interdisciplinary Journal of Feminist Scholarship, Theory, and Aesthetics* 9, no. 2 (Fall 1997): 1–12.

7. See, for example, Florence R. Krall's *Ecotone: Wayfaring on the Margins* (Albany: State University of New York Press, 1994).

8. Joni Adamson [Clarke], "Toward an Ecology of Justice: Transformative Ecological Theory and Practice," in *Reading the Earth: New Directions in the Study of Literature and the Environment,* ed. Michael P. Branch, Rochelle Johnson, Daniel Patterson, and Scott Slovic (Moscow: University of Idaho Press, 1998), 10.

9. Ibid., 15.

10. For an overview of American nature writing before Thoreau, see Michael Branch, "Indexing American Possibilities: The Natural History Writing of Bartram, Wilson, and Audubon," in *The Ecocriticism Reader: Landmarks in Literary Ecology,* ed. Cheryll Glotfelty and Harold Fromm (Athens: University of Georgia Press, 1996), 282–302. For another useful discussion of the history of canonical American nature writing, see Thomas J. Lyon, "Part I: A History," in *This Incomperable Lande: A Book of American Nature Writing,* ed. Thomas J. Lyon (Boston: Houghton Mifflin, 1989), 3–91.

11. For recent work in ecofeminist literary criticism and discussions of environmental justice literature, see *ISLE: Interdisciplinary Studies in Literature and Environment* special issue 3, no. 1 (Summer 1996), "Ecofemininist Literary Criticism," ed. Greta Gaard and Patrick Murphy (Gaard and Murphy reprinted the *ISLE* essays and additional contributions in *Ecofeminist Literary Criticism: Theory, Interpretation, Pedagogy*); Gaard, ed., *Ecofeminism: Women, Animals, Nature* (Philadelphia: Temple University Press, 1993); Vera Norwood, *Made from This Earth: American Women and Nature* (Chapel Hill: University of North Carolina Press, 1993); *Phoebe* 9, no. 1 (Spring 1997), "Ecofeminism/Ecocriticism I" and *Phoebe* 9, no. 2 (Fall 1997), "Ecofeminism/Ecocriticism II."

12. Krista Comer, "Sidestepping Environmental Justice: 'Natural' Landscapes and the Wilderness Plot," in *Breaking Boundaries: New Perspectives on Women's Regional Writing,* ed. Sherrie A. Innes and Diana Royer (Iowa City: University of Iowa Press, 1997), 225–26.

13. Ibid., 232.

14. Ibid., 230.

15. Platt, "Ecocritical Chicana Literature," 150. Platt further suggests "three possible and predominant readings for *So Far From God:* a religious reading that sees the novel as a New Mexican Catholic text; an aesthetic reading that sees the text as an example of magical realism; and the third—the kind [she] will attempt—a 'political' reading of social justice issues" (146).

16. Ibid.

17. See the SWOP Web site at <http://www.igc.org/trac/feature/hitech/swopinfo.html>, 3.

18. I am indebted to Kamala Platt's article "Ecocritical Chicana Literature" for information about these organizations (155, n. 2).

19. Adamson, "Toward an Ecology of Justice," 16.

Contributors

Cheryl Birkelo is an instructor of English and a lab-field technician in the Wildlife-Fisheries Co-op Unit at South Dakota State University. Her research interests include ecofeminist criticism, studies in ecology and literature, and Western regionalism. She received her M.A. degree from South Dakota State University. Currently, she is a technical writer in Woodland Park, Colorado.

Matthew Bolinder is a Ph.D. candidate studying American literature at Boston College.

Michael P. Branch is Associate Professor of Literature and Environment at the University of Nevada, Reno. Branch is co-editor of *The Height of Our Mountains: Nature Writing from Virginia's Blue Ridge Mountains and Shenandoah Valley* (1998) and *Reading the Earth: New Directions in the Study of Literature and Environment* (1998).

Karen Cole is Visiting Research Professor at Stetson University, DeLand, Florida, where she has taught American women's nature writing and other women and gender studies courses.

Jenny Emery Davidson is a former journalist and firefighter with the Bureau of Land Management wildland fire crew in Shoshone, Idaho. She is currently pursuing a Ph.D. in American Studies at the University of Utah.

Elizabeth A. De Wolfe is Assistant Professor of American Studies at the University of New England, where she teaches courses in women's history, communal societies, and book history.

Carol E. Dickson teaches literature and writing at Goddard College. Her teaching and research interests include American literary regionalism, environmental literature, African-American literature, women's literature, and service learning. She earned her Ph.D. in English at the University of Wisconsin–Madison.

Thomas S. Edwards is Dean of Academic Affairs at Thomas College in Waterville, Maine. Together with Karen L. Kilcup, he is the editor of *Jewett and Her Contemporaries: Reshaping the Canon* (1999).

Tamara Fritze is an Assistant Professor at Utah Valley State College, where she teaches a variety of American literature, American Studies, and Folklore classes. She holds her Ph.D. in American Studies from Washington State University.

Jen Hill is editor of *An Exhilaration of Wings: The Literature of Birdwatching* (1999). She is Assistant Professor of English at University of Nevada, Reno.

Annie Merrill Ingram is Assistant Professor of English at Davidson College, where she teaches courses about the literatures of nineteenth-century America, the environment, and Native Americans. Her scholarly work focuses on women writers and ecocriticism.

Valerie Levy is a Ph.D. candidate in American literature at the University of Georgia, Athens. Her specialties include nineteenth-century American literature and multicultural American literature.

Marcia B. Littenberg is an Assistant Professor at the State University of New York at Farmingdale. She has published on Jewett, Thaxter, and American regionalism. Other publications include a section in *Jewish American Women Writers,* edited by Ann Shapiro, an article on Jewish comedy in the Catskill resorts (in *Jewish Folklore and Ethnology Review,* 1997), and on George Burns and Gracie Allen and Jewish vaudeville comedy (in *Fools and Jesters in Art, Literature and History,* ed. Vicki Janik).

Tiffany Ana López is Assistant Professor of English at the University of California, Riverside. She is editor of *Growing Up Chicana/o* (1993). Recent articles include "Violent Inscriptions: Writing the Body and Making Community in Four Plays by Midgalia Cruz" (*Theatre Journal* 52, 2000) and "A Tolerance for Contradictions: The Short Stories of María Christina Mena," in *Nineteenth-Century American Writers: A Critical Anthology,* ed. Karen L. Kilcup (1998). Currently she is completing her book *Bodily Inscriptions: Representations of the Body in U.S. Latina Drama,* with a second manuscript in progress on questions of genre and Latina/o children's literature.

Richard M. Magee is working on his dissertation, called "From the Sentimental to the Sublime: Susan Fenimore Cooper's Nature Writing," at Fordham University. A chapter of this, "'An Artist or a Merchant's Clerk': Cooper and Landscape Painting," has been accepted for publication in a book on Cooper edited by Rochelle Johnson and Daniel Patterson. Magee is also an adjunct instructor in English at SUNY Maritime College.

Vera Norwood is Professor and Chair of American Studies at the University of New Mexico. She is the author of *Made from This Earth: American Woman and Nature* (1993) and co-author of *The Desert Is No Lady: Southwestern Landscapes in Women's Writing and Art* (1987).

Daniel J. Philippon is Assistant Professor of Rhetoric at the University of Minnesota, Twin Cities, where he teaches courses in environmental literature, history and ethics. He is the editor of *The Friendship of Nature: A New England Chronicle of Birds and Flowers,* by Mabel Osgood Wright (1999), and co-editor of *The Height of Our Mountains: Nature Writing from Virginia's Blue Ridge Mountains and Shenandoah Valley* (1998).

Kelly L. Richardson is a Ph.D. student at the University of North Carolina at Greensboro. She received her B.A. and M.A. at Winthrop University. Her primary interest is nineteenth-century American literature.

Suzanne Ross is Professor of English at St. Cloud State University in Minnesota. She teaches composition and linguistics in addition to nature literature and nonfiction nature writing. Her research interests include American and British nature writing, especially literary representations of other animals.

Mary R. Ryder is Professor and Director of Graduate Studies in English at South Dakota State University. Her book *Willa Cather and Classical Myth* (1990) earned the Classical and Modern Literature Incentive Award and the Mildred Bennett Award for distinguished Cather scholarship. Her work has appeared in journals such as *Style, Western American Literature,* and *American Literary Realism.*

Rena Sanderson is Associate Professor of English at Boise State University, where she teaches American literature. Several of her publications focus on the depiction of the American woman in such writers as Hemingway (*Cambridge Companion to Hemingway,* 1996), the psychologist Hugo Münsterberg (*Prospects: An Annual of American Cultural Studies,* 1998), and F. Scott Fitzgerald (*Cambridge Companion to Fitzgerald,* forthcoming).

Phillip Serrato is a graduate student in the Department of English at the University of California, Riverside, where he is completing his dissertation on issues of masculinity in U.S. Latino literature.

Karen E. Waldron is Professor of Literature and Associate Dean for Academic Affairs at the College of the Atlantic in Bar Harbor, Maine. She is the author of *The Power of Feminine Consciousness: Authority, Voice and Myth in* Their Eyes Were Watching God (1993) and *Coming to Consciousness, Coming to Voice: The Reinvention of Eve in American Women's Writings* (1994), as well as numerous essays on authors such as Elizabeth Bishop, Kim Chernin, and William Faulkner.

Index